NONLINEAR PROBLEMS

IN APPLIED MATHEMATICS

SIAM PROCEEDINGS SERIES LIST

Neustadt, L. W., Proceedings of the First International Congress on Programming and Control (1966)

Hull, T. E., Studies in Optimization (1970)

Day, R. H. and Robinson, S. M., Mathematical Topics in Economic Theory and Computation (1972)

Proschan, F. and Serfling, R. J., Reliability and Biometry: Statistical Analysis of Lifelength (1974)

Barlow, R. E., Reliability & Fault Tree Analysis: Theoretical & Applied Aspects of System Reliability & Safety Assessment (1975)

Fussell, J. B. and Burdick, G. R., Nuclear Systems Reliability Engineering and Risk Assessment (1977)

Erisman, A. M., Neves, K. W., and Dwarakanath, M. H., Electric Power Problems: The Mathematical Challenge (1981)

Bednar, J. B., Redner, R., Robinson, E., and Weglein, A., Conference on Inverse Scattering: Theory and Application (1983)

Santosa, Fadil, Symes, William W., Pao, Yih-Hsing, and Holland, Charles, Inverse Problems of Acoustic and Elastic Waves (1984)

Gross, Kenneth I., Mathematical Methods in Energy Research (1984)

Babuska, I., Chandra, J., and Flaherty, J., Adaptive Computational Methods for Partial Differential Equations (1984)

Boggs, Paul T., Byrd, Richard H., and Schnabel, Robert B., Numerical Optimization 1984 (1985)

Angrand, F., Dervieux, A., Desideri, J. A., and Glowinski, R., Numerical Methods for Euler Equations of Fluid Dynamics (1985)

Wouk, Arthur, New Computing Environments: Parallel, Vector and Systolic (1986)

Fitzgibbon, William E., Mathematical and Computational Methods in Seismic Exploration and Reservoir Modeling (1986)

Drew, Donald A. and Flaherty, Joseph E., Mathematics Applied to Fluid Mechanics and Stability: Proceedings of a Conference Dedicated to R.C. DiPrima (1986)

Heath, Michael T., Hypercube Multiprocessors 1986 (1986)

Papanicolaou, George, Advances in Multiphase Flow and Related Problems (1987)

Wouk, Arthur, New Computing Environments: Microcomputers in Large-Scale Computing (1987)

Chandra, Jagdish and Srivastav, Ram, Constitutive Models of Deformation (1987)

Heath, Michael T., Hypercube Multiprocessors 1987 (1987)

Glowinski, R., Golub, G. H., Meurant, G. A., and Periaux, J., First International Conference on Domain Decomposition Methods for Partial Differential Equations (1988)

Salam, Fathi M. A. and Levi, Mark L., Dynamical Systems Approaches to Nonlinear Problems in Systems and Circuits (1988)

Datta, B., Johnson, C., Kaashoek, M., Plemmons, R., and Sontag, E., Linear Algebra in Signals, Systems and Control (1988)

Ringeisen, Richard D. and Roberts, Fred S., Applications of Discrete Mathematics (1988)

McKenna, James and Temam, Roger, ICIAM '87: Proceedings of the First International Conference on Industrial and Applied Mathematics (1988)

Rodrigue, Garry, Parallel Processing for Scientific Computing (1989)

Caflish, Russel E., Mathematical Aspects of Vortex Dynamics (1989)

Wouk, Arthur, Parallel Processing and Medium-Scale Multiprocessors (1989)

Flaherty, Joseph E., Paslow, Pamela J., Shephard, Mark S., and Vasilakis, John D., Adaptive Methods for Partial Differential Equations (1989)

Kohn, Robert V. and Milton, Graeme W., Random Media and Composites (1989)

Mandel, Jan, McCormick, S. F., Dendy, J. E., Jr., Farhat, Charbel, Lonsdale, Guy, Parter, Seymour V., Ruge, John W., and Stüben, Klaus, Proceedings of the Fourth Copper Mountain Conference on Multigrid Methods (1989)

Colton, David, Ewing, Richard, and Rundell, William, Inverse Problems in Partial Differential Equations (1990)

Chan, Tony F., Glowinski, Roland, Periaux, Jacques, and Widlund, Olof B., Third International Symposium on Domain Decomposition Methods for Partial Differential Equations (1990)

Dongarra, Jack, Messina, Paul, Sorensen, Danny C., and Voigt, Robert G., Proceedings of the Fourth SIAM Conference on Parallel Processing for Scientific Computing (1990)

Glowinski, Roland and Lichnewsky, Alain, Computing Methods in Applied Sciences and Engineering (1990)

Coleman, Thomas F. and Li, Yuying, Large-Scale Numerical Optimization (1990)

Aggarwal, Alok, Borodin, Allan, Gabow, Harold, N., Galil, Zvi, Karp, Richard M., Kleitman, Daniel J., Odlyzko, Andrew M., Pulleyblank, William R., Tardos, Éva, and Vishkin, Uzi, Proceedings of the Second Annual ACM-SIAM Symposium on Discrete Algorithms (1990)

Cohen, Gary, Halpern, Laurence, and Joly, Patrick, Mathematical and Numerical Aspects of Wave Propagation Phenomena (1991)

Gómez, S., Hennart, J. P., and Tapia, R. A., Advances in Numerical Partial Differential Equations and Optimization: Proceedings of the Fifth Mexico-United States Workshop (1991)

Glowinski, Roland, Kuznetsov, Yuri A., Meurant, Gérard, Périaux, Jacques, and Widlund, Olof B., Fourth International Symposium on Domain Decomposition Methods for Partial Differential Equations (1991)

Alavi, Y., Chung, F. R. K., Graham, R. L., and Hsu, D. F., Graph Theory, Combinatorics, Algorithms, and Applications (1991)

Wu, Julian J., Ting, T. C. T., and Barnett, David M., Modern Theory of Anisotropic Elasticity and Applications (1991)

Shearer, Michael, Viscous Profiles and Numerical Methods for Shock Waves (1991)

Griewank, Andreas and Corliss, George F., Automatic Differentiation of Algorithms: Theory, Implementation, and Application (1991)

Frederickson, Greg, Graham, Ron, Hochbaum, Dorit S., Johnson, Ellis, Kosaraju, S. Rao, Luby, Michael, Megiddo, Nimrod, Schieber, Baruch, Vaidya, Pravin, and Yao, Frances, Proceedings of the Third Annual ACM-SIAM Symposium on Discrete Algorithms (1992)

Field, David A. and Komkov, Vadim, Theoretical Aspects of Industrial Design (1992)

Field, David A. and Komkov, Vadim, Geometric Aspects of Industrial Design (1992)

Bednar, J. Bee, Lines, L. R., Stolt, R. H., and Weglein, A. B., Geophysical Inversion (1992)

O'Malley, Robert E. Jr., ICIAM 91: Proceedings of the Second International Conference on Industrial and Applied Mathematics (1992)

Keyes, David E., Chan, Tony F., Meurant, Gérard, Scroggs, Jeffrey S., and Voigt, Robert G., Fifth International Symposium on Domain Decomposition Methods for Partial Differential Equations (1992)

Dongarra, Jack, Messina, Paul, Kennedy, Ken, Sorensen, Danny C., and Voigt, Robert G., Proceedings of the Fifth SIAM Conference on Parallel Processing for Scientific Computing (1992)

Corones, James P., Kristensson, Gerhard, Nelson, Paul, and Seth, Daniel L., Invariant Imbedding and Inverse Problems (1992)

Ramachandran, Vijaya, Bentley, Jon, Cole, Richard, Cunningham, William H., Guibas, Leo, King, Valerie, Lawler, Eugene, Lenstra, Arjen, Mulmuley, Ketan, Sleator, Daniel D., and Yannakakis, Mihalis, Proceedings of the Fourth Annual ACM-SIAM Symposium on Discrete Algorithms (1993)

Kleinman, Ralph, Angell, Thomas, Colton, David, Santosa, Fadil, and Stakgold, Ivar, Second International Conference on Mathematical and Numerical Aspects of Wave Propagation (1993)

Banks, H. T., Fabiano, R. H., and Ito, K., Identification and Control in Systems Governed by Partial Differential Equations (1993)

Sleator, Daniel D., Bern, Marshall W., Clarkson, Kenneth L., Cook, William J., Karlin, Anna, Klein, Philip N., Lagarias, Jeffrey C., Lawler, Eugene L., Maggs, Bruce, Milenkovic, Victor J., and Winkler, Peter, Proceedings of the Fifth Annual ACM-SIAM Symposium on Discrete Algorithms (1994)

Lewis, John G., Proceedings of the Fifth SIAM Conference on Applied Linear Algebra (1994)

Brown, J. David, Chu, Moody T., Ellison, Donald C., and Plemmons, Robert J., Proceedings of the Cornelius Lanczos International Centenary Conference (1994)

Dongarra, Jack J. and Tourancheau, B., Proceedings of the Second Workshop on Environments and Tools for Parallel Scientific Computing (1994)

Bailey, David H., Bjørstad, Petter E., Gilbert, John R., Mascagni, Michael V., Schreiber, Robert S., Simon, Horst D., Torczon, Virginia J., and Watson, Layne T., Proceedings of the Seventh SIAM Conference on Parallel Processing for Scientific Computing (1995)

Clarkson, Kenneth, Agarwal, Pankaj K., Atallah, Mikhail, Frieze, Alan, Goldberg, Andrew, Karloff, Howard, Manber, Udi, Munro, Ian, Raghavan, Prabhakar, Schmidt, Jeanette, and Young, Moti, Proceedings of the Sixth Annual ACM-SIAM Symposium on Discrete Algorithms (1995)

Cohen, Gary, Third International Conference on Mathematical and Numerical Aspects of Wave Propagation (1995)

Engl, Heinz W., and Rundell, W., GAMM–SIAM Proceedings on Inverse Problems in Diffusion Processes (1995)

Angell, T. S., Cook, Pamela L., Kleinman, R. E., and Olmstead, W. E., Nonlinear Problems in Applied Mathematics (1995)

NONLINEAR PROBLEMS

IN APPLIED MATHEMATICS

IN HONOR OF IVAR STAKGOLD ON HIS 70TH BIRTHDAY

Edited by T. S. Angell
University of Delaware
Newark, Delaware

L. Pamela Cook
University of Delaware
Newark, Delaware

R. E. Kleinman
University of Delaware
Newark, Delaware

W. E. Olmstead
Northwestern University
Evanston, Illinois

Society for Industrial and Applied Mathematics
Philadelphia

DEDICATION

This volume is dedicated to Ivar Stakgold on the occasion of his 70th birthday.

It is traditional, not to say expected, in a volume dedicated as this one is to a friend and colleague, that the path he has followed to his present station be reviewed. Whether such a *curriculum vitae* is seen as inspirational (or cautionary) by younger members of the profession, or as a celebration of a successful career, depends upon the reader, as we are constantly reminded by the deconstructionists.

Fortunately we, as editors, have the opportunity to state "up front" what our aims are: they are unabashedly celebratory. In almost a half century of service in the mathematical profession Ivar has won our admiration and respect. We are proud to be counted among his many friends. In recognition of this friendship as well as the impact he has had, we have invited a group of co-workers and colleagues to contribute work which reflects Ivar's influence on the mathematics that we all do.

For the readers of this collection, then, we record that Ivar's early life was peripatetic, possibly providing the impetus for his life-long love of travel. Born in Oslo, Norway in 1925, his family moved to Belgium, where he spent his early childhood. With the onset of the Second World War, he embarked with them on a tortuous trail through France, Spain, and Portugal to Santo Domingo before the family was able to immigrate to the United States in 1941. While for many this journey was marked by peril, bitter disappointment, and worse, Ivar treated the adventure with characteristic insouciance. After graduating from Horace Mann High School in Manhattan, he enrolled at Cornell University, earning both Bachelor's (1945) and Master's (1946) degrees. Although his studies were in engineering, he developed keen interests in mathematics and bridge and pursued both at Harvard where, under the direction of Leon Brillouin, he received a Ph.D. in Applied Mathematics in 1949. His thirty-four page thesis remains a beacon of brevity in times of verbal overkill.

His teaching career began at Harvard, where he served, first as an Instructor and then Assistant Professor, from 1949 to 1956. He then left Cambridge to become Head of the Mathematics and Logistics Branches of the Office of

Naval Research in Washington, DC. It was during his Washington years that, in addition to developing his now superb administrative skills, he served as a member of the United States Bridge team in the World Championships of 1959 and 1960.

When Ivar left his position at ONR in 1960 to join the faculty at Northwestern University, he was appointed as an Associate Professor jointly in the Departments of Engineering Sciences and Mathematics and promoted to full professor in 1964. He remained at Northwestern for fifteen years, except for a tour as Liaison Scientist at the London Branch of ONR during the 1968 academic year, accompanied by his wife Alice. It was during their stay in London that their daughter Alissa was born. He returned to Northwestern the following year to assume duties as chairman of Engineering Sciences.

Soon after arriving at Northwestern, Ivar began to offer the graduate course which spawned his two-volume set of textbooks, *Boundary Value Problems of Mathematical Physics*. From an initial enrollment of barely ten students, his popularity as an instructor soon spread widely. Within a few years, his course attracted over seventy graduate students from the various engineering disciplines as well as from mathematics and physics. More than thirty years later his course is still being taught at Northwestern. As for the textbooks, the influence of this masterful presentation of the mathematical foundations of differential equation models of physical problems is incalculable. These books became essential references in the libraries of a generation of applied mathematicians and engineers.

It was during his tenure at Northwestern that Ivar's research interests began to turn to nonlinear problems, particularly in the area of bifurcation phenomena. Ivar was a central figure in this field as it opened up in the late 1960s and early 1970s. He played a key role in several summer conferences on bifurcation problems, most notably at Colorado State (1969), Oberwolfach (1971), and Battelle (1972). His well-received survey paper "Branching of Solutions of Nonlinear Equations," which appeared in *SIAM Review* in 1971 did much to clarify the basic concepts of bifurcation theory for the community of applied mathematicians.

While at Northwestern, Ivar played a pivotal role in forming a Committee on Applied Mathematics through which graduate degrees in Applied Mathematics were awarded.

In 1975, Ivar and his family moved to Delaware, where he was appointed Professor and Chair of the Department of Mathematical Sciences and Professor of Mechanical Engineering at the University of Delaware. Learning to operate in a new administrative environment takes time and effort. Characteristically, he lost no time. Even before he arrived in Newark, he had arranged for the department to occupy new quarters, persuading the university to add another floor to a building then under construction.

Within a year of his arrival he had not only embarked on the project of building the research program in the department, but had also managed to complete the project of revising his textbook. Constantly looking for ways to improve on his earlier work, the project to rewrite an old book became the task of writing a new one. His new text *Green's Functions and Boundary Value Problems* was published in 1979, and is well-known and widely used throughout the world.

From his fifth-floor office, he encouraged the growth of the small graduate program, the increase in external funding, and the development of the gracious atmosphere which became well known and appreciated by visitors to the department. Indeed, his encouragement and support led to a "visitors program" that became the most active in the university. None of the visitors could fail to be impressed by Ivar's ability to greet them in their native tongue, however obscure, and were even more impressed by his ability to comment intelligently on their mathematics, however obscure.

Under Ivar's stewardship, the Mathematical Sciences department at Delaware gained significantly in recognition by the world-wide mathematical community. His impressive presence, quick mind, acerbic wit, and keen analytical powers not only helped him maintain relative peace among mathematicians, a group notorious for its ungovernability, but also made him the bête noire in the eyes of higher administrators, a group notorious for its tumidity.

Soon after arriving in Delaware, Ivar became active in SIAM and was elected to the SIAM Board of Trustees in 1976. He quickly became known among board members for his statesman-like thinking in the many matters brought before the board, so much so that the board elected him chair in 1979, and annually thereafter, until 1985, when he was restricted from running for reelection by SIAM bylaws.

Ivar demonstrated great skill in conducting meetings of the board, especially in its considerations of contentious issues. He was able to promote constructive dialogue among board members in addressing difficult problems. Often, his keen sense of humor proved to be instrumental in diffusing tension.

During his three terms on the board, Ivar also was one of SIAM's three representatives on the AMS–MAA–SIAM Joint Policy Board for Mathematics and he served, *ex officio*, on SIAM's Financial Management Committee.

In 1988, Ivar was elected President-Elect of SIAM, serving subsequently as President in 1989 and 1990, and Past President in 1991. When he took office in 1988, he recognized that the opportunities for SIAM members to shape the future directions of education and research were greater than ever before and encouraged the membership throughout his term to become more involved in the educational aspects of applied and industrial mathematics. He served as Chair, between 1990 and 1992, of the Conference Board of the Mathematical Sciences, an umbrella organization of mathematics and statistics societies whose total membership exceeds 150,000.

Indeed, his commitment to the profession continued after he stepped down as Chair at Delaware. In 1994 he agreed to serve one year in the Washington, DC office of the American Mathematical Society as Head of that office, ably representing the general mathematical community to the federal government.

No dedication to Ivar would be complete without mention of his efforts in promoting international cooperation in applied mathematics. Starting with his tour in ONR London, and continuing with long stays in Italy, Spain, Switzerland, England and New Zealand and shorter visits to many other places, he has left an indelible mark. His influence in shaping mathematical policy not only at home, but also abroad, is significant. The highly successful Industrial Mathematics Program at Oxford gratefully acknowledges a debt to Ivar, to name but one instance.

Indeed Ivar Stakgold is a man for all mathematical seasons. His fame precedes him. Upon meeting someone for the first time, Ivar does not know whether he will be recognized as the textbook author, the researcher, the brilliant expositor, the policy maker, or the bridge player. He is all of these and more. He is the urbane ambassador of applied mathematics. And he is our friend.

It is to this last personification that this book is warmly and affectionately dedicated.

Tom Angell
Ed Block
Pam Cook
Ralph Kleinman
Ed Olmstead

CONTENTS

IVAR STAKGOLD

Symmetry and non–symmetry
for the overdetermined Stekloff eigenvalue problem II*

G. Alessandrini†   R. Magnanini‡

Abstract
We continue the study of the overdetermined Stekloff eigenvalue problem (1.1)–(1.3) below. In [1], we constructed a variety of non–symmetric planar domains for which a solution of (1.1)–(1.3) exists. Here, we consider the problem in dimension $n \geq 3$, and prove that if there is a solution of (1.1)–(1.3) that satisfies an additional integral condition, then the domain Ω must be a ball.

1 Introduction

This article is the continuation of the research [1] originated by a paper of Payne and Philippin [14] concerning the Stekloff eigenvalue problem:

$$(1.1) \qquad\qquad \Delta u = 0 \quad \text{in } \Omega,$$

$$(1.2) \qquad\qquad \frac{\partial u}{\partial \nu} = pu \quad \text{on } \partial\Omega.$$

Here, $\Omega \subset \mathbb{R}^n$ is a bounded domain with sufficiently smooth boundary $\partial\Omega$, and ν denotes the exterior normal unit vector to $\partial\Omega$. It is well-known that this problem has infinitely many eigenvalues $0 = p_1 < p_2 \leq p_3 \leq \ldots$ (see [16]).

In [14], the authors proved, for $n = 2$, that if there is an eigenfunction u of (1.1), (1.2), corresponding to the second eigenvalue p_2, which also satisfies the overdetermined condition:

$$(1.3) \qquad\qquad |Du| = 1 \quad \text{on } \partial\Omega,$$

then u is linear and Ω must be a disk. They also pointed out that this result does not hold if $\partial\Omega$ is not of class C^2.

A natural question arises: if $\partial\Omega \in C^2$, suppose that for some $p > 0$ there exists a solution of (1.1), (1.2) satisfying (1.3); does this imply that Ω is a ball?

In [1], we examined the two–dimensional case and constructed a variety of non–symmetric domains for which a solution of (1.1)–(1.3) exists.

In the present paper, we are concerned with the case $n \geq 3$. The problem shows quite different features; in order to understand this, it is worth looking at solutions of (1.1)–(1.3) in the unit ball B_n of $\mathbb{R}^n$.

In this case, $\nu(x) = x$ on ∂B_n; by (1.2), since $x \cdot Du(x) - pu(x)$ is harmonic in B_n, we have that $x \cdot Du(x) = pu(x)$, $x \in B_n$, that is u must be a homogeneous harmonic polynomial

---

* Work partially supported by MURST 40 % and MURST 60 %.

† Università di Trieste, Dipartimento di Scienze Matematiche, Piazzale Europa 1, 34100 Trieste.

‡ Università di Firenze, Dipartimento di Matematica "U. Dini", Viale Morgagni 67/A, I-50134 Firenze.

of degree p. Therefore, (1.1)–(1.3) can be transformed into the problem:

$$(1.4) \qquad\qquad\qquad\qquad \tilde{\Delta} u = g(u),$$

$$(1.5) \qquad\qquad\qquad\qquad |\nabla u|^2 = f(u),$$

on $S^{n-1} = \partial B_n$, where $g(u) = -p(p + n - 2)u$, $f(u) = 1 - p^2 u^2$. Here, $\tilde{\Delta}$ and ∇ denote the Laplace–Beltrami operator and tangential gradient on S^{n-1}, respectively.

When $n = 2$, all solutions of (1.4), (1.5) are given by $\{\frac{1}{p} \cos(ps + s_0)\}_{p=1,2,\ldots}$, where s is the arclength parameter on S^1. Note that the above set is complete in the space $\{v \in L^2(S^1) : \int_{S^1} v ds = 0\}$, and describes all the traces of the Stekloff eigenfunctions in the disk.

If $n \geq 3$, solutions of a system of type (1.4), (1.5), with f and g smooth, are well-known in the literature as *isoparametric functions*. Their level surfaces at regular values, the *isoparametric surfaces*, enjoy the nice geometric property of having all their principal curvatures constant.

Up to this date, a complete classification of these surfaces on the sphere is not available. Here, we want to stress the fact that they seem to be very rare. When $n = 3$, for example, it can be shown that the solutions of (1.4), (1.5) are just the restrictions to S^2 of linear functions on $\mathbb{R}^3$. More results and examples in this direction are contained in the works of E. Cartan [2], [3], who first considered the isoparametric surfaces on the sphere, Nomizu [12], [13], Munzner [11], Ferus–Karcher–Munzner [7], and Wang Q. M. [17], [18], who examined them on a complete Riemannian manifold. We refer the reader to [18] for a survey on the subject.

The main result of this paper is an analogue to Payne and Philippin's theorem.

THEOREM 1.1. *Let $\Omega \subset \mathbb{R}^n$ be a contractible bounded domain with boundary $\partial\Omega \in C^2$. Suppose that there exists a solution u of (1.1)–(1.3) which also satisfies:*

$$(1.6) \qquad\qquad\qquad\qquad \int_{\partial\Omega} (u - x \cdot Du)\, u\, d\sigma = 0.$$

Then, Ω is a ball.

The proof of this result is based on Theorem 2.1, that essentially asserts that if u is a solution of equation (1.5) on a Riemannian manifold M, then the sets $\{x \in M : u(x) = c\}$, at critical values c, are smooth submanifolds of M. This quite surprising result is proved in [17], in a slightly different setting. In §2, we produce an alternative proof, based on some elementary arguments and with a more analytical flavour.

This paper is organized as follows. In §2, we state and prove Theorem 2.1. Section 3 is devoted to the proof of Theorem 1.1, as a consequence of Proposition 3.1 and Theorem 3.2.

2   Equation (1.5) on manifolds

We start with some preliminary notations. We consider a C^2 manifold M, without boundary, of dimension m, endowed with a Riemannian metric, which is represented by $\{g_{ij}(x)\}_{i,j=1,\ldots,m}$ in the local coordinates $x = (x^1, \ldots, x^m)$. If $v = (v^1, \ldots, v^m)$ and $w = (w^1, \ldots, w^m)$ are tangent vector fields on M, we define:

$$< v, w > = g_{ij}(x) v^i w^j, \qquad |v| = < v, v >^{\frac{1}{2}}.$$

Here, we adopt the usual assumption on the sum over repeated indices.

Given a C^1 function on M, we introduce the gradient of u on M as

$$\nabla u = (\nabla_1 u, \ldots, \nabla_m u), \qquad \nabla_i u = g^{ij}(x) u_{x^j}, \quad i = 1, \ldots, m;$$

here $\{g^{ij}(x)\}_{i,j=1,\ldots,m}$ is the inverse of the matrix $\{g_{ij}(x)\}_{i,j=1,\ldots,m}$.

We denote by $d : M \times M \to \mathbb{R}$ the geodetic distance on M; moreover, for any $x \in M$ and any closed subset $C \subset M$, it is well defined the number:

$$d(x, C) = \min\{d(x, y) : y \in C\}.$$

Let D be a bounded domain in M. We shall look at solutions of the following boundary value problem:

(2.1)
$$|\nabla \phi|^2 = f(\phi) \quad \text{in } D,$$
$$0 < \phi \leq \Phi \quad \text{in } D,$$

(2.2)
$$\phi = 0 \quad \text{on } \partial D.$$

Here $f \in C^1((0, \Phi])$ is a function satisfying

(2.3)
$$f > 0 \quad \text{on } (0, \Phi),$$
$$f(\Phi) = 0, \quad f'(\Phi) < 0.$$

Notice that (2.1) and (2.3) easily imply that

(2.4)
$$\max_{\overline{D}} \phi = \Phi.$$

THEOREM 2.1. *Let $D \subset M$ be a bounded domain with boundary $\partial D \in C^2$. Let f be a $C^1((0, \Phi])$ function satisfying (2.3).*

If $\phi \in C(\overline{D}) \cap C^2(D)$ is a solution of (2.1)–(2.2), then for some integer h, $0 \leq h \leq m - 1$, the extremal level set

(2.5)
$$D_\Phi = \{x \in D : \phi(x) = \Phi\}$$

is an h–dimensional C^1 connected compact submanifold without boundary of M.

Moreover, D satisfies:

(2.6)
$$D = \{x \in M : d(x, D_\Phi) < L\},$$

where

(2.7)
$$L = \int_0^\Phi \frac{ds}{\sqrt{f(s)}}.$$

COROLLARY 2.2. *If D is contractible, then D_Φ consists of a single point, and D is a geodetic ball centered at D_Φ.*

REMARK. The thesis of Theorem 2.1 provides us with a nearly sufficient condition on D for the existence of a solution of (2.1)–(2.3). In fact, let $0 \leq h \leq m - 1$; given any C^2 h–dimensional compact submanifold without boundary D_Φ in M, let K be the maximum of the absolute value of all its principal curvatures. Then, for any $L < 1/K$, the domain D defined by (2.6) is such that a solution of (2.1)–(2.3) exists; take for instance

$$(2.8) \qquad \phi(x) = \sqrt{L^2 - d(x, D_\Phi)^2};$$

in this case we have $\Phi = L$ and $f(\phi) = L^2\phi^{-2} - 1$.

LEMMA 2.3. *The closed set D_Φ defined in (2.5) has no interior points.*

Proof. Suppose by contradiction that $int(D_\Phi)$ is not empty.

Pick a point $x_0 \in \overline{int(D_\Phi)} \setminus int(D_\Phi)$; then $x_0 \in \partial D_\Phi$, since $\overline{int(D_\Phi)} \subseteq D_\Phi$. Let U be a coordinate neighborhood of x_0 and let $x \in (D \setminus D_\Phi) \cap U$. By (2.1), for every $k = 1, \ldots, m$, we have:

$$f'(\phi(x))\phi_{x^k}(x) = \partial_{x^k}|\nabla\phi|^2 = 2\phi_{x^ix^k}(x)g^{ij}(x)\phi_{x^j} + g^{ij}_{x^k}(x)\phi_{x^i}(x)\phi_{x^j}(x).$$

Since $\{g^{ij}(x)\}$ is uniformly positive definite, by the boundedness of the $g^{ij}_{x^k}(x)$'s and the Schwarz inequality, we may find positive constants c_1 and c_2 such that

$$[\sum_{i,j=1}^{m} \phi_{x^ix^j}(x)^2]^{\frac{1}{2}} \geq c_1|f'(\phi(x))| - c_2|\nabla\phi(x)|,$$

for all $x \in (D \setminus D_\Phi) \cap U$. Hence,

$$\lim_{D \setminus D_\Phi \ni x \to x_0} [\sum_{i,j=1}^{m} \phi_{x^ix^j}(x)^2]^{\frac{1}{2}} \geq c_1|f'(\Phi)| > 0,$$

whereas, obviously

$$\lim_{int(D_\Phi) \ni x \to x_0} [\sum_{i,j=1}^{m} \phi_{x^ix^j}(x)^2]^{\frac{1}{2}} = 0.$$

This is a contradiction. □

Let us set now:

$$(2.9) \qquad \begin{aligned} F(t) &= \int_0^t \frac{ds}{\sqrt{f(s)}}, \quad t \in [0, \Phi], \\ \delta(x) &= F(\phi(x)), \quad x \in M. \end{aligned}$$

Notice that $\delta \in C(\overline{D}) \cap C^2(D \setminus D_\Phi)$, and also

$$(2.10) \qquad \begin{aligned} |\nabla\delta| &= 1 \quad \text{in } D \setminus D_\Phi, \\ \delta &= 0 \quad \text{on } \partial D, \quad \delta = L \quad \text{on } D_\Phi, \\ 0 &< \delta < L \quad \text{in } D \setminus D_\Phi, \end{aligned}$$

where L is given by (2.7).

The next lemma shows the relationship between δ and the distance function. We shall use the following definition.

DEFINITION 2.4. Let $x \in D \setminus D_\Phi$. The *stream line* $\gamma(x; \cdot)$ *of δ passing through* x is the maximal solution of the initial value problem:

$$(2.11) \qquad \gamma'(x; t) = \nabla\delta(\gamma(x; t)), \quad \gamma(x; 0) = x.$$

We denote by $(\alpha(x), \beta(x))$, $\alpha(x) < 0 < \beta(x)$, the maximal existence interval for $\gamma(x; t)$.

LEMMA 2.4. *For any $x \in D \setminus D_\Phi$, we have:*

$$(2.12) \qquad \alpha(x) = -\delta(x), \quad \beta(x) = L - \delta(x);$$

moreover

$$(2.13) \qquad \lim_{t \to -\delta(x)+} \gamma(x;t) = x_0, \qquad \lim_{t \to [L-\delta(x)]+} \gamma(x;t) = x_L,$$

for some $x_0 \in \partial D, x_L \in D_\Phi$, and

$$(2.14) \qquad \lim_{t \to -\delta(x)+} \gamma'(x;t) = \xi_0, \qquad \lim_{t \to [L-\delta(x)]+} \gamma'(x;t) = \xi_L,$$

for some ξ_0, ξ_L such that $|\xi_0|, |\xi_L| = 1$.

The stream line $\gamma(x;t)$, $t \in [-\delta(x), 0]$ (resp. $t \in [0, L - \delta(x)]$) is the unique minimal geodesic joining x_0 to x (resp. x to x_L).

Finally, we have:

$$(2.15) \qquad \delta(x) = d(x, \partial D), \quad L - \delta(x) = d(x, D_\Phi), \quad x \in \overline{D}.$$

REMARK. Solutions of the eikonal equation in (2.10) have been studied by several authors and from different viewpoints (see [9], and the references therein). For instance, it is not difficult to prove (2.15) in a small one–sided neighborhood of a smooth hyprsurface (see e. g. [4], II.9). The above lemma gives a global version of this type of result adapted to the specific problem (2.10). Observe that, in this case, no smoothness is required on ∂D or D_Φ.

Proof. By (2.10) and (2.11), we have $\frac{d}{dt}\delta(\gamma(x;t)) = 1$, for any $t \in (\alpha(x), \beta(x))$, that is

$$(2.16) \qquad \delta(\gamma(x;t)) = \delta(x) + t, \quad t \in (\alpha(x), \beta(x)).$$

Since (2.11) is an autonomous system with bounded right–hand side, $\gamma(x;t)$ is defined for all (possible) t's, that is for all $-\delta(x) < t < L - \delta(x)$, since $0 < \delta < L$ in $D \setminus D_\Phi$. This implies (2.12).

The existence of the limits in (2.13) is a consequence of the fact that $\gamma(x;t)$ is uniformly Lipschitz continuous in t, by (2.10) and (2.11). Therefore, $\gamma(x;t)$ can be extended continuously to $[-\delta(x), L - \delta(x)]$, and also, by (2.16), we obtain that $x_0 \in \partial D$ and $x_L \in D_\Phi$, since $\delta = 0$ only on ∂D and $\delta = L$ only on D_Φ.

Let $r = d(x, \partial D)$ and let $x_0^* \in \partial D$ be such that $d(x, x_0^*) = r$. Since $\gamma(x;t) \to x_0$ as $t \to -\delta(x)+$, we obtain that $d(\gamma(x;t), \partial D) \to 0$ as $t \to -\delta(x)+$; thus,

$$\liminf_{t \to -\delta(x)+} d(x, \gamma(x;t)) \geq d(x, \partial D) = r.$$

Notice that (2.10) and (2.11) imply that $\gamma(x; \cdot)$ is parametrized by arclength, and hence $|t| \geq d(x, \gamma(x;t))$, for every $t \in (-\delta(x), L - \delta(x))$, thus, letting $t \to -\delta(x)$, we obtain $\delta(x) \geq r$.

Let $\tilde{\gamma}$ be the minimal geodesic joining x to x_0^*, parametrized by arclength as follows: $\tilde{\gamma} = \tilde{\gamma}(t)$, $-r \leq t \leq 0$, $\tilde{\gamma}(-r) = x_0^*$, $\tilde{\gamma}(0) = x$. We have:

$$r \leq \delta(x) = \int_{-r}^0 <\nabla\delta(\tilde{\gamma}(t), \tilde{\gamma}'(t)> \, dt \leq \int_{-r}^0 |\tilde{\gamma}'(t)| \, dt = r.$$

Consequently, $\delta(x) = r$ and also $\tilde{\gamma}'(t) = \nabla\delta(\tilde{\gamma}(t))$, for all $t \in (-r, 0)$. hence, the first formula in (2.15) holds for $x \in D \setminus D_\Phi$, and by lemma 2.3, for all $x \in \overline{D}$. The geodesic $\tilde{\gamma}(t)$ is uniquely determined and coincides with $\gamma(x; t)$ when $t \in [-\delta(x), 0]$. This also implies that $x_0^* = x_0$ and that $\gamma(x; t)$, with $t \in [-\delta(x), 0]$, is the minimal geodesic joining x_0 to x. Thus, $\gamma(x; t)$, $t \in [-\delta(x), 0]$, solves the second order differential equation for geodesics (see e.g. [5], Ch. 3), which has continuous coefficients, M being C^2-smooth. We deduce that $\gamma'(x; t)$ is uniformly Lipschitz continuous and (2.14) follows. Likewise, we obtain the latter formulas in (2.12), (2.13), (2.14), and (2.15).

LEMMA 2.5. *The set D_Φ is a deformation retract of D.*

Proof. It suffices to verify that the mapping $\tau : D \times [0, 1] \to D$, defined by

$$\tau(x, r) = \begin{cases} \gamma(x; r[L - \delta(x)]), & \text{for } (x, r) \in (D \setminus D_\Phi) \times [0, 1], \\ x, & \text{for } (x, r) \in D_\Phi \times [0, 1], \end{cases}$$

is continuous. In fact, since $\tau(\cdot, 0) = \mathrm{id}_D$, $\tau(D, r) = D_\Phi$, and $\tau(\cdot, r) = \mathrm{id}_{D_\Phi}$, for all $r \in [0, 1]$, we have that $\tau(\cdot, 1) : D \to D_\Phi$ is retraction homotopic to id_D.

REMARK. By the above lemma and [8], Ch. 1, we obtain:

(i) D_Φ is connected,

(ii) if D is contractible, then also D_Φ is contractible.

Proof of Corollary 2.2. By Theorem 2.1 and the above lemma and remark, if D is contractible, then D_Φ is contractible compact manifold without boundary. Classical results imply that $h = \dim D_\Phi = 0$ (see [10], theorem 4.1 and example p. 21, and [6], corollary 17.6.1).

Proof of Theorem 2.1. It is enough to prove that D_Φ is locally a submanifold of M, since D_Φ is connected by the above remark and is compact by (2.10).

Let $P \in D_\Phi$ and fix local coordinates $x^1, \ldots, x^m$ near P such that $P = (0, \ldots, 0)$ and $g_{ij}(P) = \delta_{ij}$, the Krönecker delta.

Observe that, by (2.15), and by rephrasing the arguments of lemma 2.5, if we choose $Q \in \partial D$ such that $d(P, Q) = d(P, \partial D)$, then there exists a unique minimal geodesic γ joining P to Q which is a stream line of δ (and also of ϕ, by (2.9)).

Let $T_P(M)$ be the tangent space to M at P and define:

$$\Xi(P) = \{\xi \in T_P(M) : |\xi| = 1,$$
$$\text{and } \exists \text{ a stream line } \gamma(t), t \in [0, L] \text{ of } \delta : \gamma(L) = P, \gamma'(L) = \xi\};$$

this set is not empty by (2.14). Let us continue each $\xi \in \Xi(P)$ as a constant vector field in a neighborhood V of Q, with respect to the chosen coordinates $x^1, \ldots, x^m$.

Pick $\xi \in \Xi(P)$ and differentiate (2.1) along ξ; in a neighborhood of P, we have:

$$2g^{ij}\phi_{x^i\xi}\phi_{x^j} + g_\xi^{ij}\phi_{x^i}\phi_{x^j} = f'(\phi)\phi_\xi.$$

We obtain the same formula (with ξ replaced by η) by differentiating (2.1) along any direction η orthogonal to ξ.

Restricting these formulas to a stream line $\gamma(t)$ of δ through P with $\gamma'(L) = \xi$, dividing by $|\nabla\phi(\gamma(t))|$, and letting $t \to L$, yield:

(2.17) $$\phi_{\xi\xi}(P) = \frac{1}{2}f'(\Phi) < 0, \quad \forall \xi \in \Xi(P),$$

(2.18) $$\phi_{\xi\eta}(P) = 0, \qquad \forall \xi \in \Xi(P), \forall \eta, <\eta, \xi> = 0.$$

Let $k = k(P)$ be the maximum number of linearly indipendent elements in $\Xi(P)$ and let $\xi_1, \ldots, \xi_k \in \Xi(P)$ be a choice of such elements. By (2.17) and (2.18), we get for $i, j = 1, \ldots, k$

$$(2.19) \qquad \phi_{\xi_i \xi_j}(P) = <\xi_i, \xi_j> \phi_{\xi_i \xi_i}(P) = \frac{1}{2} <\xi_i, \xi_j> f'(\Phi).$$

Therefore, the C^1 mapping $\psi : V \to \mathbb{R}^k$ defined by $\psi(x) = (\phi_{\xi_1}(x), \ldots, \phi_{\xi_k}(x))$ has rank k at P, by (2.19) and (2.18), and also $\psi(P) = (0, \ldots, 0)$ since $P \in D_\Phi$. By the implicit function theorem, we may find a neighborhood $U \subseteq V$ of P such that $N = \{x \in U : \psi(x) = (0, \ldots, 0)\}$ is a C^1 submanifold. Furthermore, the normal space to N at P is spanned by $\xi_1, \ldots, \xi_k$.

Obviously $D_\Phi \cap U \subseteq N$, since $|\nabla \phi| = 0$ on D_Φ. In order to conclude the proof, we need to show that $D_\Phi \cap U = N$, by possibly restricting U. The number h in the statement of the theorem will be given by $m - k$.

Let us denote by $B_\varepsilon(P)$ the geodetic ball in M centered at P and of radius $\varepsilon > 0$. Suppose by contradiction that for any $\varepsilon > 0$ there exists $P_\varepsilon \in (N \setminus D_\Phi) \cap B_\varepsilon(P)$. Let $P_\varepsilon^* \in D_\Phi$ be such that $d(P_\varepsilon, P_\varepsilon^*) \leq d(P_\varepsilon, P) < \varepsilon$, and hence $d(P, P_\varepsilon^*) < 2\varepsilon$.

Let γ_ε be the stream line of δ through P_ε; by lemma 2.5 we have that its endpoint on D_Φ is P_ε^*. Let us parametrize γ_ε by arclength in such a way that $\gamma_\varepsilon(L) = P_\varepsilon^*$, and let $\xi_\varepsilon = \gamma_\varepsilon'(L)$; ξ_ε is a unit vector in $T_{P_\varepsilon^*}(M)$. By possibly passing to subsequences, P_ε and $P_\varepsilon^* \to P$, and $\xi_\varepsilon \to \xi$ as $\varepsilon \to 0$, where $\xi \in T_P(M)$ is a unit vector. By the continuous dependence on the Cauchy data, for all $t \in [0, L]$, $\gamma_\varepsilon(t)$ converges to $\gamma(t)$, where γ is a geodesic such that $\gamma(L) = P$ and $\gamma'(L) = \xi$. Thus, γ is a stream line of δ, since all γ_ε's are stream lines of δ. In particular, we deduce $\xi \in \Xi(P)$.

We will show now that $\xi \in T_P(N)$, contradicting the fact that $\Xi(P)$ is contained in the normal space to N at P. We choose a local coordinate system near P such that N is represented by the equations $x^{h+1} = \cdots = x^m = 0$. In this system $P_\varepsilon = (x_\varepsilon^1, \ldots, x_\varepsilon^h, 0, \ldots, 0)$, and analogously for P_ε^*. Since by Taylor's formula, we have:

$$P_\varepsilon^* = P_\varepsilon + d(P_\varepsilon^*, P_\varepsilon)\xi_\varepsilon + o(d(P_\varepsilon^*, P_\varepsilon)), \quad \text{as } \varepsilon \to 0,$$

we obtain that $\xi_\varepsilon^i \to 0$ as $\varepsilon \to 0$, $\forall i = h+1, \ldots, m$.

This means that $\xi^i = 0, i = h+1, \ldots, m$, that is $\xi \in T_P(N)$.

Finally (2.6) and (2.7) follow from (2.10) and (2.15).

3 Overdetermined Stekloff eigenfunctions

In the sequel, we will denote by $x = (x', x^n)$ a point of $\mathbb{R}^n$, where $x' \in \mathbb{R}^{n-1}$ has coordinates $(x^1, \ldots, x^{n-1})$; $\nu = (\nu_1, \ldots, \nu_n)$ will indicate the exterior normal unit vector to $\partial\Omega$.

We begin with the following result, which has its own interest.

PROPOSITION 3.1 *Let $u \in C^2(\Omega) \cap C^1(\overline{\Omega})$ satisfy (1.1) and (1.3). If*

$$(3.1) \qquad \int_{\partial\Omega} [u - x \cdot Du] \frac{\partial u}{\partial \nu} \, d\sigma = 0,$$

then u is linear.

Proof. By Rellich's identity (see [15]),

$$\int_{\partial\Omega} \{2(x \cdot Du)\frac{\partial u}{\partial \nu} - |Du|^2(x \cdot \nu)\} \, d\sigma = \int_\Omega \{2(x \cdot Du)\Delta u + (2 - N)|Du|^2\} \, dx.$$

By (1.1) and (1.3), we obtain via the divergence theorem:

$$\int_{\partial\Omega} \{2(x \cdot Du)\frac{\partial u}{\partial \nu} - (x \cdot \nu)\}\, d\sigma = 2\int_{\partial\Omega} u\,\frac{\partial u}{\partial \nu}\, d\sigma - N\int_{\Omega} |Du|^2\, dx.$$

Thus, (3.1) yields:

$$(3.2) \qquad N\int_{\Omega} |Du|^2\, dx = \int_{\partial\Omega} x \cdot \nu\, d\sigma = N|\Omega|,$$

where $|\Omega|$ is the Lebesgue measure of Ω.

Since $|Du|^2$ is subharmonic in Ω, by (1.3), we have $|Du| \leq 1$ in Ω, so that (2.2) implies $|Du| \equiv 1$ in Ω. Therefore, $2\sum_{i,j=1}^{n} u_{ij}^2 = \Delta|Du|^2 \equiv 0$ in Ω, and hence u is linear in Ω.

Theorem 1.1 will be a consequence of the following more general result.

THEOREM 3.2 *Let $\Omega \subset \mathbb{R}^n$ be a bounded domain with boundary $\partial\Omega \in C^2$. Suppose that u is a linear solution of (1.1)–(1.3).*

Then, up to a rigid change of coordinates, for some $h = 0, 1, \ldots, n-2$, there exists a C^1 h-dimensional submanifold $D_\Phi \subset \{x \in \mathbb{R}^n : x^n = 0\}$, such that

$$(3.3) \qquad \Omega = \{x \in \mathbb{R}^n : \operatorname{dist}(x, D_\Phi) < \frac{1}{p}\}.$$

Furthermore, if Ω is contractible, then Ω is a ball.

Proof. Up to a rigid change of coordinates, we may assume that $u(x) = x^n$. By (1.2), we have:

$$(3.4) \qquad \nu_n = px^n \quad \text{on } \partial\Omega.$$

If we consider $\Gamma = \{x \in \partial\Omega : x^n > 0\}$, we have that $\nu_n > 0$ on Γ, hence Γ is the graph of a function $\phi = \phi(x')$, where x' ranges over $D = \{x \in \Omega : x^n = 0\}$. The vector ν is then given by $\{1 + |\nabla\phi|^2\}^{-1/2}(-\nabla\phi, 1)$ on D, where ∇ denotes the gradient in the variable $x' \in D$.

Therefore, (3.4) yields $\{1 + |\nabla\phi|^2\}^{-1/2} = px^n = p\phi$, that is

$$(3.5) \qquad \begin{aligned} |\nabla\phi|^2 &= \frac{1}{p^2\phi^2} - 1 \quad \text{in } D, \\ \phi &= 0 \quad \text{on } \partial D. \end{aligned}$$

Since $\partial\Omega \in C^2$, we also have that $\phi \in C(\overline{D}) \cap C^2(D)$; hence, by setting $m = n - 1$, theorem 3.1 applies to the (flat) domain D. Note that $D_\Phi = \{x' \in D : \phi(x') = \frac{1}{p}\}$ and that

$$(3.6) \qquad \phi(x') = \sqrt{\frac{1}{p^2} - \operatorname{dist}(x', D_\Phi)^2}.$$

Therefore,

$$\{x \in \Omega : x^n > 0\} = \{x \in \mathbb{R}^n : x^n > 0, \operatorname{dist}(x, D_\Phi) < \frac{1}{p}\}$$

and, by the same argument,

$$\{x \in \Omega : x^n < 0\} = \{x \in \mathbb{R}^n : x^n < 0, \operatorname{dist}(x, D_\Phi) < \frac{1}{p}\}.$$

Consequently, we obtain (3.3).

Finally, one easily sees that D is a deformation retract of Ω, and hence D is contractible, if Ω is so. Corollary 2.2 implies that D is an $(n-1)$–dimensional ball, that is, by (3.6), Ω is an n–dimensional ball

Proof of Theorem 1.1. By (1.1)–(1.3), (1.6), and proposition 3.1, we have that u is linear, and hence Theorem 3.2 applies.

References

[1] G. Alessandrini and R. Magnanini, *Symmetry and non–symmetry for the overdetermined Stekloff eigenvalue problem*, Journ. Appl. Math. Phys.(ZAMP), 45 (1994), pp. 44–52.

[2] E. Cartan, *Familles de surfaces isoparamétriques dans les espaces à courbure constante*, Ann. Mat., 17 (1938), pp. 177–191.

[3] ———, *Sur des familles remarquables d'hypersurfaces isoparamétriques dans les espaces sphériques*, Math. Z., 45 (1939), pp. 335–367.

[4] R. Courant and D. Hilbert, *Methods of Mathematical Physics*, Wiley Interscience, New York, 1962.

[5] M. DoCarmo, *Differential Geometry of Curves and Surfaces*, Prentice Hall, Englewood Cliffs, NJ, 1976.

[6] B. A. Dubrovin, A. T. Fomenko, and S. P. Novikov, *Modern Geometry-Methods and Applications. Part II. The Geometry and Topology of Manifolds*, Springer Verlag, New York, 1984.

[7] D. Ferus, H. Karcher, and H. F. Munzner, *Cliffordalgebren und neue isoparametrische Hyperflachen*, Math. Z., 177 (1981), pp. 479–502.

[8] S. T. Hu, *Theory of Retracts*, Wayne State University Press, Detroit, 1964.

[9] P. L. Lions,*Generalized Solutions of Hamilton–Jacobi Equations*, Pitman, New York, 1982.

[10] W. S. Massey, *Singular Homology Theory*, Springer-Verlag, New York, 1980.

[11] H. F. Munzner, *Isoparametrische Hyperflachen in Spharen*, Math. Ann., 251 (1980), pp. 57–71.

[12] K. Nomizu, *Some results in E. Cartan's theory of isoparametric families of hypersurfaces* , Bull. AMS 79 (1973), pp. 1184–1188.

[13] ———, *Elie Cartan's work on isoparametric families of hypersurfaces* , Proc. Symp. Pure Math. 27 (1975), pp. 191–200.

[14] L. E. Payne and G. A. Philippin, *Some overdetermined boundary value problems for harmonic functions*, Journ. Appl. Math. Phys.(ZAMP), 42 (1991), pp. 864–873.

[15] F. Rellich, *Darstellung der eigenwerte $\Delta u + \lambda u$ durch ein randintegral*, Math. Z., 46 (1940), pp. 635–646.

[16] M. W. Stekloff, *Sur les problèmes fondamentaux en physique mathématique*, Ann. Sci. Ecole Norm. Sup., 19 (1902), pp. 455–490.

[17] Q. M. Wang, *Isoparametric functions on Riemannian manifolds. I*, Math. Ann., 277 (1987), pp. 639–646.

[18] ———, *Isoparametric maps of Riemanniann manifolds and their applications*, Advances in science of China. Mathematics, 2 (1986),pp. 79–103.

OPTIMAL ASYMPTOTICS FOR SOLUTIONS TO THE INITIAL VALUE PROBLEM FOR THE POROUS MEDIUM EQUATION

S. B. ANGENENT AND D. G. ARONSON

For Ivar Stakgold on his 70th birthday.

ABSTRACT. The solution to the initial value problem for the equation of heat conduction in $\mathbf{R} \times \mathbf{R}^+$ behaves asymptotically as $t \to \infty$ like a multiple of the fundamental solution on the scale $\sqrt{t}$. An analogous result holds for the porous medium equation with the Barenblatt solution playing the rôle of the fundamental solution. In this paper we use a rigorous asymptotic representation of solutions to the porous medium equation to obtain the optimal approximation, both in one space dimension and, for radially symmetric flows, in any space dimension.

1. INTRODUCTION

Many partial differential equations and systems of equations possess self-similar solutions [11]. A classical example is the fundamental solution

$$(1.1) \qquad g(x,t) = \frac{1}{(4\pi t)^{\frac{d}{2}}} e^{\frac{-|x|^2}{4t}},$$

of the equation of heat conduction

$$(1.2) \qquad \frac{\partial u}{\partial t} = \Delta u$$

in $\mathbf{R}^d \times \mathbf{R}^+$ for $d \geq 1$. Note that g actually depends on t and the self-similar variable $\frac{|x|}{2\sqrt{t}}$ which reflects a group invariance of the equation (1.2). Self-similar solutions are important since they provide explicit or, at least, semi-explicit solutions to complicated problems, but even more so since they often describe the detailed behavior of some class of solutions in some crucial limit or at some crucial transition. For example, let $u(x,t)$ denote the solution to the initial value problem for (1.2) in $\mathbf{R}^d \times \mathbf{R}^+$ with $u(\cdot,0) = u_0$, where u_0 is a given compact support function on $\mathbf{R}^d$. Define

$$M = \int_{\mathbf{R}^d} u_0(x)\,dx.$$

1991 *Mathematics Subject Classification.* 35K65, 76S05, 35B40.
Key words and phrases. Asymptotics, porous medium, self-similarity.
Supported, in part, by NSF PYI Grant DMS-9058492 (S.B.A.) and by NSF Grant DMS-9503392 (D.G.A.).

It is well known that, on the appropriate scale, $Mg(x,t)$ approximates $u(x,t)$ as $t \to \infty$. Indeed, the appropriate scale is $t^{\frac{d}{2}}$ and the error in the approximation is, in general, of order t^{-1}, i.e.,

$$(1.3) \qquad t^{\frac{d}{2}} |u(x,t) - Mg(x,t)| = O(t^{-1}) \text{ as } t \to \infty.$$

If $d = 1$ the error in (1.3) can be reduced to order $t^{-\frac{3}{2}}$ by judicious placement of the singularity in the fundamental solution [9]. Specifically, there exist numbers x_0 and θ such that

$$(1.4) \qquad t^{\frac{1}{2}} |u(x,t) - Mg(x - x_0, t + \theta)| = O(t^{-\frac{3}{2}}) \text{ as } t \to \infty.$$

Moreover, if u_0 is an even function, then $x_0 = 0$ and the error term in (1.4) is actually $O(t^{-2})$. This sharper result continues to hold for radially symmetric solutions in all space dimensions d. Thus if $u(x,0) = u_0(|x|)$ then there exists a number θ such that

$$(1.5) \qquad t^{\frac{d}{2}} |u(x,t) - Mg(x, t + \theta)| = O(t^{-2}) \text{ as } t \to \infty.$$

We will discuss these results briefly in Section 2 below.

Our main concern in this paper is with the equation

$$(1.6) \qquad \frac{\partial u}{\partial t} = \Delta(u^m)$$

for constant $m \geq 1$. For $m = 1$, (1.6) is just the equation of heat conduction (1.2). For $m > 1$ it is known as the porous medium equation since it can be viewed as describing the evolution of the density $u(x,t)$ of an ideal gas flowing isentropically in a homogeneous porous medium [2]. Note that for $m > 1$, (1.6) is a nonlinear degenerate parabolic equation. In contrast to the infinite speed of propagation of disturbances from rest in classical heat conduction, the porous medium equation possesses a finite speed of propagation. In particular, if the initial datum has compact support, the solution to the corresponding initial value problem will also have compact support for all positive times.

The fundamental solution (1.1) of the heat conduction equation (1.2) is the solution for positive time whose initial datum is a Dirac measure of unit mass concentrated at the origin. The corresponding solution to the porous medium equation (1.6) is known generally as the Barenblatt solution ([10],[4]) and is easiest to describe in terms of pressure v rather than density u. These are related, via the ideal gas law and a convenient scaling, by

$$v = mu^{m-1}.$$

The evolution of v is governed by the porous medium pressure equation

$$(1.7) \qquad \frac{\partial v}{\partial t} = v \, \Delta v + \gamma |\nabla v|^2 \,,$$

where $\gamma = \frac{1}{m-1}$. The Barenblatt pressure $V(r, t; M)$ in $\mathbf{R}^d \times \mathbf{R}^+$ corresponding to an initial Dirac measure of mass M concentrated at the origin is the (generalized) solution to (1.7) which satisfies

$$m^\gamma M = \int_{\mathbf{R}^d} V^\gamma(|x|, t; M) \, dx$$

for all $t \in \mathbf{R}^+$. It can be written in the form

$$(1.8) \qquad V(r,t;M) = \frac{K}{t^\alpha}(1 - \varsigma^2)_+,$$

where $(\cdot)_+ = \max(0, \cdot)$,

$$\varsigma = \frac{r}{Rt^\beta}, \quad \alpha = \frac{d(m-1)}{d(m-1)+2}, \quad \text{and} \quad \beta = \frac{1}{d(m-1)+2}.$$

The constants K and R are related by

$$(1.9) \qquad m^\gamma M = K^\gamma R^d \omega_d \int_0^1 (1 - \varsigma^2)^\gamma \varsigma^{d-1} d\varsigma$$

and

$$R = \sqrt{2K(d + 2\gamma)},$$

where ω_d denotes the volume of the d-dimensional unit ball. Observe that for each $t > 0$ the support of the Barenblatt solution is the interval $0 \le \varsigma \le 1$, i.e.,

$$0 \le r \le Rt^\beta.$$

The curve $r = Rt^\beta$ is called the interface.

Let $v(x, t)$ denote the solution to the initial value problem for (1.7) with $v(\cdot, 0) = v_0$, where v_0 is a compact support function on $\mathbf{R}^d$ and let

$$M = \int_{\mathbf{R}^d} v_0^{\frac{1}{m-1}}(x) dx.$$

Then the Barenblatt solution $V(|x|, t; M)$ approximates $v(x, t)$ as $t \to \infty$ and the analogue of (1.3) is

$$(1.10) \qquad t^\alpha |v(x, t) - V(|x|, t; M)| \to 0 \text{ as } t \to \infty.$$

This result was first proved in the special case $d = 1$ by Kamenomostkaya (Kamin) [6] and subsequently for all d by Friedman and Kamin [5]. Two very elegant proofs can be found in [7]. Vazquez [8] has shown in the case $d = 1$, that by centering the Barenblatt solution at the center of mass

$$x_0 = \frac{1}{M} \int_{-\infty}^\infty x v_0^{\frac{1}{m-1}}(x) dx$$

of the initial distribution one can improve (1.10). Specifically, he shows that, in general,

$$(1.11) \qquad t^{\frac{m}{m+1}} |v(x, t) - V(x - x_0, t; M)| \to 0 \text{ as } t \to \infty,$$

and if $v_0(x) = v_0(-x)$ then

$$(1.12) \qquad |v(x, t) - V(x, t; M)| = O\left(t^{-\frac{2m}{m+1}}\right) \text{ as } t \to \infty.$$

The best convergence rate for $d = 1$ was obtained by Zel'dovich and Barenblatt [9] by means of a formal calculation. Their result is that a number θ can be chosen so that

$$(1.13) \qquad |v(x, t) - V(x - x_0, t + \theta; M)| = o\left(t^{-\frac{2m}{m+1}}\right) \text{ as } t \to \infty.$$

In Section 3 of this paper we use a rigorous asymptotic expansion for the porous medium pressure derived in [1] to obtain a more precise version of (1.13). In particular, we show that, for an appropriately chosen θ,

(1.14)

$$t^{\frac{m-1}{m+1}} |v(x,t) - V(x - x_0, t - \theta, M)| = \begin{cases} O\left(t^{-2}\right) \text{ if } m > 2 \\ O\left(t^{-\frac{3m}{m+1}}\right) \text{ if } m < 2 \\ O\left(t^{-2} \log t\right) \text{ if } m = 2 \end{cases} \quad \text{as } t \to \infty.$$

Finally, in Section 4 we discuss the extension to radially symmetric flows of the expansion derived in [1].

2. LINEAR HEAT CONDUCTION

In this Section we sketch the derivations of (1.3) and (1.4). We first consider the case $d = 1$. Let $u(x,t)$ denote the solution to (1.2) with $d = 1$ and initial values $u(\cdot, t) = u_0$ with compact support. Define

$$M_j = \int_{-\infty}^{\infty} x^j u_0(x) dx \text{ for } j = 0, 1, 2, \ldots.$$

Then

$$u(x,t) = \int_{-\infty}^{\infty} g(x - y, t) u_0(y) dy,$$

where g is the fundamental solution given by (1.1) with $d = 1$. Expand $g(x - y, t)$ as a power series in y to obtain

(2.1)

$$u(x,t) = \frac{e^{-\xi^2}}{2(\pi t)^{\frac{1}{2}}} \left\{ M_0 + \frac{\xi}{t^{\frac{1}{2}}} M_1 + \frac{2\xi^2 - 1}{4t} M_2 + \frac{(2\xi^3 - 3\xi)}{12t^{\frac{3}{2}}} M_3 + O\left(t^{-2}\right) \right\},$$

where

$$\xi = \frac{x}{2\sqrt{t}}.$$

On the other hand

(2.2)

$$g(x + \varsigma, t + \theta) = g(x,t) \left\{ 1 - \frac{\xi\varsigma}{t^{\frac{1}{2}}} + \frac{(2\xi^2 - 1)\left(\theta + \frac{\varsigma^2}{2}\right)}{2t} - \frac{(2\xi^3 - 3\xi)\left(\theta\varsigma + \frac{\varsigma^3}{6}\right)}{2t^{\frac{3}{2}}} + O\left(t^{-2}\right) \right\}$$

Thus $u(x,t)$ and $M_0 g(x + \varsigma, t + \theta)$ will agree up to terms of $O\left(t^{-\frac{3}{2}}\right)$ provided that

(2.3)

$$\varsigma = -\frac{M_1}{M_0} \quad \text{and} \quad \theta = \frac{1}{2M_0}\left(M_2 - \frac{M_1^2}{M_0}\right).$$

Therefore (1.4) holds. If u_0 is an even function then $M_1 = M_3 = 0$ so that $\varsigma = 0$. Hence there are no $O(t^{-\frac{3}{2}})$ terms in either (2.1) or (2.2), and (1.5) holds for θ given by (2.3).

We now consider radially symmetric heat conduction in $\mathbf{R}^d$. Let $u(x.t)$ denote the solution to (1.2) with initial values $u(x,0) = u_0(|x|)$, where u_0 has compact support. Then u is given by the formula

$$(2.4) \qquad u(x,t) = \int_{\mathbf{R}^d} g(x-y,t) u_0(|y|) dy.$$

Rotate the coordinates so that the positive y_1-axis is parallel to the vector x, and introduce "spherical" coordinates

$$
\begin{aligned}
y_1 &= s\cos\theta_1 \\
y_2 &= s\sin\theta_1\cos\theta_2 \\
&\;\;\vdots \\
y_{d-1} &= s\sin\theta_1\sin\theta_2\cdots\sin\theta_{d-2}\cos\theta_{d-1} \\
y_d &= s\sin\theta_1\sin\theta_2\cdots\sin\theta_{d-2}\sin\theta_{d-1},
\end{aligned}
$$

where $s = |y|$. Then (2.4) becomes

$$u(x,t) = g(x,t)\int_0^\infty s^{d-1} u_0(s) e^{-\frac{s^2}{4t}} J\, ds,$$

where

$$J = \int_0^{2\pi} d\theta_{d-1} \int_0^\pi \sin\theta_{d-2} d\theta_{d-2} \cdots \int_0^\pi \sin^{d-3}\theta_2 d\theta_2 \int_0^\pi e^{\frac{rs}{2t}\cos\theta_1} \sin^{d-2}\theta_1 d\theta_1$$

and $r = |x|$. Using the power series for the exponential we find that

$$\int_0^\pi e^{\frac{rs}{2t}\cos\theta_1} \sin^{d-2}\theta_1 d\theta_1 = \frac{\sqrt{\pi}\Gamma(\frac{d-1}{2})}{\Gamma(\frac{d}{2})}\left(1 + \frac{s^2\xi^2}{2dt}\right) + O(t^{-2}),$$

where

$$\xi = \frac{r}{2\sqrt{t}}.$$

Thus

$$J = \omega_d\left(1 + \frac{s^2\xi^2}{2dt}\right) + O(t^{-2}),$$

where ω_d denotes the volume of the unit ball in $\mathbf{R}^d$, and

$$u(x,t) = g(x,t)\int_0^\infty s^{d-1}u_0(s)e^{-\frac{s^2}{4t}}\left\{\omega_d\left(1 + \frac{s^2\xi^2}{2dt}\right) + O(t^{-2})\right\} ds.$$

Finally, replacing the remaining exponential by its power series, we obtain

$$u(x,t) = g(x,t)\left\{M + \frac{\omega_d}{2t}\left(\frac{\xi^2}{d} - \frac{1}{2}\right)\int_0^\infty s^{d+1}u_0(s)ds + O(t^{-2})\right\}.$$

Since

$$g(x,t+\theta) = g(x,t)\left\{1 + \frac{\theta d}{t}\left(\frac{\xi^2}{d} - \frac{1}{2}\right) + O(t^{-2})\right\}$$

it is clear that $u(x,t)$ and $Mg(x,t+\theta)$ agree up to $O(t^{-2})$ provided that

$$\theta = \frac{\omega_d}{2Md}\int_0^\infty s^{d+1}u_0(s)ds.$$

For $d > 1$ without assuming radial symmetry one can approximate $t^{\frac{d}{2}}u(x,t)$ up to order t^{-1} by $t^{\frac{d}{2}}Mg(x - x_0, t)$, where x_0 is the center of mass of u_0. Moreover, this convergence rate cannot be improved by displacing g in time.

3. POROUS MEDIUM FLOW: $d = 1$

Let $v(x,t)$ denote the solution to the initial value problem

$$(3.1) \qquad \frac{\partial v}{\partial t} = v\frac{\partial^2 v}{\partial x^2} + \gamma\left(\frac{\partial v}{\partial x}\right)^2 \text{ in } \mathbf{R} \times \mathbf{R}^+$$
$$v(\cdot, 1) = v_0 \text{ in } \mathbf{R},$$

where v_0 has compact support, and we have chosen $t = 1$ as the initial time for later convenience. We assume that v corresponds to a density with total mass M, i.e.,

$$M = m^{-\gamma}\int_{-\infty}^{\infty}v^{\gamma}(x,t)dx = m^{-\gamma}\int_{-\infty}^{\infty}v_0^{\gamma}(x)dx.$$

We seek conditions which guarantee that v is close to the Barenblatt solution (1.8) and can be written in the form

$$(3.2) \qquad v(x,t) = W(\xi, t)$$

with

$$(3.3) \qquad \xi \equiv \frac{x - \mu(t)}{\rho(t)},$$

where the functions μ and ρ are to be determined.

Assume that the initial datum $v(x,1) = v_0(x)$ satisfies

$$v_0 = 0 \text{ on } (-\infty, a_1] \cup [a_2, \infty), \quad v_0 > 0 \text{ on } (a_1, a_2),$$

$v_0(a_j) = 0$, and $v_0'(a_j) \neq 0$ for $j = 1, 2$. Here $-\infty < a_1 < a_2 < \infty$. Let $\mu_0 = \frac{1}{2}(a_1 + a_2)$ and $\rho_0 = \frac{1}{2}(a_2 - a_1)$. For $\xi = \frac{x - \mu_0}{\rho_0}$ define W_0 by $v_0(x) \equiv W_0(\xi)$. Note that $W_0 > 0$ on $(-1,1)$, $W_0(\pm 1) = 0$, and $W_0'(\pm 1) > 0$. It is shown in [1] that if W_0 is $C^{2+\alpha}$ close to $1 - \xi^2$ then there exist smooth functions $W(\xi, t)$, $\mu(t)$, and $\rho(t)$ with initial values $W(\cdot, 1) = W_0$, $\mu(1) = \mu_0$, and $\rho(1) = \rho_0$ such that $W(\pm 1, \cdot) = 0$ and $W_\xi(\pm 1, \cdot) \neq 0$ on $[1, \infty)$, and the solution v to (3.1) is given by (3.2) for all $t \geq 1$. In addition, it is proved in [1] that

$$(3.4) \qquad W(\xi, t) \sim t^{-\frac{m-1}{m+1}}\left\{K\left(1 - \xi^2\right) + \sum_{\alpha \in \mathcal{A}}p_\alpha(\log t, \xi)t^{\frac{\alpha(m-1)}{m+1}}\right\},$$

where K is the constant in (1.8), the p_α are polynomials in $\log t$ and ξ, and

$$\mathcal{A} = \left\{\alpha : \alpha = -(\lambda_{n_1} + \cdots + \lambda_{n_k}) \text{ for } n_1, ..., n_k \geq 2 \text{ and } \lambda_n = \frac{n}{2}\left(n + \frac{2}{m-1} - 1\right)\right\}.$$

Actually p_α will depend on $\log t$ only when a resonance occurs, i.e., only when one of the λ_k can be written as

$$(3.5) \qquad \lambda_k = \lambda_{n_1} + \cdots + \lambda_{n_j}.$$

In particular, the p_α will all be independent of t if and only if γ is irrational. The condition that $W(\xi, 1)$ be $C^{2+\alpha}$ close to $1 - \xi^2$ is no essential restriction, since it is proved in [3] that this occurs in finite time for any solution of (1.7) whose initial datum has compact support.

Asymptotic representations for μ and ρ can be derived from (3.4) and the invariants M and

$$M x_0 = m^{-\gamma} \int_{-\infty}^{\infty} x v^\gamma(x,t)dx = m^{-\gamma} \int_{-\infty}^{\infty} x v_0^\gamma(x)dx.$$

In particular,

(3.6)
$$\rho(t) = \frac{m^\gamma M}{\int_{-1}^1 W^\gamma(\xi,t)d\xi},$$

and

(3.7)
$$\mu(t) = x_0 - \frac{\rho^2(t)\int_{-1}^1 \xi W^\gamma(\xi,t)d\xi}{m^\gamma M}.$$

The analysis in [1] shows that there exists a constant c such that

$$p_{-\lambda_2}(\log t, \xi) = c(1-\xi^2).$$

Thus we can rewrite (3.4) as

(3.8)

$$W(\xi,t) \sim t^{-\frac{m-1}{m+1}} \left\{ \left(K + \frac{c}{t}\right)(1-\xi^2) + \sum_{\substack{\alpha \in \mathcal{A} \\ \alpha \neq -\lambda_2}} p_\alpha(\log t, \xi) t^{\frac{\alpha(m-1)}{m+1}} \right\}.$$

The largest number $\beta \in \mathcal{A}\backslash\{-\lambda_2\}$ is either $-2\lambda_2$ or $-\lambda_3$ depending upon the value of m. It is shown in [1] that

$$p_\beta(\log t, \xi) = \begin{cases} f(\xi) \text{ if } \beta = -2\lambda_2, \text{ i.e., if } m > 2 \\ \xi(1-\xi^2) \text{ if } \beta = -\lambda_3, \text{ i.e., if } m > 2 \\ \{f(\xi) + g(\xi)\log t\} \text{ if } \beta = -2\lambda_2 = -\lambda_3, \text{ i.e., if } m = 2 \end{cases},$$

where f and g are polynomials which can be computed. It follows that (3.8) can be written in the form

(3.9)
$$W(\xi,t) \sim t^{-\frac{m-1}{m+1}} \left\{ \left(K + \frac{c}{t}\right)(1-\xi^2) + \mathcal{E} \right\},$$

where

(3.10)
$$\mathcal{E} = \begin{cases} O(t^{-2}) \text{ if } m > 2 \\ O(t^{-\frac{3m}{m+1}}) \text{ if } m < 2 \\ O(t^{-2}\log t) \text{ if } m = 2 \end{cases}.$$

In general, we will use the symbol $\mathcal{E}$ to denote any function which satisfies the estimate (3.10).

If we write (3.9) with t replaced by $t + \theta$ and then expand in powers of θ the result is

$$W(\xi,t+\theta) \sim t^{-\frac{m-1}{m+1}} \left\{ \left(K + \frac{c}{t} - \frac{K\theta(m-1)}{(m+1)t}\right)(1-\xi^2) + \mathcal{E} \right\},$$

where the error terms which are of $O(t^{-2})$ have been included in $\mathcal{E}$. Thus if

$$\theta = \frac{c(m+1)}{K(m-1)},$$

then

(3.11) $$W(\xi, t + \theta) \sim t^{-\frac{m-1}{m+1}} \left\{ K \left(1 - \xi^2 \right) + \mathcal{E} \right\}.$$

The function ξ in (3.11) is given by (??) with t replaced by $t + \theta$.

If we use (3.11) in (3.6) we obtain

$$\rho(t + \theta) = t^{\frac{1}{m+1}} (R + \mathcal{E}).$$

where R is given by (1.9). Similarly, using (3.11) in (3.7) together with the fact that

$$\int_{-1}^{1} \xi(1 - \xi^2)^\gamma d\xi = 0,$$

we obtain

$$\mu(t + \theta) = x_0 + t^{\frac{1}{m+1}} \mathcal{E},$$

Hence it follows from (??) that

$$\xi = \varsigma + \mathcal{E},$$

where

$$\varsigma = \frac{x - x_0}{R t^{\frac{1}{m+1}}}.$$

Therefore we can rewrite (3.11) in its final form

$$v(x, t + \theta) \sim t^{-\frac{m-1}{m+1}} \left\{ K \left(1 - \varsigma^2 \right) + \mathcal{E} \right\}$$

which is equivalent to (1.14). The interface for v is given by

$$x = x_0 \pm t^{\frac{1}{m+1}} R + t^{\frac{1}{m+1}} \mathcal{E}.$$

4. Porous Medium Flow: $d > 1$

To extend the results of the previous Section to flows in $\mathbf{R}^d$ for $d > 1$ we restrict our attention to radially symmetric flows. Let $v(r, t)$ denote the solution to the initial value problem

(4.1)
$$\frac{\partial v}{\partial t} = v \left(\frac{\partial^2 v}{\partial r^2} + \frac{d-1}{r} \frac{\partial v}{\partial r} \right) + \gamma \left(\frac{\partial v}{\partial r} \right)^2 \quad \text{in } \mathbf{R}^+ \times \mathbf{R}^+$$
$$v(\cdot, 1) = v_0 \quad \text{in } \mathbf{R}^+,$$

where v_0 has compact support. We assume that the density corresponding to v has total mass M, i.e.,

$$m^\gamma M = \int_{\mathbf{R}^d} v^\gamma(|x|, t) dx = \omega_d \int_0^\infty v^\gamma(r, t) r^{d-1} dr = \omega_d \int_0^\infty v_0^\gamma(r) r^{d-1} dr.$$

We seek conditions which guarantee that v is close to the Barenblatt solution (1.8) corresponding to mass M, and can be written in the form

(4.2)
$$v(r, t) = W \left(\frac{r}{R(t)}, t \right),$$

where the function $R(t)$ will be determined by the moving boundary condition

(4.3)
$$W(1, t) = 0 \text{ for all } t \geq 0.$$

Let

$$\rho = \rho(t) = \frac{r}{R(t)}.$$

If we substitute W in equation (4.1) we find that W must satisfy

(4.4) $$W_t = \frac{W}{R^2}\left(W_{\rho\rho} + \frac{d-1}{\rho}W_\rho\right) + \frac{\rho R'}{R}W_\rho + \frac{\gamma}{R^2}W_\rho^2,$$

where $' = \frac{d}{dt}$. From the moving boundary condition (4.3) we conclude that

$$R' = -\frac{\gamma}{R}W_\rho(1,t).$$

Define

$$\mathcal{P}(f)(\rho) \equiv f(\rho) - \rho f(1).$$

Then (4.4) can be written in the form

$$W_t = R^{-2}\left(WW_{\rho\rho} + \frac{d-1}{\rho}WW_\rho + \gamma\mathcal{P}(W_\rho)W_\rho\right).$$

To eliminate R we use the invariant M to obtain

$$m^\gamma M = \int_{\mathbf{R}^d} W^\gamma\left(\frac{r}{R(t)},t\right)dx = \omega_d R^d(t)\int_0^1 W^\gamma(\rho,t)\,\rho^{d-1}d\rho.$$

Thus

$$R^d(t) = \frac{m^\gamma M}{\omega_d \int_0^1 W^\gamma(\rho,t)\,\rho^{d-1}d\rho}$$

and W satisfies the (nonlocal) equation

(4.5)

$$W_t = \left(\frac{m^\gamma M}{\omega_d \int_0^1 W^\gamma(\rho,t)\,\rho^{d-1}d\rho}\right)^{-\frac{d}{2}}\left(WW_{\rho\rho} + \frac{d-1}{\rho}WW_\rho + \gamma\mathcal{P}(W_\rho)W_\rho\right).$$

By a slight extension of the argument in [1] it can be shown that if $W(\rho,1)$ is $C^{2+\alpha}$ close to $1 - \rho^2$ then there exists a function $R(t)$ such that (4.2) holds. Moreover

$$W(\rho,t) \sim t^{-\alpha}\left\{K(1-\rho^2) + \sum_{\sigma\in\mathcal{S}}p_\sigma(\log t,\rho)t^{\alpha\sigma}\right\},$$

where K is the constant in (1.8),

$$\alpha = \frac{d(m-1)}{d(m-1)+2},$$

the p_σ are polynomials in $\log t$, and

$$\mathcal{S} = \left\{\sigma : \sigma = -(\lambda_{n_1} + \cdots + \lambda_{n_k}) \text{ for } n_j \geq 1 \text{ and } \lambda_n = \frac{n}{d}(2n + 2\gamma + d - 2)\right\}.$$

As in the one-dimensional case, the p_σ depend on $\log t$ only when a resonance occurs, i.e., only when (3.5) holds. Otherwise the p_σ are the eigenfunctions of the linearization of (4.5) about $K(1 - \rho^2)$, i.e., nontrivial solutions to $Lp = -\sigma p$ for $\sigma \in \mathcal{S}$, where

$$Lz \equiv \frac{1}{2d}(1 - \rho^2)\left(z_{\rho\rho} + \frac{d-1}{\rho}z_\rho\right) - \frac{\gamma\rho}{d}\mathcal{P}(z_\rho) - \frac{2\gamma\int_0^1(1-\rho^2)^{\gamma-1}z\rho^{d-1}d\rho}{d\int_0^1(1-\rho^2)^\gamma\rho^{d-1}d\rho}(1 - \rho^2).$$

Since

$$L(1 - \rho^2) = -\frac{d+2\gamma}{d}(1 - \rho^2) = -\lambda_1(1 - \rho^2)$$

it follows that

$$p_{-\lambda_1}(\log t, \rho) = c(1 - \rho^2)$$

for some constant $c \neq 0$. Moreover $-2\lambda_1$ is the largest element in $\mathcal{S}\backslash\{-\lambda_1\}$. Hence

$$W(\rho, t) \sim t^{-\alpha}\left\{(1 - \rho^2)(K + \frac{c}{t}) + O(t^{-2})\right\}.$$

Now one argues exactly as in the previous Section to show that

$$t^\alpha|v(r, t) - V(r, t; M)| = O(t^{-2}) \text{ as } t \to \infty.$$

References

[1] S.B.Angenent, *Large time asymptotics for the porous medium equation*, in: Nonlinear Diffusion Equations and their Equilibrium States I, edited by W.-M.Ni, L.A.Peletier, and J.Serrin, Springer, Berlin, 1988.

[2] D.G.Aronson, *The porous medium equation*, in: Some Problems in Nonlinear Diffusion, edited by A.Fasano and M.Primicerio, Lecture Notes in Mathematics 1224, Springer, Berlin, 1986.

[3] D.G.Aronson and J.L.Vazquez, *Eventual C^∞-regularity and concavity for flows in one-dimensional porous media*, Arch. Rat. Mech. Math., 99(1987), 329-348.

[4] G.I.Barenblatt, *On self-similar motions of compressible fluids in porous media*, Prikl. Mat. Mech., 16(1952), 679-698. (Russian)

[5] A.Friedman and S.Kamin, *The asymptotic behavior of a gas in an n-dimensional porous medium*, Trans. Amer. Math. Soc., 262(1980), 551-563.

[6] S.Kamenomostskaya, *The asymptotic behavior of the solution of the filtration equation*, Israel J. Math., 14(1973), 76-87.

[7] S.Kamin and J.L.Vazquez, *Fundamental solutions and the asymptotic behaviour for the p-Laplacian equation*, Rev. Mat. Iberoamericana, 4(1988), 339-354.

[8] J.L.Vazquez, *Asymptotic behaviour and propagation of the one-dimensional flow of a gas in a porous medium*, Trans. Amer. Math. Soc., 277(1983), 507-527.

[9] Ya.B.Zel'dovich and G.I.Barenblatt, *The asymptotic properties of self-modelling solutions of the nonstationary gas filtration equations*, Sov. Phys. Dokl., 3(1958), 44-47.

[10] Ya.B.Zel'dovich and A.S.Kompaneets, *On the theory of propagation of heat with thermal conductivity depending on temperature*, in: Collection of Papers Dedicated to the 70th Birthday of A. F. Ioffe, Izd. Acad. Nauk. SSSR, Moscow, 1950. (Russian)

[11] Ya. B. Zel'dovich and Yu. P. Raizer, *Physics of Shock-waves and High-temperature Phenomena*, Academic Press, New York, 1966.

Department of Mathematics, University of Wisconsin, Madison WI 53706, USA
E-mail address: angenent@math.wisc.edu

School of Mathematics, University of Minnesota, Minneapolis MN 55455, USA
E-mail address: don@i2.math.umn.edu

The Shearing of Nonlinearly Viscoplastic Slabs

Stuart S. Antman[*]   Frank Klaus[†]

1 Introduction

In this paper we study an initial-boundary-value problem for a general, properly invariant theory of nonlinear viscoplasticity. The governing equations form a quasilinear parabolic-hyperbolic system of partial-functional differential equations, with the memory induced by the evolution of internal variables. Our emphasis is on the novel yet reasonable constitutive assumptions that support a proof that our problem has globally defined regular solutions. Our methods can handle far more complicated problems for this class of materials.

Since the 1960's there has been a sustained effort, now reaching fruition, to develop properly invariant theories of large-strain plasticity and viscoplasticity (see [9, 10, 14, 16, 17, 18, 20]). It now seems possible not only to treat these theories numerically (cf. [8, 11, 17, 19]), but also to subject them to mathematical analysis in a way that extends the studies of nonlinear problems with small strains (cf. [1, 12, 15]). The treatment of problems for bodies with specific geometry and general material response may help illuminate the central question of determining what reasonable constitutive equations must really look like.

Notation. Vectors, which are elements of a Euclidean space, and vector-valued functions are denoted by lower-case, bold-face symbols. The differential of $\mathbf{u} \mapsto \mathbf{f}(\mathbf{u})$ at $\mathbf{v}$ in the direction $\mathbf{h}$ is $\frac{d}{dt}\mathbf{f}(\mathbf{v}+t\mathbf{h})\big|_{t=0}$. When it is linear in $\mathbf{h}$, it is denoted by either $\frac{\partial \mathbf{f}}{\partial \mathbf{u}}(\mathbf{v})\cdot\mathbf{h}$ or $\mathbf{f}_{\mathbf{u}}(\mathbf{v})\cdot\mathbf{h}$. The partial derivative of a function f with respect to a scalar argument t is denoted by either f_t or $\partial_t f$. Obvious analogs of these notations will also be used. We denote the norm on $\mathrm{L}^2(0,1)$ by $\|\cdot\|$.

We let c and C denote typical positive constants that are supplied as data or that can be estimated in terms of data. Their meanings typically change with each appearance (even in the same equation or inequality. C may be regarded as increasing and c as decreasing with each appearance). Similarly, Γ denotes a typical positive-valued continuous function of t. Tacit in an inequality of the form $\|u\| \leq C$ is an assertion that there exists a positive number C such that this estimate holds. In the sequel we use without comment the Hölder inequality and such standard inequalities as $ab \leq (\varepsilon a)^p/p + (b/\varepsilon)^q/q$ for $a, b, \varepsilon > 0$, $p > 1$, $q = p/(p-1)$. As in this inequality, ε always denotes a small positive number at our disposal.

---

[*] Department of Mathematics and Institute for Physical Science and Technology, University of Maryland, College Park, MD 20742. This research was supported in part by NSF Grant DMS 9302726.

[†] Department of Mathematics, University of Maryland, College Park, MD 20742 and Fachbereich Mathematik, Technische Hochschule Darmstadt, Germany. This research was supported by a grant from the DFG.

2 Formulation of the Governing Equations

Let $\{\mathbf{i}_1, \mathbf{i}_2, \mathbf{k}\}$ be a fixed orthonormal basis for Euclidean 3-space. We study the motion of an incompressible viscoplastic material whose reference configuration is the slab consisting of all material points $x_1\mathbf{i}_1 + x_2\mathbf{i}_2 + s\mathbf{k}$ with $x_1, x_2 \in \mathbf{R}$, $0 \leq s \leq 1$. We limit our attention to shearing motions in which the displacement of the material point $x_1\mathbf{i}_1 + x_2\mathbf{i}_2 + s\mathbf{k}$ has the form

$$(1) \qquad \mathbf{u}(s,t) \equiv u_1(s,t)\mathbf{i}_1 + u_2(s,t)\mathbf{i}_2.$$

To construct our one-dimensional problem, we must assume that the material properties and the projection of the body force onto the $\{\mathbf{i}_1, \mathbf{i}_2\}$-plane are independent of the material coordinates x_1 and x_2. For notational simplicity, we make the stronger assumption that the material is homogeneous and that the body forces and surface tractions are zero. The equations of motion in the $\{\mathbf{i}_1, \mathbf{i}_2\}$-plane then have the vectorial form

$$(2) \qquad \rho\mathbf{u}_{tt} = \mathbf{n}_s$$

where ρ is the given constant density at any material point, and where $\mathbf{n} = n_1\mathbf{i}_1 + n_2\mathbf{i}_2$ is the vector whose components n_1 and n_2 are the shear components of the Piola-Kirchhoff stress tensor on the k-face in the $\mathbf{i}_1$- and $\mathbf{i}_2$-directions. (The third equation of motion, in the k-direction, which we do not exhibit, is used to determine the pressure field.)

We assume that the material has a viscous response of differential type and that the plastic response is described by the evolution of an internal variable $\boldsymbol{\pi}$. (The variable $\boldsymbol{\pi}$ is a collection of real-valued functions. Two of its components could be the permanent plastic deformation, i.e., the value of $\mathbf{u}_s$ when the stress is zero.)

We accordingly assume that there is a continuously differentiable constitutive function $(\mathbf{y}, \mathbf{z}, \boldsymbol{\pi}) \mapsto \hat{\mathbf{n}}(\mathbf{y}, \mathbf{z}, \boldsymbol{\pi})$ such that the stress at (s,t) is given by

$$(3) \qquad \mathbf{n}(s,t) = \hat{\mathbf{n}}(\mathbf{u}_s(s,t), \mathbf{u}_{st}(s,t), \boldsymbol{\pi}(s,t))$$

where the evolution of $\boldsymbol{\pi}$ is dictated by a continuously differentiable *yield* function $(\mathbf{y}, \boldsymbol{\pi}) \mapsto \gamma(\mathbf{y}, \boldsymbol{\pi})$ with the property that its gradient vanishes nowhere: The material behaves *elastically* where $\gamma(\mathbf{u}_s(s,t), \boldsymbol{\pi}(s,t)) < 0$ and it behaves *plastically* where $\gamma(\mathbf{u}_s(s,t), \boldsymbol{\pi}(s,t)) = 0$. (In our theory, we suppose that the fields $\mathbf{u}_s, \boldsymbol{\pi}$ can never make γ positive, but this assumption is not essential for the analysis; cf. [11].) The material point at (s,t) undergoes *plastic loading* if it is behaving plastically and if $\gamma_{\mathbf{y}}(\mathbf{u}_s(s,t), \boldsymbol{\pi}(s,t)) \cdot \mathbf{u}_{st}(s,t) > 0$. We assume that there is another continuously differentiable constitutive function $(\mathbf{y}, \mathbf{z}, \boldsymbol{\pi}) \mapsto \boldsymbol{\zeta}(\mathbf{y}, \mathbf{z}, \boldsymbol{\pi})$ defined on a neigbourhood of

$$(4) \qquad \{(\mathbf{y}, \mathbf{z}, \boldsymbol{\pi}) : \mathbf{z} \in \mathbf{R}^2,\ \gamma(\mathbf{y}, \boldsymbol{\pi}) = 0\}$$

such that $\boldsymbol{\pi}$ satisfies an ordinary differential equation, the *flow rule*:

$$(5) \qquad \boldsymbol{\pi}_t = \begin{cases} \boldsymbol{\zeta}(\mathbf{u}_s(s,t), \mathbf{u}_{st}(s,t), \boldsymbol{\pi}(s,t)) & \text{during plastic loading,} \\ \mathbf{0} & \text{otherwise.} \end{cases}$$

We assume that $\boldsymbol{\zeta} = \mathbf{0}$ when $\gamma(\mathbf{y}, \boldsymbol{\pi}) = 0$ and $\mathbf{z} = \mathbf{0}$.

We assume that the *elastic part* $\hat{\mathbf{n}}(\mathbf{y}, \mathbf{0}, \mathbf{0})$ of $\mathbf{n}$ is the derivative of a stored-energy function $\varphi \geq 0$, so that $\hat{\mathbf{n}}(\mathbf{y}, \mathbf{0}, \mathbf{0}) = \varphi_{\mathbf{y}}(\mathbf{y})$. We decompose $\hat{\mathbf{n}}$ into its elastic and inelastic parts:

$$(6) \qquad \hat{\mathbf{n}}(\mathbf{y}, \mathbf{z}, \boldsymbol{\pi}) = \varphi_{\mathbf{y}}(\mathbf{y}) + \hat{\mathbf{m}}(\mathbf{y}, \mathbf{z}, \boldsymbol{\pi}).$$

Thus $\hat{\mathbf{m}}(\mathbf{y}, 0, 0) = \mathbf{0}$. We decompose $\hat{\mathbf{m}}$ into its *dissipative part* $\mathbf{m}^D$ and its plastic part $\mathbf{m}^P$ by

(7)
$$\hat{\mathbf{m}}(\mathbf{u}_s, \mathbf{u}_{st}, \boldsymbol{\pi}) = \mathbf{m}^D(\mathbf{y}, \mathbf{z}, \boldsymbol{\pi}) + \mathbf{m}^P(\mathbf{y}, \mathbf{z}, \boldsymbol{\pi})$$

where

(8)
$$\mathbf{m}^D(\mathbf{y}, \mathbf{z}, \boldsymbol{\pi}) \equiv \left[\int_0^1 \frac{\partial \hat{\mathbf{n}}}{\partial \mathbf{z}}(\mathbf{y}, \lambda\mathbf{z}, \lambda\boldsymbol{\pi}) \, d\lambda \right] \cdot \mathbf{z},$$

$$\mathbf{m}^P(\mathbf{y}, \mathbf{z}, \boldsymbol{\pi}) \equiv \left[\int_0^1 \frac{\partial \hat{\mathbf{n}}}{\partial \boldsymbol{\pi}}(\mathbf{y}, \lambda\mathbf{z}, \lambda\boldsymbol{\pi}) \, d\lambda \right] \cdot \boldsymbol{\pi}.$$

In lieu of imposing any convexity conditions on φ, we merely require that there be a constant $c > 0$ such that

(9)
$$\varphi(\mathbf{y}) \geq c\,|\mathbf{y}|.$$

To ensure that our material is uniformly dissipative, we require that there exist a positive constant c such that

(10)
$$\mathbf{a} \cdot \frac{\partial \hat{\mathbf{n}}}{\partial \mathbf{z}}(\mathbf{y}, \mathbf{z}, \boldsymbol{\pi}) \cdot \mathbf{a} \equiv \mathbf{a} \cdot \frac{\partial \hat{\mathbf{m}}}{\partial \mathbf{z}}(\mathbf{y}, \mathbf{z}, \boldsymbol{\pi}) \cdot \mathbf{a} \geq c\,|\mathbf{a}|^2$$

for all $\mathbf{y}, \mathbf{z}, \boldsymbol{\pi}$ and all $\mathbf{a} \in \mathbf{R}^2$. We ensure that the viscosity is mildly strain-dependent (as in the constitutive equation for a Navier-Stokes fluid in the material formulation) by assuming that there is a continuously differentiable function ψ with $\psi(y) \to \infty$ as $y \to \infty$ and that there are non-negative numbers c and C such that

(11)
$$\hat{\mathbf{n}}(\mathbf{y}, \mathbf{z}, \boldsymbol{\pi}) \cdot \frac{\mathbf{y}}{|\mathbf{y}|} \geq \psi'(|\mathbf{y}|)\frac{\mathbf{y}}{|\mathbf{y}|} \cdot \mathbf{z} - C\psi(|\mathbf{y}|) + c\,|\mathbf{y}||\mathbf{z}|$$

for all $\mathbf{y}, \mathbf{z}, \boldsymbol{\pi}$. We ensure that the dependence of our constitutive functions ζ on $(\mathbf{y}, \mathbf{z})$ and $\mathbf{m}^P$ on $\boldsymbol{\pi}$ is mild by assuming that there are positive constants $\alpha < 1$ and C such that

(12)
$$|\zeta(\mathbf{y}_1, \mathbf{z}_1, \boldsymbol{\pi}_1) - \zeta(\mathbf{y}_2, \mathbf{z}_2, \boldsymbol{\pi}_2)| \leq C\left(|\mathbf{y}_1 - \mathbf{y}_2|^\alpha + |\mathbf{z}_1 - \mathbf{z}_2|^\alpha + |\boldsymbol{\pi}_1 - \boldsymbol{\pi}_2|\right),$$

(13)
$$\mathbf{m}^P(\mathbf{y}, \mathbf{z}, \boldsymbol{\pi}) \cdot \mathbf{z} \leq C(1 + \varphi(\mathbf{y}) + |\mathbf{z}||\boldsymbol{\pi}|)$$

for all possible values of the arguments.

Our *initial-boundary-value problem* consists of the equations of motion (2), the constitutive equation (3), the flow rule (5), boundary conditions for complementary combinations of components of $\mathbf{u}_t$ and $\mathbf{n}$ at $s = 0$ and $s = 1$, and compatible initial conditions for $\mathbf{u}$, $\mathbf{u}_t$, and $\boldsymbol{\pi}$. For details on the mechanical principles underlying the formulation of our equations, see [2, 4, 6, 8].

3 Energy Estimates

In this and the next section we assume that a sufficiently regular solution of our initial-boundary-value problem is given. We obtain bounds on the solution that tell us where to seek solutions and that thereby enable us to find solutions with the requisite regularity.

We substitute (3) and (6) into (2), take the dot product of the resulting equation with $\mathbf{u}_t(s, t)$, and integrate the resulting dot product by parts over $[0, 1] \times [0, t]$ to obtain the *energy equation*:

(14)
$$K(t) + \Phi(t) + W^D(t) = K(0) + \Phi(0) - W^P(t) + \left[\int_0^t \mathbf{n}(s, \tau) \cdot \mathbf{u}_t(s, \tau) \, d\tau \right]_{s=0}^{s=1},$$

where

$$K(t) = \frac{1}{2} \int_0^1 \rho |\mathbf{u}_t(s,t)|^2 \, ds, \quad \Phi(t) = \int_0^1 \varphi(\mathbf{u}_s(s,t)) \, ds,$$

(15)
$$W^{\mathrm{D}}(t) = \int_0^t \int_0^1 \mathbf{m}^{\mathrm{D}}(\mathbf{u}_s, \mathbf{u}_{st}, \boldsymbol{\pi}) \cdot \mathbf{u}_{st} \, ds \, d\tau,$$

$$W^{\mathrm{P}}(t) = \int_0^t \int_0^1 \mathbf{m}^{\mathrm{P}}(\mathbf{u}_s, \mathbf{u}_{st}, \boldsymbol{\pi}) \cdot \mathbf{u}_{st} \, ds \, d\tau$$

are the kinetic energy, the stored energy, the dissipative work, and the plastic work at time t.

We now estimate the time-dependent terms on the right-hand side of (14). Let

(16) $\mathcal{P}(t) \equiv \{s \in [0,1] : \gamma(\mathbf{u}_s(s,t), \boldsymbol{\pi}(s,t)) = 0, \gamma_{\mathbf{y}}(\mathbf{u}_s(s,t), \boldsymbol{\pi}(s,t)) \cdot \mathbf{u}_{st}(s,t) > 0\}.$

Then from (5) and (12) we get

(17)
$$\frac{d}{dt} \|\boldsymbol{\pi}(\cdot, t)\|^2 = 2 \int_0^1 \boldsymbol{\pi}(s,t) \cdot \boldsymbol{\pi}_t(s,t) \, ds$$

$$= 2 \int_{\mathcal{P}(t)} \boldsymbol{\pi}(s,t) \cdot \boldsymbol{\zeta}(\mathbf{u}_s(s,t), \mathbf{u}_{st}(s,t), \boldsymbol{\pi}(s,t)) \, ds$$

$$\leq C \int_0^1 (1 + |\mathbf{u}_s(s,t)|^\alpha + |\mathbf{u}_{st}(s,t)|^\alpha + |\boldsymbol{\pi}(s,t)|) \, |\boldsymbol{\pi}(s,t)| \, ds$$

$$\leq C \left(1 + \int_0^1 |\mathbf{u}_s(s,t)|^\alpha |\boldsymbol{\pi}(s,t)| \, ds + \|\mathbf{u}_{st}(\cdot, t)\|^{2\alpha} + \|\boldsymbol{\pi}(\cdot, t)\|^2 \right).$$

Now

(18) $\displaystyle \int_0^1 |\mathbf{u}_s(s,t)|^\alpha |\boldsymbol{\pi}(s,t)| \, ds \leq \frac{1}{2} \int_0^1 |\mathbf{u}_s(s,t)|^{2\alpha} \, ds + \frac{1}{2} \int_0^1 |\boldsymbol{\pi}(s,t)|^2 \, ds.$

If $\alpha < \frac{1}{2}$, then (9) implies that

(19)
$$\int_0^1 |\mathbf{u}_s(s,t)|^{2\alpha} \, ds \leq C \, (1 + \Phi(t)).$$

If $\frac{1}{2} < \alpha < 1$, then

(20) $\displaystyle \int_0^1 |\mathbf{u}_s(s,t)|^{2\alpha} \, ds \leq \|\mathbf{u}_s(\cdot, t)\|^{2\alpha} \leq \Gamma(t) \left[\|\mathbf{u}_s(\cdot, 0)\|^{2\alpha} + \int_0^t \|\mathbf{u}_{st}(\cdot, \tau)\|^{2\alpha} \, d\tau \right].$

Substituting (18), (19), (20) into (17), integrating the result, and using the Gronwall inequality, we obtain

(21)
$$\|\boldsymbol{\pi}(\cdot, t)\|^2 \leq \Gamma(t) \left[1 + \int_0^t \|\mathbf{u}_{st}(\cdot, \tau)\|^{2\alpha} \, d\tau + \int_0^t \Phi(\tau) \, d\tau \right].$$

From (13) and (15) we estimate the plastic work:

(22)
$$\left| W^{\mathrm{P}}(t) \right| \leq C \int_0^t \left(1 + \Phi(\tau) + \|\mathbf{u}_{st}(\cdot, \tau)\| \, \|\boldsymbol{\pi}(\cdot, \tau)\| \right) \, d\tau.$$

24

From (21) we get

(23) $\int_0^t \|\mathbf{u}_{st}(\cdot, \tau)\| \, \|\pi(\cdot, \tau)\| \, d\tau$

$$\leq \Gamma(t) \int_0^t \|\mathbf{u}_{st}(\cdot, \tau)\| \left[1 + \int_0^\tau \|\mathbf{u}_{st}(\cdot, \eta)\|^{2\alpha} \, d\eta + \int_0^\tau \Phi(\eta) \, d\eta\right]^{1/2} d\tau$$

$$\leq \Gamma(t) \left\{ \left[\int_0^t \|\mathbf{u}_{st}(\cdot, \tau)\|^2 \, d\tau\right]^{1/2} \left[1 + \int_0^t \Phi(\tau) \, d\tau\right]^{1/2} + \left[\int_0^t \|\mathbf{u}_{st}(\cdot, \tau)\|^2 \, d\tau\right]^{(1+\alpha)/2} \right\}.$$

Thus (22), (10), and (23) yield

(24) $$\left|W^{\mathrm{P}}(t)\right| \leq \Gamma(t) + \Gamma(t) \int_0^t \Phi(\tau) \, d\tau + \frac{1}{2} W^{\mathrm{D}}(t).$$

The substitution of (24) into (14) yields

(25) $$K(t) + \Phi(t) + \frac{1}{2} W^{\mathrm{D}}(t) \leq \Gamma(t) \left[1 + \int_0^t \Phi(\tau) \, d\tau\right] + \left[\int_0^t \mathbf{n}(s, \tau) \cdot \mathbf{u}_t(s, \tau) \, d\tau\right]_{s=0}^{s=1}.$$

We now treat the boundary terms in (25), limiting our attention to the representative case that $\mathbf{u}(0, t)$ and $\mathbf{n}(1, t)$ are prescribed with $\mathbf{u}_{tt}(0, \cdot), \mathbf{n}(1, \cdot) \in \mathrm{L}^2_{\mathrm{loc}}(0, \infty)$. Since

(26) $$|\mathbf{u}_t(s, t)| \leq |\mathbf{u}_t(0, t)| + \int_0^1 |\mathbf{u}_{st}(s, t)| \, ds \leq \Gamma(t) + \|\mathbf{u}_{st}(\cdot, t)\|$$

for any $s \in [0, 1]$, we obtain from (10) that

(27) $$\left|\int_0^t \mathbf{n}(1, \tau) \cdot \mathbf{u}_t(1, \tau) \, d\tau\right| \leq \Gamma(t) + \varepsilon W^{\mathrm{D}}(t).$$

From (2) we obtain

(28) $$\int_0^t \mathbf{n}(0, \tau) \cdot \mathbf{u}_t(0, \tau) \, d\tau = \int_0^t \left[\mathbf{n}(1, \tau) - \int_0^1 \rho \mathbf{u}_{tt}(s, \tau) \, ds\right] \cdot \mathbf{u}_t(0, \tau) \, d\tau$$

$$= \int_0^t \mathbf{n}(1, \tau) \cdot \mathbf{u}_t(0, \tau) \, d\tau - \left[\int_0^1 \rho \mathbf{u}_t(s, \tau) \, ds\right] \cdot \mathbf{u}_t(0, \tau) \Big|_{\tau=0}^{\tau=t}$$

$$+ \int_0^t \left[\int_0^1 \rho \mathbf{u}_t(s, \tau) \, ds\right] \cdot \mathbf{u}_{tt}(0, \tau) \, d\tau,$$

so that

(29) $$\left|\int_0^t \mathbf{n}(0, \tau) \cdot \mathbf{u}_t(0, \tau) \, d\tau\right| \leq \Gamma(t) + \frac{1}{2} K(t) + \int_0^t K(\tau) \, d\tau.$$

Substituting (27) and (29) into (25), and choosing ε sufficiently small, we deduce that

(30) $$K(t) + \Phi(t) \leq \Gamma(t) + \Gamma(t) \int_0^t [K(\tau) + \Phi(\tau)] \, d\tau.$$

From Gronwall's inequality we obtain $K(t) + \Phi(t) \leq \Gamma(t)$, which implies that (29) is bounded. In view of (27) with ε sufficiently small, we thus obtain from (25) the *energy inequality*:

(31) $$K(t) + \Phi(t) + W^{\mathrm{D}}(t) \leq \Gamma(t).$$

For the treatment of other boundary conditions, see [5, 6, 7].

4 Bounds on the strain

We set $\kappa \equiv \sqrt{\mathbf{u}_s \cdot \mathbf{u}_s}$ and write

$$\mathbf{u}_s = \kappa \, \mathbf{e}_1 \tag{32}$$

with

$$\mathbf{e}_1 = \cos\theta \, \mathbf{i}_1 + \sin\theta \, \mathbf{i}_2, \quad \mathbf{e}_2 = -\sin\theta \, \mathbf{i}_1 + \cos\theta \, \mathbf{i}_2. \tag{33}$$

Thus

$$\partial_t \mathbf{e}_1 = \theta_t \, \mathbf{e}_2, \quad \mathbf{u}_{st} = \kappa_t \, \mathbf{e}_1 + \kappa \, \theta_t \, \mathbf{e}_2. \tag{34}$$

Let us suppose that $\mathbf{n}(1,t)$ is prescribed. For $\nu = 1, 2$ we define

$$g_\nu(s,t) = \mathbf{e}_\nu(s,t) \cdot \int_s^1 \rho \mathbf{u}_t(\xi, t) \, d\xi. \tag{35}$$

The integration of (2) yields

$$\mathbf{n}(s,t) = \mathbf{n}(1,t) - \int_s^1 \rho \mathbf{u}_{tt}(\xi, t) \, d\xi, \tag{36}$$

from which we obtain

$$\mathbf{e}_1(s,t) \cdot \mathbf{n}(s,t) = \mathbf{e}_1(s,t) \cdot \mathbf{n}(1,t) - \partial_t g_1(s,t) + \theta_t(s,t) \, g_2(s,t). \tag{37}$$

From (11) and (37) we obtain

$$\psi'(\kappa) \, \kappa_t - C\psi(\kappa) \leq \Gamma - \partial_t g_1 + |\theta_t \, g_2| - c \, \kappa \, |\mathbf{u}_{st}|. \tag{38}$$

From (31) it follows that there are functions $\Gamma_\nu(t)$ depending on the data such that

$$|g_\nu(s,t)| \leq \Gamma_\nu(t), \quad \nu = 1, 2. \tag{39}$$

Since $|\mathbf{u}_{st}| = \sqrt{\kappa_t^2 + \kappa^2 \theta_t^2}$, we obtain

$$|\theta_t(s,t) \, g_2(s,t)| - c \, \kappa(s,t) \, |\mathbf{u}_{st}(s,t)| \leq |\theta_t(s,t)| \, [\Gamma_2(t) - c \, \kappa(s,t)^2]. \tag{40}$$

Thus, wherever $\kappa(s,t)^2 \geq \Gamma_2(t)/c$, (38) reduces to

$$\psi'(\kappa) \, \kappa_t - C\psi(\kappa) \leq \Gamma - \partial_t g_1. \tag{41}$$

Let $\kappa(\cdot, 0)$ be bounded and let Γ_3 be a continuous function bounded below by $\sqrt{\Gamma_2/c} + \max_s \kappa(s, 0)$. We now show that κ is bounded by a continuous function of t. Suppose that there is a ξ in $[0, 1]$ and a $\tau_2 > 0$ such that $\kappa(\xi, \tau_2) > \Gamma_3(\tau_2)$. Were there no such (ξ, τ_2), there would be nothing to prove. Since κ is continuous and since $\kappa(\xi, 0) < \Gamma_3(\tau_2)$, there exists a last time $\tau_1 < \tau_2$ at which $\kappa(\xi, \tau_1) = \Gamma_3(\tau_2)$. We integrate (41) for $s = \xi$ from τ_1 to $t \leq \tau_2$ and use (39) to obtain

$$\psi(\kappa(\xi, t)) - \psi(\Gamma_3(\tau_2)) - C \int_{\tau_1}^t \psi(\kappa(\xi, \tau)) \, d\tau \leq \Gamma(t) - g_1(\xi, t) + g_1(\xi, \tau_1) \leq \Gamma(t). \tag{42}$$

Gronwall's inequality then implies that $\psi(\kappa(\xi, t)) \leq \Gamma(t)$, so that

$$|\mathbf{u}_s(\xi, t)| = \kappa(\xi, t) \leq \Gamma(t), \quad \text{whence} \quad |\mathbf{u}_s| \leq \Gamma(t). \tag{43}$$

5 Bounds on u_{st} and π

We now supplement the pointwise bonds obtained in Section 4 with further pointwise bounds on u_{st} and π, so that all the arguments of the constitutive functions are pointwise bounded. Formally differentiating (2) with respect to t, taking the dot product of the differentiated equation with u_{tt}, and integrating the resulting equation by parts with respect to s over $(0, 1)$ yields

$$(44) \qquad \frac{d}{dt} \int_0^1 \rho |u_{tt}(s,t)|^2 \, ds + 2 \int_0^1 u_{stt}(s,t) \cdot n_t(s,t) \, ds = 2 \Big[u_{tt} \cdot n_t \Big]_{s=0}^{s=1}.$$

The substitution of (3) into the integrand of the second integral of (44) converts this integrand to

$$(45) \qquad u_{stt} \cdot \left[\frac{\partial \hat{n}}{\partial y}(u_s, u_{st}, \pi) \cdot u_{st} + \frac{\partial \hat{n}}{\partial z}(u_s, u_{st}, \pi) \cdot u_{stt} + \frac{\partial \hat{n}}{\partial \pi}(u_s, u_{st}, \pi) \cdot \pi_t \right].$$

where we suppressed the arguments (s, t). The second bilinear form in (45) is a positive-definite quadratic form, by the monotonicity condition (10).

To estimate the other two bilinear forms in (45), we introduce a further constitutive assumption, which says that when the strain is bounded, the elastic and plastic responses are controlled by the viscous response: There is a continuous function λ such that

$$(46) \qquad \left| a \cdot \frac{\partial \hat{n}}{\partial y}(y, z, \pi) \right|^2 + \left| a \cdot \frac{\partial \hat{n}}{\partial \pi}(y, z, \pi) \right|^2 \leq \lambda(M) \, a \cdot \frac{\partial \hat{n}}{\partial z}(y, z, \pi) \cdot a$$

for all positive M and all a, y, z, π with $|y| \leq M$. Thus (43) implies that

$$(47) \qquad \left| \int_0^1 u_{stt} \cdot \left[\frac{\partial \hat{n}}{\partial y} \cdot u_{st} + \frac{\partial \hat{n}}{\partial \pi} \cdot \pi_t \right] ds \right|$$

$$\leq \frac{1}{2} \int_0^1 u_{stt} \cdot \frac{\partial \hat{n}}{\partial z} \cdot u_{stt} \, ds + \lambda(\Gamma(T)) \int_0^1 \left(|u_{st}|^2 + |\pi_t|^2 \right) ds$$

for $0 \leq t \leq T$. To estimate the last term in (47) we use (5) and (12) to get

$$(48) \qquad \int_0^1 |\pi_t|^2 \, ds = \int_{\mathcal{P}(t)} |\zeta(u_s, u_{st}, \pi)|^2 \, ds \leq C \int_0^1 \left(1 + |u_s|^2 + |u_{st}|^2 + |\pi|^2 \right) ds.$$

Substituting (45), (47), (48) into (44), we obtain

$$(49) \qquad \frac{d}{dt} \int_0^1 \rho |u_{tt}|^2 \, ds + \int_0^1 u_{stt} \cdot \frac{\partial \hat{n}}{\partial z} \cdot u_{stt} \, ds$$

$$\leq 2 \Big[u_{tt} \cdot n_t \Big]_{s=0}^{s=1} + \Gamma(T) \int_0^1 \left(1 + |u_s|^2 + |u_{st}|^2 + |\pi|^2 \right) ds$$

for $0 \leq t \leq T$. From estimates like (21) and from the energy estimate (31) we then get

$$(50) \qquad \int_0^1 \rho |u_{tt}|^2 ds + \int_0^t \int_0^1 u_{stt} \cdot \frac{\partial \hat{n}}{\partial z} \cdot u_{stt} \, ds \, d\tau \leq 2 \int_0^t \Big[u_{tt} \cdot n_t \Big]_{s=0}^{s=1} d\tau + \Gamma(T)$$

for $0 \leq t \leq T$. Suppose that $u(0, \cdot)$ and $n(1, \cdot)$ are prescribed with $u_{ttt}(0, \cdot), n_t(1, \cdot) \in L^2_{loc}(0, \infty)$. Using (2) we write the bracketed boundary term of (50) as

$$(51) \qquad [u_{tt}(1,t) - u_{tt}(0,t)] \cdot n_t(1,t)$$

$$+ \partial_t \left[u_{tt}(0,t) \cdot \int_0^1 \rho u_{tt}(s,t) \, ds \right] - u_{ttt}(0,t) \cdot \int_0^1 \rho u_{tt}(s,t) \, ds$$

and replace $\mathbf{u}_{tt}(1,t) - \mathbf{u}_{tt}(0,t)$ with $\int_0^1 \mathbf{u}_{stt}(s,t)\,\mathrm{d}s$. Substituting (51) into (50) and using our standard estimates, we deduce from (50) that

$$
(52) \qquad \int_0^1 \rho|\mathbf{u}_{tt}|^2\,\mathrm{d}s + \int_0^t \int_0^1 |\mathbf{u}_{stt}|^2\,\mathrm{d}s\,\mathrm{d}\tau \leq \Gamma(T)
$$

for $0 \leq t \leq T$.

From (2), (10), and (46) we obtain

$$
(53) \quad \int_0^1 \mathbf{u}_{sst} \cdot \frac{\partial \hat{\mathbf{n}}}{\partial \mathbf{z}} \cdot \mathbf{u}_{sst}\,\mathrm{d}s = \int_0^1 \mathbf{u}_{sst} \cdot \left(\rho \mathbf{u}_{tt} - \frac{\partial \hat{\mathbf{n}}}{\partial \mathbf{y}} \cdot \mathbf{u}_{ss} - \frac{\partial \hat{\mathbf{n}}}{\partial \boldsymbol{\pi}} \cdot \boldsymbol{\pi}_s \right)\,\mathrm{d}s
$$
$$
\leq \varepsilon \int_0^1 \mathbf{u}_{sst} \cdot \frac{\partial \hat{\mathbf{n}}}{\partial \mathbf{z}} \cdot \mathbf{u}_{sst}\,\mathrm{d}s + \Gamma(t) \int_0^1 \left(\rho|\mathbf{u}_{tt}|^2 + |\mathbf{u}_{ss}|^2 + |\boldsymbol{\pi}_s|^2 \right)\,\mathrm{d}s.
$$

Let us estimate the last term of (53), which is the only source of difficulty. From (5) we find that $\boldsymbol{\pi}_{st} = \boldsymbol{\zeta}_\mathbf{y} \cdot \mathbf{u}_{ss} + \boldsymbol{\zeta}_\mathbf{z} \cdot \mathbf{u}_{sst} + \boldsymbol{\zeta}_{\boldsymbol{\pi}} \cdot \boldsymbol{\pi}_s$, where $\gamma_\mathbf{y} \cdot \mathbf{u}_{ss} + \gamma_{\boldsymbol{\pi}} \cdot \boldsymbol{\pi}_s = 0$ for s in any interval on which there is plastic loading. We therefore obtain

$$
(54) \qquad \frac{\mathrm{d}}{\mathrm{d}t} \|\boldsymbol{\pi}_s(\cdot,t)\|^2 = 2 \int_0^1 \boldsymbol{\pi}_s(s,t) \cdot \boldsymbol{\pi}_{st}(s,t)\,\mathrm{d}s
$$
$$
\leq 2 \int_0^1 \left| \left(\frac{\partial \boldsymbol{\zeta}}{\partial \mathbf{y}} \cdot \mathbf{u}_{ss} + \frac{\partial \boldsymbol{\zeta}}{\partial \mathbf{z}} \cdot \mathbf{u}_{sst} + \frac{\partial \boldsymbol{\zeta}}{\partial \boldsymbol{\pi}} \cdot \boldsymbol{\pi}_s \right) \cdot \boldsymbol{\pi}_s \right|\,\mathrm{d}s
$$
$$
\leq \Gamma(T) \left(\|\mathbf{u}_{ss}(\cdot,t)\|^2 + \|\mathbf{u}_{sst}(\cdot,t)\|^2 + \|\boldsymbol{\pi}_s(\cdot,t)\|^2 \right).
$$

We solve (54) à la Gronwall for $\|\boldsymbol{\pi}_s(\cdot,t)\|^2$ as a time-integral of $\|\mathbf{u}_{ss}(\cdot,t)\|^2 + \|\mathbf{u}_{sst}(\cdot,t)\|^2$, substitute the result into (53), note that the first term in the second integral of (53) is bounded by virtue of (52), and use

$$
(55) \qquad \|\mathbf{u}_{ss}(\cdot,t)\|^2 \leq 2\|\mathbf{u}_{ss}(0)\|^2 + \Gamma(T) \int_0^t \int_0^1 \mathbf{u}_{sst} \cdot \frac{\partial \hat{\mathbf{n}}}{\partial \mathbf{z}} \cdot \mathbf{u}_{sst}\,\mathrm{d}s\,\mathrm{d}\tau,
$$

which is obtained from (10). Thus, by using the Gronwall inequality, we obtain

$$
(56) \qquad \int_0^1 |\mathbf{u}_{sst}(s,t)|^2\,\mathrm{d}s \leq \Gamma(t).
$$

We now integrate the inequality $|\mathbf{u}_{st}(s,t)| \leq |\mathbf{u}_{st}(\xi,t)| + \int_0^1 |\mathbf{u}_{sst}(\eta,t)|\,\mathrm{d}\eta$ with respect to ξ over $[0,1]$ and use (52), (56), and the Gronwall inequality to obtain

$$
(57) \qquad |\mathbf{u}_{st}(s,t)| \leq \Gamma(t).
$$

The boundedness of $\mathbf{u}_s$ and $\mathbf{u}_{st}$, ensured by (43) and (57), and the Lipschitz continuity of $\boldsymbol{\zeta}$, ensured by (12), enable us to use the Gronwall inequality to deduce from (5) that

$$
(58) \qquad |\boldsymbol{\pi}(s,t)| \leq \Gamma(t).
$$

6 Comments on the existence theory

Estimates (43), (57), and (58) ensure that all the arguments of our constitutive functions are pointwise bounded for $0 \leq t \leq T$. We could therefore consider the restriction of our constitutive functions to arguments satisfying these bounds with $\Gamma(T)$ replaced by $2\Gamma(T)$. We could then extend these restricted constitutive functions to the whole space in a way that preserves the uniformity of behavior of the restriction on the bounded set. The virtue of these new constitutive functions is that they behave nicely everywhere, whereas the original constitutive functions need not. (The only difficulty in effecting this extension is showing that the extension enjoys the monotonicity condition (10). The proof is not as obvious as it might seem; see [7].) Let us denote these extended constitutive functions with superposed tildes. The analog of (2) and (3) is

$$(59) \qquad \rho(s)\,\mathbf{v}_t = \tilde{\mathbf{n}}(s, \mathbf{u}_s, \mathbf{v}_s, \boldsymbol{\pi})_s,$$

where $\mathbf{v} = \mathbf{u}_t$. If we regard $\mathbf{u}_s$ as frozen, the properties of $\tilde{\mathbf{n}}$ ensure that (59) is a well-behaved parabolic system, which can be analyzed by exploiting the monotonicity condition (10) in the Galerkin method [13]. To handle our actual system for $\mathbf{u}$, $\mathbf{v}$ and $\boldsymbol{\pi}$, we supplement (59) with

$$(60) \qquad \mathbf{u}_s(s, t) = \mathbf{u}_s(s, 0) + \int_0^t \mathbf{v}_s(s, \tau)\,\mathrm{d}\tau$$

and with the integral version of (5). The Galerkin method is readily applied to the whole system (cf. [5]).

We could get the existence theory by an alternative route: The energy estimate together with the monotonicity condition (10) ensures the existence of a weak solution. The estimates on the strains of Section 4 then have the virtue of making the constitutive hypotheses milder than they would otherwise have to be. The work in Section 5 then constitutes the regularity theory for the weak solutions.

The proof of uniqueness is straightforward.

7   Conclusion

This work indicates that in the presence of strong dissipation, characterized by our hypotheses (10) and (11), initial-boundary-value problems are well-posed. (On the other hand, the work of [4] indicates that in the absence of other dissipative mechanisms, we would not get a well-defined shock structure in the limit as viscous effects evanesce.) We see no obstacle preventing the extension of our methods to the treatment of much more complicated problems in one independent spatial variable, like those of [7]. Our results also suggest that concrete problems, such as the stability problems of [3], are tractable for this class of materials.

References

[1] H. D. Alber, *On a system of equations from the theory of nonlinear visco-elasticity*, SIAM J. Math. Anal., to appear.

[2] S. S. Antman, *Nonlinear Problems of Elasticity*, Springer, 1995.

[3] S. S. Antman and H. Koch, *Self-sustained oscillations of nonlinearly viscoelastic bodies*, to appear.

[4] S. S. Antman and R. Malek-Madani, *Travelling waves in nonlinearly viscoelastic media and shock structure in elastic media*, Quart. Appl. Math., 46 (1988), 77–93.

[5] S. S. Antman and T. I. Seidman, *Quasilinear hyperbolic-parabolic equations of nonlinear viscoelasticity*, J. Diff. Eqs., to appear.

[6] S. S. Antman and T. I. Seidman, *Large shearing motions of nonlinearly viscoelastic slabs*, Bull. Tech. Univ. Istanbul, 47 (1994), 41–56.

[7] S. S. Antman and T. I. Seidman, *Hyperbolic-parabolic systems governing the spatial motion of nonlinearly viscoelastic rods*, in preparation.

[8] S. S. Antman and W. G. Szymczak, *Nonlinear elastoplastic Waves*, in Current Progress in Riemann Problems, edited by W B. Lindquist, Contemp. Math., 100 (1989), 27–54.

[9] J. Casey and P. M. Naghdi, *Constitutive results for finitely deforming elastic-plastic materials* in Constitutive Equations: Macro and Computational Aspects, edited by K. J. Willem, Amer. Soc. Mech. Engg., 1984, 53–71.

[10] A. E. Green and P. M. Naghdi, *A general theory of an elastic-plastic continuum*, Arch. Rational Mech. Anal., 18 (1965), 251–281.

[11] J. M. Greenberg, *Models of elastic-perfectly plastic materials*, Euro. J. Appl. Math., 1 (1990), 131–150.

[12] F. Klaus, *Lokaler Existenz- und Eindeutigkeitssatz für die Millerschen Gleichungen zur nichtlinearen Viskoelastizität mit Härtung*, Dissertation, Technische Hochschule, Darmstadt, 1994.

[13] J.-L. Lions, *Quelques méthodes de résolution des problèmes aux limites non linéaires*, Dunod, Gauthier-Villars, 1969.

[14] M. Lucchesi, D. R. Owen, and P. Podio-Guidugli, *Materials with elastic range: A theory with a view toward applications, III, Approximate constitutive relations*, Arch. Rational Mech. Anal., 117 (1992), 53–96.

[15] A. Nouri and M. Rascle, *A global existence and uniqueness theorem for a model problem in dynamic elasto-plasticity with isotropic strain hardening*, SIAM J. Math. Anal., 26 (1995), 850–868.

[16] D. R. Owen, *A mechanical theory of materials with elastic range*, Arch. Rational Mech. Anal., 37 (1970), 85–110.

[17] J. C. Simo, *Topics on the numerical analysis and simulation of plasticity*, to appear in Handbook of Numerical Analysis, edited by P. G. Ciarlet and J.-L. Lions, Elsevier.

[18] J. C. Simo and M. Ortiz, *A unified approach to finite deformation elastoplastic analysis based on the use of hyperelastic constitutive equations*, Comput. Methods Appl. Mech. Engrg., 49 (1985), 221–245.

[19] J.A. Trangenstein and P. Colella, *A Godunov method for elastic-plastic deformations*, Comm. Pure Appl. Math., 44 (1991), 41–100.

[20] C. Truesdell, *Hypo-elasticity*, J. Rational Mech. Anal., 4 (1955), 83–133, 1019-1020.

Bounds for Ratios of the First, Second, and Third Membrane Eigenvalues

Mark S. Ashbaugh [*]   Rafael D. Benguria [†]

Dedicated to Ivar Stakgold on the occasion of his seventieth birthday.

Abstract

In this paper the method of H. Yang is incorporated into that of P. Marcellini to obtain an improved bound for λ_3/λ_1 in terms of λ_2/λ_1, where λ_1, λ_2, and λ_3 are the first three Dirichlet eigenvalues of $-\Delta$ on an arbitrary bounded domain $\Omega \subset I\!\!R^2$. In this context, Yang's approach can be distilled down to the application of the following principle: "Make optimal use of the Cauchy–Schwarz inequality." Based on this approach, the best bounds yet known for λ_3/λ_1 and $(\lambda_2 + \lambda_3)/\lambda_1$ become $\lambda_3/\lambda_1 \leq 3.83103^-$ and $(\lambda_2 + \lambda_3)/\lambda_1 \leq 5.50661^-$. These improve upon the values 3.89804^+ and 5.52485^+, which had been established earlier by Yang and ourselves, respectively. Similar advances can also be made for the analogous problem in n dimensions.

1991 Mathematics Subject Classification (MSC) numbers: 35P15, 49Rxx

1 Introduction

We consider the Dirichlet eigenvalues of the Laplacian on a bounded domain Ω in $I\!\!R^2$. That is, we consider the eigenvalue problem

$$(1) \qquad\qquad -\Delta u = \lambda u \qquad \text{in } \Omega$$

$$(2) \qquad\qquad u = 0 \qquad \text{on } \partial\Omega.$$

This is the mathematical formulation of the so–called fixed membrane problem. It is well-known that the spectrum of this problem consists of a discrete set of positive real eigenvalues whose only accumulation point is infinity. Thus, we denote the eigenvalues by $\lambda_1, \lambda_2, \lambda_3, \ldots$ where $0 < \lambda_1 < \lambda_2 \leq \lambda_3 \leq \ldots$ and $\lambda_k \to \infty$ as $k \to \infty$ and we let $\{u_k\}_{k=1}^\infty$ be an associated orthonormal basis of real eigenfunctions (where each eigenvalue is repeated according to its multiplicity).

Our work here improves upon the inequalities of P. Marcellini [13] and H. Yang [16] for the allowed range of λ_2/λ_1 and λ_3/λ_1 by combining their approaches. By tailoring the

---

[*]Department of Mathematics, University of Missouri, Columbia, Missouri 65211 (mark @ash-baugh.math.missouri.edu). This author's work was partially supported by National Science Foundation grants DMS-9114162, INT-9123481, and DMS-9500968.

[†]Facultad de Física, P. Universidad Católica de Chile, Avenida Vicuña Mackenna 4860, Casilla 306, Santiago 22, Chile (rbenguri@lascar.puc.cl). This author's work was partially supported by Fondo Nacional de Ciencia y Tecnología (Chile) project number 193–0561.

approach of Yang to our purposes, we simplify it to the point where it amounts to the addition of one key idea to Marcellini's approach. This allows us to present a good portion of our own work in parallel with the paper of Marcellini. Indeed, so as to make our account self-contained we present most of Marcellini's arguments. We do this briefly, in a way which fits into our overall scheme, only commenting in more detail when we introduce the improvements that are made possible by the incorporation of Yang's idea.

Our new inequalities for λ_2/λ_1 and λ_3/λ_1 allow us to obtain the best bounds yet found for λ_3/λ_1 and $(\lambda_2 + \lambda_3)/\lambda_1$:

$$(3) \qquad \lambda_3/\lambda_1 \leq 3.83103^-$$

and

$$(4) \qquad (\lambda_2 + \lambda_3)/\lambda_1 \leq 5.50661^-.$$

Previously the best bounds known for these quantities were $\lambda_3/\lambda_1 \leq 3.89804^+$ and $(\lambda_2 + \lambda_3)/\lambda_1 \leq 5.52485^+$. The first of these was established by Yang [16], while the second is found in our paper [7]. Indeed, in that paper we had derived bounds of 3.90514^+ and 5.52485^+ for λ_3/λ_1 and $(\lambda_2 + \lambda_3)/\lambda_1$, respectively, based on Marcellini's inequality (see (42) below) and our inequality (46) which is an improved version of an earlier inequality of de Vries [10]. Similarly, Yang had obtained $\lambda_3/\lambda_1 \leq 3.89804^+$ and $(\lambda_2 + \lambda_3)/\lambda_1 \leq 5.59076^+$ by making use of his inequality (44) and the inequality $(\lambda_2 + \lambda_3)/\lambda_1 \leq 5 + \lambda_1/\lambda_2$ of Brands [8]. Already by combining the inequalities (44) and (46) we could obtain better values (cf. Figure 1) than either Yang or we had obtained previously. However, we can do even better by combining the methods of Yang and Marcellini to obtain the new inequality (45) (see also (48) and (40)). This new inequality gets us the bounds (3) and (4) (for (4), (45) is applied in conjunction with (46); see Figure 1).

Payne, Pólya, and Weinberger [15] had conjectured that $(\lambda_2 + \lambda_3)/\lambda_1$ attains its maximum iff Ω is a disk. If this were true then the best bound for $(\lambda_2 + \lambda_3)/\lambda_1$ in two dimensions would be $\approx 5.077^+$ (cf. our bound 5.507^- above). On the other hand, there is no obvious candidate for where λ_3/λ_1 is maximized (the disk is definitely not the answer, since for it $\lambda_3/\lambda_1 = \lambda_2/\lambda_1 \approx 2.5387$). We do know that the best upper bound must be greater than or equal to $35/11 \approx 3.1818$, since λ_3/λ_1 attains this value for a $\sqrt{8}$ by $\sqrt{3}$ rectangle.

Because of certain alternatives that arise, we find it convenient to organize our development around obtaining various upper and lower bounds for the quantity $(\int_\Omega x u_1 u_2 \, dA)^2$, where the Cartesian variables x and y have been set up so that $\int_\Omega x u_1^2 \, dA$, $\int_\Omega y u_1^2 \, dA$, and $\int_\Omega y u_1 u_2 \, dA$ all vanish (see below for details; here dA denotes standard Lebesgue measure in the plane). This allows us to decompose the arguments into their constituent parts and to combine them in a variety of interesting ways. It also brings order to comparisons between the various bounds available and allows us to focus our efforts on those showing the most promise.

In addition, $(\int_\Omega x u_1 u_2 \, dA)^2$ is a physical quantity of interest in its own right. If one had information about this quantity for a given problem, then our bounds below would yield even stronger bounds on the allowed range of values of λ_2/λ_1 and λ_3/λ_1. From a slightly different point of view, one can think of adding a third axis, one for $\lambda_1(\int_\Omega x u_1 u_2 \, dA)^2$ (the factor of λ_1 makes this a suitable dimensionless quantity), to our range of values of λ_2/λ_1 and λ_3/λ_1 graph (see Figure 1) to obtain a 3-dimensional region containing the allowed values of $(\lambda_2/\lambda_1, \lambda_3/\lambda_1, \lambda_1(\int_\Omega x u_1 u_2 \, dA)^2)$. Above many points on the perimeter of the region identified in Figure 1 (for the range of λ_2/λ_1 and λ_3/λ_1) there is only one value of $\lambda_1(\int_\Omega x u_1 u_2 \, dA)^2$ possibly allowed. This includes all bounds developed in this paper and, in particular, the special points that we identify in Section 4. Thus, if one could show that some such value leads to a contradiction, then our bounds for the allowed region for λ_2/λ_1 and λ_3/λ_1 could be tightened.

We also want to mention that while the idea of making better ("optimal") use of the Cauchy–Schwarz inequality is a key idea of Yang, it is not the only one. Indeed, using this idea together with his method of introducing factors and summing he is able to establish the general bound

$$(5) \qquad \sum_{i=1}^{k} (\lambda_{k+1} - \lambda_i)(n\lambda_{k+1} - (n+4)\lambda_i) \le 0$$

for all $k = 1, 2, 3, \ldots$, which contains the bound (44) that we refer to in this paper as "Yang's bound" as a special case (the case where $k = 2$; n here denotes the dimension, i.e., (5) holds for $\Omega \subset I\!R^n$). This bound also turns out to be better than the corresponding bounds of Payne, Pólya, and Weinberger [15], Hile and Protter [12], and others for eigenvalues of large index, and specifically in the limit as k goes to infinity. This was called attention to by Yang himself [16], with further investigations now being carried out by Harrell and Stubbe [11].

The literature of bounds for ratios of eigenvalues of the Laplacian begins with Payne, Pólya, and Weinberger [14],[15]. Further advances were made by Brands [8], de Vries [10], Hile and Protter [12], Marcellini [13], and Chiti [9], and, most recently by Yu [17], Yang [16], and ourselves [1]–[5], [7]. For other related work, see the references in [2], [5] and in our review paper [6].

2 Improving Upon Marcellini's Argument

Following Marcellini, we choose Cartesian coordinates x, y so that

$$(6) \qquad \int_{\Omega} x u_1^2 \, dA = 0 = \int_{\Omega} y u_1^2 \, dA = \int_{\Omega} y u_1 u_2 \, dA.$$

The first two of these amount to choosing the origin to be at the center of mass of Ω with mass density u_1^2, while the last can be arranged via rotation, if necessary, by a continuity argument.

We now introduce trial functions for $i = 1, 2$ (as in [15], [13])

$$(7) \qquad \phi_i = x u_i - \sum_{j=1}^{2} a_{ij} u_j \qquad \text{where} \qquad a_{ij} = \int_{\Omega} x u_i u_j \, dA = a_{ji}.$$

Clearly $\phi_i \perp u_1, u_2$ and

$$(8) \qquad \int_{\Omega} \phi_i^2 \, dA = \int_{\Omega} x u_i \phi_i \, dA = \int_{\Omega} x^2 u_i^2 \, dA - \sum_{j=1}^{2} a_{ij}^2.$$

Also

$$(9) \qquad -\Delta \phi_i = \lambda_i x u_i - 2 u_{ix} - \sum_{j=1}^{2} \lambda_j a_{ij} u_j$$

and hence

$$(10) \qquad \int_{\Omega} \phi_i(-\Delta \phi_i) \, dA = \lambda_i \int_{\Omega} x u_i \phi_i \, dA - 2 \int_{\Omega} \phi_i u_{ix} \, dA = \lambda_i \int_{\Omega} \phi_i^2 \, dA - 2 \int_{\Omega} \phi_i u_{ix} \, dA.$$

By the Rayleigh–Ritz inequality (for λ_3) we have

(11)
$$(\lambda_3 - \lambda_i) \int_\Omega \phi_i^2 \, dA \le -2 \int_\Omega \phi_i u_{ix} \, dA.$$

Now

$$
\begin{aligned}
-2 \int_\Omega \phi_i u_{ix} \, dA &= -2 \int_\Omega [x u_i - \sum_{j=1}^{2} a_{ij} u_j] u_{ix} \, dA \\
&= -\int_\Omega x(u_i^2)_x \, dA + 2 \sum_{j=1}^{2} a_{ij} \int_\Omega u_{ix} u_j \, dA \\
&= 1 + 2 \sum_{j=1}^{2} a_{ij} b_{ij}.
\end{aligned}
$$

(12)

where we have integrated the first term by parts and defined

(13)
$$b_{ij} = \int_\Omega u_{ix} u_j \, dA.$$

Integration by parts shows $b_{ji} = -b_{ij}$ and, even better, we can relate b_{ij} to a_{ij} via

(14)
$$2 b_{ij} = (\lambda_i - \lambda_j) a_{ij}$$

by making use of the identity

$$
\begin{aligned}
\lambda_j a_{ij} &= \lambda_j \int_\Omega x u_i u_j \, dA = \int_\Omega x u_i (-\Delta u_j) \, dA = \int_\Omega u_j [-\Delta(x u_i)] \, dA \\
&= \int_\Omega u_j [\lambda_i x u_i - 2 u_{ix}] \, dA = \lambda_i a_{ij} - 2 \int_\Omega u_{ix} u_j \, dA = \lambda_i a_{ij} - 2 b_{ij}.
\end{aligned}
$$

(15)

We therefore obtain (from (12) and (14))

(16)
$$-2 \int_\Omega \phi_i u_{ix} \, dA = 1 + \sum_{j=1}^{2} (\lambda_i - \lambda_j) a_{ij}^2.$$

We now turn to the use of the Cauchy–Schwarz inequality to bound $-2 \int_\Omega \phi_i u_{ix} \, dA$. It is here that, following Yang's lead, we can improve upon what Marcellini did. Because ϕ_i is orthogonal to u_1 and u_2 it does not change the inner product $\int \phi_i u_{ix} \, dA$ if we replace u_{ix} by u_{ix} less its components along u_1 and u_2, i.e., by

(17)
$$u_{ix} - \sum_{j=1}^{2} b_{ij} u_j.$$

On the other hand, this is very advantageous from the point of view of what results from applying the Cauchy–Schwarz inequality, since we obtain

$$
\begin{aligned}
\left(-2 \int_\Omega \phi_i u_{ix} \, dA\right)^2 &= 4\left(\int_\Omega \phi_i [u_{ix} - \sum_{j=1}^{2} b_{ij} u_j] \, dA\right)^2 \\
&\le 4\left(\int_\Omega \phi_i^2 \, dA\right)\left(\int_\Omega [u_{ix} - \sum_{j=1}^{2} b_{ij} u_j]^2\right) = 4\left(\int_\Omega \phi_i^2 \, dA\right)\left[\int_\Omega u_{ix}^2 - \sum_{j=1}^{2} b_{ij}^2\right].
\end{aligned}
$$

(18)

34

In Marcellini's version of this argument he did not have the terms in b_{ij}^2 on the right and hence he had a weaker inequality.

Since (11) implies that $-2 \int_\Omega \phi_i u_{ix} \, dA \geq 0$, we can conclude from (18) that

$$(19) \qquad \frac{-2 \int_\Omega \phi_i u_{ix} \, dA}{\int_\Omega \phi_i^2 \, dA} \leq \frac{4 \int_\Omega u_{ix}^2 - 4 \sum_{j=1}^2 b_{ij}^2}{-2 \int_\Omega \phi_i u_{ix} \, dA}$$

(where we are prepared to treat the right–hand side as infinity if $-2 \int_\Omega \phi_i u_{ix} \, dA = 0$; note, too, that by (16) this is impossible if $i = 2$). Putting this together with the Rayleigh–Ritz inequality (11) and using (12) yields

$$(20) \qquad \lambda_3 - \lambda_i \leq \frac{4 \int_\Omega u_{ix}^2 - 4 \sum_{j=1}^2 b_{ij}^2}{1 + 2 \sum_{j=1}^2 a_{ij} b_{ij}}.$$

Finally, we use the fact that $b_{11} = b_{22} = 0$ to reduce (20) to

$$(21) \qquad (\lambda_3 - \lambda_1)[1 - (\lambda_2 - \lambda_1)a_{12}^2] \leq 4 \int_\Omega u_{1x}^2 \, dA - (\lambda_2 - \lambda_1)^2 a_{12}^2$$

in the case where $i = 1$ or

$$(22) \qquad (\lambda_3 - \lambda_2)[1 + (\lambda_2 - \lambda_1)a_{12}^2] \leq 4 \int_\Omega u_{2x}^2 \, dA - (\lambda_2 - \lambda_1)^2 a_{12}^2$$

in the case where $i = 2$.

A similar derivation, but starting from (7) with x replaced by y, yields the simpler analogs of (21) and (22)

$$(23) \qquad \lambda_3 - \lambda_1 \leq 4 \int_\Omega u_{1y}^2 \, dA$$

and

$$(24) \qquad \lambda_3 - \lambda_2 \leq 4 \int_\Omega u_{2y}^2 \, dA,$$

because in that case $\tilde{a}_{12} \equiv \int_\Omega y u_1 u_2 \, dA = 0$, by virtue of (6). Since

$$(25) \qquad \int_\Omega [u_{ix}^2 + u_{iy}^2] \, dA = \int_\Omega |\nabla u_i|^2 \, dA = \lambda_i,$$

summing (21) and (23) yields

$$(26) \qquad (\lambda_3 - \lambda_1)[2 - (\lambda_2 - \lambda_1)a_{12}^2] \leq 4\lambda_1 - (\lambda_2 - \lambda_1)^2 a_{12}^2,$$

while summing (22) and (24) yields

$$(27) \qquad (\lambda_3 - \lambda_2)[2 + (\lambda_2 - \lambda_1)a_{12}^2] \leq 4\lambda_2 - (\lambda_2 - \lambda_1)^2 a_{12}^2.$$

From these it follows easily that

$$(28) \qquad a_{12}^2 \geq \frac{2(\lambda_3 - 3\lambda_1)}{(\lambda_2 - \lambda_1)(\lambda_3 - \lambda_2)}$$

and

$$(29) \qquad a_{12}^2 \leq \frac{2(3\lambda_2 - \lambda_3)}{(\lambda_2 - \lambda_1)(\lambda_3 - \lambda_1)}.$$

We note that the right–hand side of (29) is always positive since we know that $\lambda_1 < \lambda_2 \leq \lambda_3$ and, from [3], [4], $\lambda_3/\lambda_2 < K_2 \approx 2.539^-$ (indeed, (29) implies $\lambda_3/\lambda_2 \leq 3$). Also, the right–hand side of (28) can be replaced by 0 whenever $\lambda_3 < 3\lambda_1$ (since clearly $a_{12}^2 \geq 0$). This applies, in particular, if $\lambda_2 = \lambda_3$, because we know that $\lambda_2/\lambda_1 \leq K_2$.

If one follows through the arguments above without making "optimal" use of the Cauchy–Schwarz inequality as done in (18), then one arrives at the weaker variants of (28) and (29)

$$(30) \qquad a_{12}^2 \geq \frac{2(\lambda_3 - 3\lambda_1)}{(\lambda_2 - \lambda_1)(\lambda_3 - \lambda_1)}$$

and

$$(31) \qquad a_{12}^2 \leq \frac{2(3\lambda_2 - \lambda_3)}{(\lambda_2 - \lambda_1)(\lambda_3 - \lambda_2)}.$$

Marcellini obtained (31) in his paper (see the inequality following his inequality (17)). He did not exhibit or use (30) (although he surely knew of it), because it was not strong enough to surpass other bounds already known at that time (for example, (30) and (31) together imply only the bound of Hile and Protter [12], $y \leq 1 + x + \sqrt{1 - x - x^2}$, where $x \equiv \lambda_2/\lambda_1$ and $y \equiv \lambda_3/\lambda_1$). Instead, for his lower bound to a_{12}^2 he used

$$(32) \qquad a_{12}^2 \geq \frac{2\lambda_2\lambda_3 - 5\lambda_1\lambda_2 - \lambda_1^2}{(\lambda_3 - \lambda_2)(4\lambda_1\lambda_2 + \lambda_1^2 - \lambda_2\lambda_3)}.$$

This bound is derived by starting from the Rayleigh–Ritz inequality (11) for $i = 1$, but without using the Cauchy–Schwarz inequality. Instead we use (8) and (16) to rewrite (11) as

$$(33) \qquad (\lambda_3 - \lambda_1) \int_\Omega x^2 u_1^2 \, dA \leq 1 + (\lambda_3 - \lambda_2)a_{12}^2$$

and, if x is replaced by y, this reduces to

$$(34) \qquad (\lambda_3 - \lambda_1) \int_\Omega y^2 u_1^2 \, dA \leq 1$$

by virtue of (6) (note that these are Marcellini's inequalities (15) and (17)). These two inequalities are then combined with the inequality of Brands,

$$(35) \qquad \left(\int_\Omega x^2 u_1^2 \, dA\right)^{-1} + \left(\int_\Omega y^2 u_1^2 \, dA\right)^{-1} \leq \lambda_1(3 + \lambda_1/\lambda_2),$$

to yield (32). See [8] for the derivation of (35) and Section 6 of [5] for the n–dimensional version of this bound. While the derivation of (35) does make use of the Cauchy–Schwarz inequality, this part of the argument does not appear amenable to improvement using Yang's idea in the same way that the earlier parts were. We also observe that in the course of deriving (32) we find that the denominator on the right is nonnegative, i.e., $\lambda_3/\lambda_1 \leq 4 + \lambda_1/\lambda_2$ which is only slightly weaker than the bound $\lambda_3/\lambda_1 \leq 5 + \lambda_1/\lambda_2 - \lambda_2/\lambda_1$ of Brands [8]. As in (28) and (30), if the numerator of the lower bound to a_{12}^2 in (32) ever becomes negative we are better off using the trivial lower bound $a_{12}^2 \geq 0$.

Remark. There are two further bounds on a_{12}^2 worthy of note. The first is

$$(36) \qquad a_{12}^2 \leq \frac{1}{\lambda_2 - \lambda_1}$$

and follows immediately from (16) with $i = 1$ (recall that (11) implies that $-2\int_\Omega \phi_i u_{ix} \, dA \geq 0$). Alternatively, one can use Cauchy–Schwarz and (33) to see that $a_{12}^2 \leq \int_\Omega x^2 u_1^2 \, dA \leq$

$(\lambda_2 - \lambda_1)^{-1}$. The second bound follows from adding (33) and (34) and using rearrangements following Chiti [9] (see also [1], [2], [5]):

$$2 + (\lambda_3 - \lambda_2)a_{12}^2 \geq (\lambda_3 - \lambda_1)\int_\Omega r^2 u_1^2 \, dA \geq (\lambda_3 - \lambda_1)\int_{\Omega^\star} r^2 u_1^{\star 2} \, dA$$

$$\geq (\lambda_3 - \lambda_1)\frac{j_{0,1}^2 - 2}{3\lambda_1}$$

where $j_{0,1} \approx 2.4048255577$ denotes the first positive zero of the Bessel function $J_0(t)$. We thereby obtain the lower bound

(37)
$$a_{12}^2 \geq \frac{k\lambda_3 - (k+2)\lambda_1}{\lambda_1(\lambda_3 - \lambda_2)}$$

where $k = (j_{0,1}^2 - 2)/3 \approx 1.261061987$.

3 The Bounds

We are now in position to combine our inequalities for a_{12}^2 to obtain inequalities for λ_3/λ_1 in terms of λ_2/λ_1. For the purposes of this and later sections it is convenient to adopt the notation $x \equiv \lambda_2/\lambda_1$ and $y \equiv \lambda_3/\lambda_1$.

By combining (28) and (29) we obtain the inequality (cf. (5) with $k = 2$ and $n = 2$)

(38)
$$(y-1)(y-3) + (y-x)(y-3x) \leq 0,$$

or

$$2y^2 - 4(1+x)y + 3(1+x^2) \leq 0,$$

yielding Yang's bound
(39)
$$y \leq 1 + x + \sqrt{2x - (1+x^2)/2}.$$

By combining (29) and (32) we obtain the cubic inequality

(40) $$2xy^3 - 2(5x^2 + 3x + 1)y^2 + (6x^3 + 39x^2 + 2x - 1)y - (24x^3 + 11x^2 - 4x - 1) \geq 0.$$

We claim that the middle root of this cubic gives an upper bound for y, which is an improvement upon the bound of Marcellini [13]. The latter bound is obtained by combining (31) and (32), which yields

(41)
$$2xy^2 - 2(4x^2 + 3x + 1)y + (29x^2 + 2x - 1) \geq 0.$$

To see which roots of the polynomials in (40) and (41) give the upper bounds we seek, we note that when $x = 1$ (40) has $y = 1, 3, 5$ as roots and (41) has $y = 3, 5$ as roots. By substitution of $y = 4$ into the respective polynomials we find that each is negative for all x, and hence $y = 4$ can never satisfy (40) or (41). Since we already know that $y < 4$ always holds, this shows that the upper bounds that we seek are, respectively, the middle root of (40) and the smaller root of (41), i.e.,

(42)
$$y \leq \{4x^2 + 3x + 1 - [(4x^2 + 3x + 1)^2 - 2x(29x^2 + 2x - 1)]^{1/2}\}/(2x).$$

Remarks. (1) Similar arguments could also be given starting from the known upper bounds $y < 5$ and $y \leq 4 + 1/x$ (both of which follow from Brands' bound $y \leq 5 + 1/x - x$).

(2) That the upper bound based on (40) is better than (42) follows immediately from the fact that (29) is always better than (31) (for cases which are realizable).

(3) It is interesting to investigate the bounds that can be obtained by combining any of the lower bounds (28), (30), (32), or (37) for a_{12}^2 with any of the upper bounds (29), (31), or (36). The best bounds are had by combining (28) and (29) (Yang's bound (39)) and (29) and (32) (our bound (40)) but other sometimes simpler bounds result from other combinations. We have already noted the combinations (30) and (31) (Hile and Protter's bound) and (31) and (32) (Marcellini's bound (42)). Other combinations of note are (32) and (36) giving Brands' bound $y \leq 5 + 1/x - x$ and (28) and (31) (or (29) and (30)) giving $y \leq 3(1+x)/2$, a bound which is already better than the bounds $y \leq 1 + 2x$ of Payne, Pólya, and Weinberger [15] and $y \leq 1 + x + \sqrt{1 - x + x^2}$ of Hile and Protter [12]. The bound $y \leq 3(1+x)/2$ is the special case of Yang's inequality (3) in [16] with $k = 2$ and $n = 2$, which is in turn weaker than his main inequality (see (5) above, and its $k = 2$ specialization, (39)).

(4) By virtue of the bounds (28) and (29) any bound stemming from (30) or (31) can be written as a strict inequality.

4 The Best Inequalities

In this section we summarize the best upper bounds currently known for $y = \lambda_3/\lambda_1$ as a function of $x = \lambda_2/\lambda_1$ for $1 < x \leq K_2$ where $K_2 \equiv j_{1,1}^2/j_{0,1}^2 \approx 2.5387$ (here $j_{p,1}$ denotes the first positive zero of the Bessel function $J_p(t)$). That x lies in the range $1 < x \leq K_2$, with $x = K_2$ if and only if Ω is a disk, follows from our earlier work [1], [2]. The lower bound $y \geq x$ is of course immediate.

We have

$$(43) \qquad y < K_2 x \qquad \text{for} \qquad 1 < x \leq 1.396^-,$$

$$(44) \qquad y \leq 1 + x + \sqrt{2x - (1 + x^2)/2} \qquad \text{for} \qquad 1.396^- \leq x \leq 1.634^-,$$

$$(45) \qquad y \leq F(x) \qquad \text{for} \qquad 1.634^- \leq x \leq 1.676^-,$$

$$(46) \qquad y \leq H(x) - x \qquad \text{for} \qquad 1.676^- \leq x \leq 2.198^+,$$

and

$$(47) \qquad y \leq G(x) \qquad \text{for} \qquad 2.198^+ \leq x \leq 2.539^-,$$

where the function $H(x)$ is as given in [7] and the functions $F(x)$ and $G(x)$ are defined by

$$(48) \qquad F(x) = \text{middle root of the cubic}$$

$$2xy^3 - 2(5x^2 + 3x + 1)y^2 + (6x^3 + 39x^2 + 2x - 1)y - (24x^3 + 11x^2 - 4x - 1) = 0$$

and

$$(49) \qquad G(x) = \inf_{\beta > 1/2} \left[\frac{\beta^2}{2\beta - 1} + \frac{x - \beta^2/(2\beta - 1)}{C_2(\beta)(x - \beta^2/(2\beta - 1)) - 1} \right]$$

with the infimum taken over values of β satisfying $x > \beta^2/(2\beta - 1) + 1/C_2(\beta)$ and with

$$(50) \qquad C_2(\beta) \equiv \frac{2\beta - 1}{\beta} \frac{\int_0^{j_{0,1}} t^3 J_0^{2\beta}(t)\, dt}{\int_0^{j_{0,1}} t J_0^{2\beta}(t)\, dt}.$$

Here $J_0(t)$ denotes the standard Bessel function of order zero and $j_{0,1}$ is its first positive zero. For the derivation of the bound (46) and more discussion of it, see [7]. The other

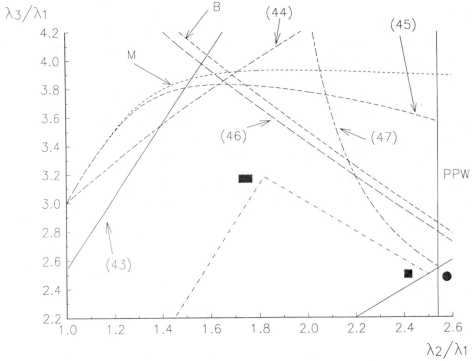

FIG. 1. *Range of values of λ_2/λ_1 and λ_3/λ_1. These curves represent the inequalities (43)–(47) as well as the Marcellini (M) and Brands (B) inequalities. The permitted region is bounded above by the curves (43)–(47), on the right by $x \leq 2.5387^+$, and below by $y \geq x$. The dot–dashed broken line gives points corresponding to all rectangles, while special symbols mark the disk, the square, and the $\sqrt{8}$ by $\sqrt{3}$ rectangle.*

bounds given above are due to H. Yang (see [16] for (44)) and ourselves (see [3], [4] for (43), [7] for (47), and this paper for (45)).

We remark that the inequalities (43)–(47) all apply on broader intervals of x–values than the intervals specified explicitly with them; the given intervals indicate the range for which the corresponding inequality gives the best bound yet found.

The various inequalities stated above, as well as some of the inequalities that these supersede (for more details on these see [7]), are graphed in Figure 1. In addition, since inequalities (45)–(47) are best handled numerically, we have included tables of their values for the relevant ranges of x. These are presented in Tables 1–3.

By noting that $F(x)$ has a maximum within the interval where it is our best bound we obtain the best upper bound (3) yet found for λ_3/λ_1. The maximum occurs at the point $(x, y) \approx (1.657279842, 3.831028110)$. On the other hand the best upper bound (4) yet found for $(\lambda_2 + \lambda_3)/\lambda_1$ occurs at the intersection point between the bounds (45) and (46). This point is found to be $(x, y) \approx (1.67581, 3.83080)$.

Other points of interest include the intersection point between (43) and (44) at $(x, y) \approx (1.395836534, 3.543657624)$ and the intersection point between (44) and (45) at $(x, y) \approx (1.633596268, 3.830623483)$. It is interesting to note that $F(x)$ varies very little over the (small) interval where it is our best bound: it has a minimum of ≈ 3.8306 at the left end, a maximum of ≈ 3.8310, and a relative minimum of ≈ 3.8308 at the right end.

x	$(46)(x)$
1.66	3.8516
1.676	3.8308
1.68	3.8253
1.70	3.7992
1.72	3.7732
1.74	3.7473
1.76	3.7217
1.78	3.6961
1.80	3.6707
1.82	3.6454
1.84	3.6202
1.86	3.5952
1.88	3.5702
1.90	3.5454
1.92	3.5207
1.94	3.4961
1.96	3.4715
1.98	3.4471
2.00	3.4227
2.02	3.3985
2.04	3.3743
2.06	3.3502
2.08	3.3262
2.10	3.3022
2.12	3.2784
2.14	3.2546
2.16	3.2308
2.18	3.2072
2.198	3.1858
2.20	3.1836

Table 2.

x	$(45)(x)$
1.62	3.83000
1.63	3.83049
1.634	3.83062
1.64	3.83081
1.65	3.83099
1.66	3.83102
1.67	3.83092
1.676	3.83080
1.68	3.83068
1.69	3.83032

Table 1.

x	$(47)(x)$
2.198	3.1858
2.20	3.1793
2.22	3.1157
2.24	3.0575
2.26	3.0043
2.28	2.9553
2.30	2.9100
2.32	2.8681
2.34	2.8292
2.36	2.7930
2.38	2.7593
2.40	2.7276
2.42	2.6980
2.44	2.6702
2.46	2.6440
2.48	2.6193
2.50	2.5960
2.52	2.5739
2.54	2.5530

Table 3.

5 N–Dimensional Bounds

All the bounds developed above have analogs for the problem (1)–(2) where Ω is a bounded domain in n dimensions. Because the proofs are identical to those for the 2–dimensional case, we jump directly to a brief summary of the results. Following Section 3 above, the analogs of (28) and (29) are

$$(51) \qquad a_{12}^2 \geq \frac{n\lambda_3 - (n+4)\lambda_1}{(\lambda_2 - \lambda_1)(\lambda_3 - \lambda_2)}$$

and

$$(52) \qquad a_{12}^2 \leq \frac{(n+4)\lambda_2 - n\lambda_3}{(\lambda_2 - \lambda_1)(\lambda_3 - \lambda_1)}.$$

Also,

$$(53) \qquad a_{12}^2 \geq \frac{n\lambda_2\lambda_3 - (n+3)\lambda_1\lambda_2 - \lambda_1^2}{(\lambda_3 - \lambda_2)[(n+2)\lambda_1\lambda_2 + \lambda_1^2 - (n-1)\lambda_2\lambda_3]}.$$

is the n–dimensional analog of (32). Combining (51) and (52) yields the special case of Yang's bound (5) with $k = 2$, or, explicitly,

$$(54) \qquad y \leq \frac{1}{2}\left(1 + \frac{2}{n}\right)(1 + x) + \frac{1}{2}\sqrt{2\left(1 + \frac{2}{n}\right)^2 x - \left(1 + \frac{4}{n} - \frac{4}{n^2}\right)(1 + x^2)},$$

which is the n-dimensional analog of (39). Similarly, combining (52) and (53) gives

$$
\begin{aligned}
(55) \qquad n(n-1)xy^3 \quad &- \quad [(2n^2 + 3n - 4)x^2 + n(n+1)x + n]y^2 \\
&+ \quad [(n-1)(n+4)x^3 + (2n^2 + 10n + 11)x^2 + 2x - 1]y \\
&- \quad [(n+2)(n+4)x^3 + (2n+7)x^2 - (n+2)x - 1] \geq 0,
\end{aligned}
$$

the n–dimensional analog of (40). Again, the explicit upper bound for y that follows from this is given by the middle root of the cubic appearing on the left. This can be readily seen by evaluating the cubic at $y = (n+2+1/x)/(n-1)$, the right–hand side of which is known to be an upper bound on y by the n–dimensional version of Brands' bound (see inequalities (6.10) and (6.18) of [5], for example). One finds the value to be $-(1-1/x)(3x+1)^2/(n-1)^2$, and since this is negative for $x > 1$ it is clear that $y = (n+2+1/x)/(n-1)$ lies below the largest root of the given cubic and in a forbidden region. Since there surely are allowed values for (x, y), it must be that the cubic has its other two roots real (at least for some choices of $x > 1$) and that, in particular, the allowed region lies below the middle root. (For completeness we note that the lower bound determined by the lowest root is not of interest because it lies below the curve $y = x$ and we already know that $y \geq x$ is a bound.) **Remarks.** (1) We could also have made the previous argument by substituting $y = (n+3)/(n-1)$ and confirming that this lies in a forbidden region, i.e., it violates (55).

(2) Another way to make the argument above is to work with the n–dimensional analog of Marcellini's bound (given in Section 6 of [5], for example)

$$
\begin{aligned}
(56) \qquad n(n-1)xy^2 \quad &- \quad [(n^2 + 4n - 4)x^2 + n(n+1)x + n]y \\
&+ \quad [(n^2 + 7n + 11)x^2 + 2x - 1] \geq 0,
\end{aligned}
$$

and first show that the lower root of this quadratic is an upper bound to y by showing that the substitution $y = (n+2+1/x)/(n-1)$ (or $y = (n+3)/(n-1)$) makes the left–hand side of (56) negative for $x > 1$, i.e., that the curve $y = (n+2+1/x)/(n-1)$ (respectively, $y = (n+3)/(n-1)$) lies in a forbidden region. Then in turn one shows that the curve defined by the lower root of (56) lies in a forbidden region for (55) (using the fact that (55) comes from (52) and (53) while (56) comes from (53) and the weakening of (52) where the $\lambda_3 - \lambda_1$ in the denominator of the right–hand side is replaced by $\lambda_3 - \lambda_2$). As argued earlier this shows that the cubic on the left in (55) must have all its roots real for any allowed value of x (i.e., for $1 < x \leq K_n \equiv (j_{n/2,1}/j_{n/2-1,1})^2$) and its middle root must give an upper bound for y (so that the allowed region lies between the curve $y = x$ and the curve determined by this middle root).

(3) The reader might well wonder at the validity of our bounds (51), (52), and (53) above, and at the meaning of a_{12} in the n–dimensional setting. We claim that, by suitable choice of origin and rotation of axes, we can find Cartesian coordinates $x_1, \ldots, x_n$ such that $\int_\Omega x_i u_1^2 = 0$ for $i = 1, \ldots, n$, and $\int_\Omega x_i u_1 u_2 = 0$ for $i = 2, 3, \ldots, n$ (all our integrals are now to be taken with respect to n–dimensional Lebesgue measure). By a_{12} above we mean $a_{12} = \int_\Omega x_1 u_1 u_2$ with respect to this coordinate system. In the derivations of (51) (respectively, (52)) one then has one inequality (in x_1) analogous to (21) (resp., (22)) and $n - 1$ inequalities (in $x_2, \ldots, x_n$) analogous to (23) (resp., (24)). Summing, and using the n–dimensional analog of (25), yields (51) (resp., (52)). Inequality (53) follows in a similar fashion from one "copy" of (33) and $n - 1$ "copies" of (34) combined according to the n–dimensional version of Brands' bound (35) (see Section 6 of [5]):

$$(57) \qquad \sum_{i=1}^{n} \left(\int_\Omega x_i^2 u_1^2 \right)^{-1} \leq \lambda_1 (3 + \lambda_1/\lambda_2).$$

Acknowledgements

We thank Evans Harrell and Pawel Kröger for informing us of the preprint of H. Yang and providing us with a copy of it. M. S. A. gratefully acknowledges the hospitality of the Physics Department of the Catholic University of Chile in June–July 1995 when much of this work was completed.

References

[1] M. S. Ashbaugh and R. D. Benguria, Proof of the Payne-Pólya-Weinberger conjecture, Bull. Amer. Math. Soc. **25** (1991), 19–29.

[2] M. S. Ashbaugh and R. D. Benguria, A sharp bound for the ratio of the first two eigenvalues of Dirichlet Laplacians and extensions, Annals of Math. **135** (1992), 601–628.

[3] M. S. Ashbaugh and R. D. Benguria, Isoperimetric bound for λ_3/λ_2 for the membrane problem, Duke Math. J. **63** (1991), 333–341.

[4] M. S. Ashbaugh and R. D. Benguria, Isoperimetric bounds for higher eigenvalue ratios for the n-dimensional fixed membrane problem, Proc. Royal Soc. Edinburgh **123A** (1993), 977–985.

[5] M. S. Ashbaugh and R. D. Benguria, More bounds on eigenvalue ratios for Dirichlet Laplacians in n dimensions, SIAM J. Math. Anal. **24** (1993), 1622–1651.

[6] M. S. Ashbaugh and R. D. Benguria, Isoperimetric inequalities for eigenvalue ratios, in **Partial Differential Equations of Elliptic Type, Cortona, 1992**, A. Alvino, E. Fabes, and G. Talenti, editors, Symposia Mathematica, vol. 35, Cambridge University Press, Cambridge, 1994, pp. 1–36.

[7] M. S. Ashbaugh and R. D. Benguria, The range of values of λ_2/λ_1 and λ_3/λ_1 for the fixed membrane problem, Rev. in Math. Phys. **6** (1994), 999–1009.

[8] J. J. A. M. Brands, Bounds for the ratios of the first three membrane eigenvalues, Arch. Rat. Mech. Anal. **16** (1964), 265–268.

[9] G. Chiti, A bound for the ratio of the first two eigenvalues of a membrane, SIAM J. Math. Anal. **14** (1983), 1163–1167.

[10] H. L. de Vries, On the upper bound for the ratio of the first two membrane eigenvalues, Zeitschrift für Naturforschung **22A** (1967), 152–153.

[11] E. M. Harrell and J. Stubbe, On trace identities and universal eigenvalue estimates for some partial differential operators, 1995 preprint.

[12] G. N. Hile and M. H. Protter, Inequalities for eigenvalues of the Laplacian, Indiana Univ. Math. J. **29** (1980), 523–538.

[13] P. Marcellini, Bounds for the third membrane eigenvalue, J. Diff. Eqs. **37** (1980), 438–443.

[14] L. E. Payne, G. Pólya, and H. F. Weinberger, Sur le quotient de deux fréquences propres consécutives, Comptes Rendus Acad. Sci. Paris **241** (1955), 917–919.

[15] L. E. Payne, G. Pólya, and H. F. Weinberger, On the ratio of consecutive eigenvalues, J. of Math. and Phys. **35** (1956), 289–298.

[16] Hongcang Yang, Estimates of the difference between consecutive eigenvalues, 1995 preprint (revision of International Centre for Theoretical Physics preprint IC/91/60, Trieste, Italy, April, 1991).

[17] Qihuang Yu, On the first and second eigenvalues of Schrödinger operator, Chinese Ann. of Math. **14B** (1993), 85–92.

POTENTIAL REPRESENTATIONS OF INCOMPRESSIBLE VECTOR FIELDS

GILES AUCHMUTY

Dedicated to Ivar Stakgold on his 70th birthday.

ABSTRACT. Let Ω be a nice bounded domain in $\mathbb{R}^3$. We first describe an orthogonal Helmholtz decomposition of L^2-vector fields on Ω which holds on domains with arbitrary topology. Under mild regularity and boundary conditions this is also H^1-orthogonal. Then it is shown that if v is incompressible and obeys a flux condition, there is an incompressible vector potential A such that $v = \operatorname{curl} A$ on Ω. Variational characterizations of the potentials are described.

1. INTRODUCTION

Given a 3-dimensional vector field v on a domain $\Omega \subseteq \mathbb{R}^3$, there are a variety of ways to define scalar and vector potentials (φ, A) to yield a Helmholtz decomposition

$$(1.1) \qquad v = \nabla\varphi + \operatorname{curl} A.$$

It appears to be an open question whether the usual definitions of these potentials provides an orthogonal decomposition of $L^2(\Omega; \mathbb{R}^3)$ with the standard inner product. See Section 12 of [1] for a discussion of this.

Over 50 years ago, Hermann Weyl [5] showed that, assuming some regularity on the domain Ω,

$$(1.2) \qquad v = \nabla\varphi + \operatorname{curl} A + h$$

when v is an L^2-vector field on Ω. He constructed φ and A using potential-theoretic methods and described the harmonic vector field h. In section 3 of this paper, this construction will be adapted to provide an L^2-orthogonal Helmholtz decomposition

Department of Mathematics, University of Houston, Tx 77204-3476.
This research was partially supported by NSF Grant DMS 9501148.

of the form (1.1). When div v is also in $L^2(\Omega)$, this decomposition is also orthogonal in some natural Sobolev space inner-products provided that the boundary condition

$$(1.3) \qquad\qquad v \cdot \nu = 0 \quad \text{on } \partial\Omega \text{ holds.}$$

Weyl's construction in (1.2) reduces to (1.1) when $\partial\Omega$ is contractible. The construction of the orthogonal Helmholtz decomposition described in section 3 holds on domains of arbitrary differential topological complexity. In particular we specify exactly how the scalar and vector potential potentials in (1.1) should be determined. They will be described as solutions of certain natural variational principles rather than using the classical methods employed by Weyl.

A vector field is said to be incompressible if

$$(1.4) \qquad\qquad \text{div } v = \sum_{j=1}^{3} v_{j,j} = 0 \quad \text{on } \Omega.$$

In section 4, we shall show that if v is incompressible and certain flux conditions hold, then there is a vector potential A in $H^1(\Omega; \mathbb{R}^3)$ which also obeys (1.3), (1.4) and

$$(1.5) \qquad\qquad v = \text{curl } A \quad \text{on } \Omega.$$

In particular any incompressible L^2-vector field on Ω obeying (1.3) can be written as the curl of another vector field with the same properties. This does not require any differential topological restrictions on Ω.

It is a particular pleasure to dedicate this article to Ivar Stakgold in appreciation of his advice, help and friendship over the years.

2. DOMAINS AND FUNCTION SPACES

Throughout this paper Ω will be a bounded, connected, open set in $\mathbb{R}^3$ with boundary $\partial\Omega$. We shall require Ω be $C^{1,1}$ in the sense of [3] section 1.1 and that Ω be locally on one side of Ω. Let $\nu : \partial\Omega \to \mathbb{R}^3$ be the unit outward normal to $\partial\Omega$. Then ν will be Lipschitz continuous on $\partial\Omega$ and various integration-by-parts (or Gauss-Green) formulae hold. We shall assume the conventions of [1], when terms are not otherwise defined here.

We shall also need some differential topological properties of Ω. Let $\{U_0, U_1, \dots, U_K\}$ be the maximal connected open non-empty components of $\mathbb{R}^3 \setminus \bar{\Omega}$ with U_0 unbounded, so that

(2.1)
$$\mathbb{R}^3 \setminus \bar{\Omega} = \bigcup_{k=1}^K U_k.$$

K is called the second Betti number, $\beta_2(\Omega)$, of Ω and is a topological invariant of the domain. When $K = 0$, we say that $\partial\Omega$ is contractible while, when $K \geq 1$, Ω may be regarded as having K "holes." The boundary $\partial\Omega_k$ of the k-th hole will be denoted S_k for $1 \leq k \leq K$.

Let $L^2(\Omega), H^1(\Omega), H_0^1(\Omega)$ be the usual real Hilbert Sobolev spaces and

(2.2)
$$\langle \varphi, \psi \rangle = \int_\Omega \varphi\psi \, dx$$

be the usual L^2-inner product. Let $H^1(\Omega; \mathbb{R}^3)$ be the space of all real vector fields on Ω whose components (v_1, v_2, v_3) each lie in $H^1(\Omega)$. All derivatives used in this paper should be taken as weak, or distributional, derivatives.

Let $H := H(\operatorname{div}, \Omega)$ be the subspace of $L^2(\Omega; \mathbb{R}^3)$ consisting of those vector fields with $\operatorname{div} v \in L^2(\Omega)$. It is a Hilbert space under the inner product

(2.3)
$$[v, w] = \int_\Omega [v \cdot w + \operatorname{div} v \cdot \operatorname{div} w] \, dx.$$

The normal trace operator T_ν is defined for continuous vector fields on $\bar{\Omega}$ by

$$(T_\nu v)(x) = v(x) \cdot \nu(x) \quad \text{for } x \in \partial\Omega.$$

This operator has a continuous linear extension to H and the resulting map $T_\nu : H \to H^{-1/2}(\partial\Omega)$ is surjective. See theorem 1, section IX.2 of [2].

Define $H_0 := H_0(\operatorname{div}, \Omega)$ and $H_\nu := H_\nu^1(\Omega; \mathbb{R}^3)$ to be the closed subspaces of H and $H^1(\Omega; \mathbb{R}^3)$ respectively consisting of all fields obeying

(2.4)
$$T_\nu v = 0.$$

(2.4) is the weak form of (1.3).

H_ν will be a Hilbert space under the inner product

(2.5)
$$[v, w] = \int_\Omega [v \cdot w + \operatorname{div} v \cdot \operatorname{div} w + \operatorname{curl} v \cdot \operatorname{curl} w] \, dx$$

and this is equivalent to the usual inner product on $H^1(\Omega; \mathbb{R}^3)$. See theorem 3, section IX.2 of [2] for a proof.

Finally our main result will be a characterization of the closed subspace $V_0 = V_0(\Omega)$ of all incompressible vector fields in H_0. That is, a vector field v is in $V_0(\Omega)$ if v and div v are L^2 on Ω and (1.3)-(1.4) hold in a weak sense. It is the space of allowable flows for inviscid, incompressible fluid mechanics and also arises in magnetic field theory.

3. An Orthogonal Helmholtz Decomposition

In this section, one of the Hodge-Weyl decompositions will be shown to provide an orthogonal Helmholtz decomposition of L^2-vector fields on Ω.

Define $G_0(\Omega) = \{\nabla\varphi : \varphi \in H_0^1(\Omega)\}$ and $C(\Omega) = \{ \operatorname{curl} A : A \in H^1(\Omega; \mathbb{R}^3)\}$. In [1], sections 4 and 5, it was shown that $G_0(\Omega), C(\Omega)$ are closed, orthogonal subspaces of $L^2(\Omega; \mathbb{R}^3)$ and that the associated projections could be defined variationally.

Namely if $P_G : L^2(\Omega; \mathbb{R}^3) \to G_0(\Omega)$ is the projection onto $G_0(\Omega)$, then

$$(3.1) \qquad P_G v = \nabla\hat{\varphi}$$

where $\hat{\varphi}$ is the unique minimizer in $H_0^1(\Omega)$ of

$$(3.2) \qquad \mathcal{D}(\varphi) = \int_\Omega |v - \nabla\varphi|^2 \, dx.$$

In consequence, $\hat{\varphi}$ is the unique solution in $H_0^1(\Omega)$ of

$$(3.3) \qquad \Delta\varphi = \operatorname{div} v \quad \text{on } \Omega.$$

Similarly the projection P_C of $L^2(\Omega; \mathbb{R}^3)$ onto $C(\Omega)$ is given by

$$(3.4) \qquad P_C v = \operatorname{curl} \hat{A}$$

where $\hat{A}$ is any minimizer in $H^1(\Omega; \mathbb{R}^3)$ of

$$(3.5) \qquad \mathcal{C}(A) = \int_\Omega |v - \operatorname{curl} A|^2 \, dx.$$

There is an affine subspace of such minimizers and one may always find a minimizer in the subspace V_0 of incompressible fields obeying (1.3). The set of all minimizers of $\mathcal{C}$ on V_0 is given by $\hat{A} + h$ where $\hat{A}$ is a minimizer and h is a harmonic vector field on Ω which also obeys (1.3). A vector field h on Ω is said to be harmonic if

$$(3.6) \qquad \operatorname{div} h = 0 \quad \text{and} \quad \operatorname{curl} h = 0 \quad \text{on } \Omega.$$

The class of all such fields obeying (1.3) will be denoted $\mathcal{H}^1(\Omega)$ and is a finite dimensional subspace of $H_\nu(\Omega; \mathbb{R}^3)$. When Ω is simply connected, $\mathcal{H}^1(\Omega) = \{0\}$ and there will be a unique minimizer $\hat{A}$ of $\mathcal{C}$ on V_0; see Section 5 of [1] for proofs of all these statements.

Part II of Theorem 3.1 of [1] can be restated as saying that for any v in $L^2(\Omega; \mathbb{R}^3)$,

$$(3.7) \qquad v = P_G v + P_C v + P_H v$$

where $P_H v$ is the projection of v onto a class of harmonic vector fields and also that this decomposition is L^2-orthogonal.

If $\beta_2(\Omega) = 0$, then $P_H v = 0$ and (3.7) provides an orthogonal decomposition of $L^2(\Omega; \mathbb{R}^3)$. When $\beta_2(\Omega) = K \geq 1$, then Picard [4; section 3] has shown that $\dim R(P_H) = K$ and that $\{h^{(k)} : 1 \leq k \leq K\}$ is a basis of the range of P_H where $h^{(k)} = \nabla \psi_k$ and each ψ_k is defined as the solution of the Dirichlet problem

$$(3.8) \qquad \Delta \psi = 0 \text{ in } \Omega.$$

$$(3.9) \qquad \psi = \begin{cases} 0 & \text{on } \partial\Omega \setminus S_k \\ 1 & \text{on } S_k \end{cases}.$$

Under our regularity conditions on $\partial\Omega$, there is a unique H^1 function in Ω obeying (3.8), (3.9). Thus for any v in $L^2(\Omega; \mathbb{R}^3)$ one may find coefficients $c_1, \dots, c_k$ such that

$$(3.10) \qquad P_H v = \sum_{k=1}^{K} c_k \nabla \psi_k \text{ on } \Omega.$$

These results may be summarized in the following orthogonal Helmholtz representation theorem.

Theorem 1. *Suppose v is in $L^2(\Omega; \mathbb{R}^3)$, then there is a unique function $\hat{\varphi}$ in $H_0^1(\Omega)$, a vector field $\hat{A}$ in V_0, and a harmonic function ψ in $H^1(\Omega)$, such that*

$$(3.11) \qquad v = \nabla(\hat{\varphi} + \psi) + \operatorname{curl} \hat{A}.$$

Moreover the vector fields $\nabla\hat{\varphi}, \nabla\psi$ and $\operatorname{curl} \hat{A}$ are L^2-orthogonal on Ω.

Proof. Substitute in (3.7) from (3.1), (3.4) and (3.10) taking $\psi = \sum_{k=1}^{K} c_k \psi_k$. Then the result follows. $\qquad \square$

Note that (3.11) reduces to (1.1) when $\varphi := \hat{\varphi} + \psi$ and $A := \hat{A}$, so this is a Helmholtz decomposition.

Corollary. *If v is in H_0 (or $H_\nu^1(\Omega; \mathbb{R}^3)$), then $\nabla\hat{\varphi}, \nabla\psi$ and $\operatorname{curl} A$ are orthogonal with respect to the inner products (2.3) (or (2.5)).*

Proof. From (3.8) and (3.10), ψ is a harmonic function, so $\operatorname{div} \nabla\psi = 0$ on Ω. Thus

$$[\nabla\hat{\varphi}, \nabla\psi] = \langle \nabla\hat{\varphi}, \nabla\psi \rangle \quad \text{and} \quad [\nabla\hat{\varphi}, \operatorname{curl} \hat{A}] = \langle \hat{\varphi}, \operatorname{curl} \hat{A} \rangle$$

when $[,]$ is defined by (2.3). Hence L^2-orthogonality of the terms in (3.11) implies orthogonality with respect to (2.3). Similarly for the inner product (2.5) on H_ν. $\square$

4. Incompressible Vector Fields

The orthogonal Helmholtz representation described in the preceding section simplifies considerably for incompressible vector fields. Moreover, it becomes independent of the topology of Ω when the boundary condition (2.4) is also required.

Result 4.1 *Suppose v is in $H(\operatorname{div},\Omega)$, then $P_G v = 0$ if and only if v is incompressible.*

Proof. If v is incompressible, then the unique solution of (3.3) is $\hat\varphi = 0$ so $P_G v = 0$. Conversely if $\nabla\hat\varphi = 0$, then $\hat\varphi \equiv 0$ since $\hat\varphi$ is in $H_0^1(\Omega)$ and so $\operatorname{div} v \equiv 0$ on Ω. $\square$

Result 4.2 *Suppose v is in $H(\operatorname{div},\Omega)$ and $\mathbb{R}^3\backslash\bar\Omega$ has the form (2.1). Then $P_H v = 0$ whenever v is incompressible and*

$$(4.1) \qquad \int_{S_k} (v \cdot \nu) d\sigma = 0 \text{ for } 1 \le k \le K.$$

Proof. Let φ_k, $h^{(k)}$ be defined as in section 3, then

$$\langle v, h^{(k)}\rangle = \int_{\partial\Omega} \psi_k(v \cdot \nu)\, d\sigma - \int_\Omega \psi_k \operatorname{div} v\, dx = \int_{S_k} v \cdot \nu\, d\sigma$$

using Gauss's theorem, (3.8)-(3.9) and assuming v is incompressible.

From the orthogonality of the decomposition (3.7), $\langle v, h^{(k)}\rangle = \langle P_H v, h^{(k)}\rangle$. Thus $\langle P_H v, h^{(k)}\rangle = 0$ for all $1 \le k \le K$ when v is incompressible and (4.1) holds. Since the $h^{(k)}$ form a basis of the range of P_H, this implies $P_H v = 0$. $\square$

Theorem 2. *Suppose v is in $H(\operatorname{div},\Omega)$, v is incompressible and (4.1) holds. Then there is a vector field $\hat A$ in $V_0(\Omega)$ such that $v = \operatorname{curl} \hat A$.*

Proof. Under these assumptions $P_G v = 0$ and $P_H v = 0$ from results 4.1 and 4.2. Hence the result follows from theorem 1. $\square$

The condition (4.1) requires that there be no net flux of v through the boundary S_k of any "hole" U_k in the domain Ω. It is a condition that depends on the topology of Ω. When the boundary conditions (1.3) or (2.4) are required, all such topological considerations may be ignored, as we have the following corollary.

Corollary 1 *If v is in $V_0(\Omega)$, then there is a vector field $\hat{A}$ in $V_0(\Omega)$ such that $v = \operatorname{curl} \hat{A}$.*

Proof. When v is in $V_0(\Omega)$, then $v \cdot \nu = 0$ in a weak sense on $\partial\Omega$ so (4.1) holds. Then this result follows from theorem 2. □

REFERENCES

[1]   G. Auchmuty, "Orthogonal Decompositions and Bases for Three-dimensional Vector Fields," Numer. Funct. Anal. and Optim. **15** (1994), 455-488.

[2]   R. Dautray & J.L. Lions, Mathematical Analysis and Numerical Methods for Science and Technology, Vol. 3, Springer Verlag (1990), Berlin.

[3]   V. Girault and P.A. Raviart, Finite Element Methods for the Navier Stokes equations, Springer Verlag (1986), Berlin.

[4]   R. Picard, "On the Boundary Value Problems of Electro- and Magnetostatics," Proc. Roy. Soc. Edinburgh **92A** (1982), 165-174.

[5]   H. Weyl, "The Method of Orthogonal Projection in Potential Theory," Duke Math. J. **7**(1940), 411-444.

Inequalities for the Capillary Problem with Volume Constraint

Catherine BANDLE and Jesus Ildefonso DIAZ

Dedicated to our friend and colleague Ivar Stakgold

Abstract

We use **rearrangements methods** to estimate the minimal height of a liquid in a tube. We also use comparison techniques in order to give upper and lower bounds for the critical volume which gives rise to a free boundary. The main motivation comes from a conjecture posed by R. Finn in 1986.

Keywords: Capillary problem, isoperimetric inequality, rearrangements
1991 Mathematics Subject Classification: 76845, 35J60, 53A10

1 Introduction

In this paper we discuss the equilibrium surface of a liquid in a cylindrical tube of cross–section D and horizontal basis. Its height is denoted by $u(x, y)$ and is determined by means of the principle of virtual work. (see e.g. [5]). If $\kappa > 0$ is the capillarity constant, $\gamma \in [0, \pi/2)$ the wetting angle and V the prescribed volume of the liquid, the total energy is

$$J[u] = \int_D \sqrt{1 + |\nabla u|^2} \, dx \, dy + \frac{1}{2} \kappa \int_D u^2 \, dx \, dy - \cos \gamma \oint_{\partial D} u \, ds \,.$$

The height $u(x, y)$ is the solution of the variational problem

$$(1) \qquad J[v] \rightarrow \min, \quad v \in K := \{w \in BV(D) : w \geq 0, \int_D w \, dx \, dy = V\},$$

where $BV(D)$ is the space of bounded variations functions.

It was shown in [6] that (1) has a unique solution $u \in C^1(\bar{D})$ which is locally analytic in $D^+ := \{(x, y) : u(x, y) > 0\}$, has the property that $\operatorname{div} Tu \in L^\infty(D^+)$, $Tu := \frac{\nabla u}{\sqrt{1+|\nabla u|^2}}$, and satisfies the Laplace equation and the capillary boundary condition

$$(2) \qquad \begin{cases} \operatorname{div} Tu = \kappa u + \lambda & \text{in } D^+, \quad u \geq 0 \quad \text{in } D, \\ (Tu, n) = \cos \gamma & \text{on } \partial D. \end{cases}$$

Here n is the outer normal to ∂D and $(\ ,\ )$ denotes the scalar product. The constant λ is a Lagrange multiplier which is easily obtained from (2) by integration by parts. More precisely we have

$$(3) \qquad \lambda = \frac{1}{A^+}(L \cos \gamma - \kappa V),$$

50

where

$$A^+ = \text{meas } D^+ \quad \text{and} \quad L = \oint_{\partial D} ds.$$

If no volume constraint is imposed then the height u_0 is a solution of (1) with K replaced by $K_0 = \{w \in BV(D)\}$. The minimizer is positive and satisfies $\text{div}\, Tu = \kappa u$ in D. The volume V_0 of the liquid is then

(4)
$$V_0 = L \cos \gamma / \kappa.$$

The maximum principle applies and implies that u_0 takes its maximum at the boundary and its minimum, say α, at an inner point. It is well–known [6] that there exists a critical volume V^c such that $D_0 = D - D^+ = \{(x, y) \in D : u(x, y) = 0\}$ is void if $V > V^c$. It is readily seen that $D_0 \neq \phi$ for $V \leq V^c$. By shifting the solution $u_0(x, y)$ of the problem without volume constraint, it follows that

(5)
$$V^c = V_0 - \alpha A, \quad A := \text{meas } D.$$

The case $\kappa = 0$ is peculiar. It corresponds to the capillary problem without gravity. A solution exists only if $\lambda A^+ = \cos \gamma L$ [5].

Equation (2) can also be used to describe a fluid in a tube with a pressure different than the one of the surrounding media. In this case λ is determined by the difference of the pressures. The aim of this paper is to estimate u and in particular α, by means of rearrangement methods [1]. It was motivated by the following conjecture of Finn [5]: *Among all cross–sections D of given area, the disk has the minimal α.* For references on this and related problems we refer to [2], [3], [4], [5], [7], [8]. We also construct an upper solution to localize D^+.

2 Rearrangement Inequalities

Let us first introduce some notation connected with the solution $u(x, y)$ of problem (2). Given $\tilde{u} \in \mathbb{R}^+$ let

$$D(\tilde{u}) := \{(x, y) \in D : u(x, y) < \tilde{u}\}$$

$a(\tilde{u}) := \text{meas } D(\tilde{u})$ is the <u>distribution function</u>. The increasing rearrangement of u is defined by $\tilde{u}(s) := \inf\{t : a(t) \geq s\}$. The boundary of $D(\tilde{u})$ consists of the sets

$$\Gamma(\tilde{u}) := \{(x, y) \in D : u(x, y) = \tilde{u}\} \quad \text{and} \quad \Gamma_0(\tilde{u}) := \partial D \cap \partial D(\tilde{u}).$$

For $\tilde{u} \in (\alpha, \min_{\partial D} u)$, $\Gamma_0(\tilde{u})$ is empty.

The goal of this section is to derive a differential inequality for the rearrangement $\tilde{u}(a)$. The divergence theorem yields

(6)
$$\int_{D(\tilde{u})} \text{div}\, Tu \, dx\, dy = \int_{\Gamma(\tilde{u})} \frac{|\nabla u|}{\sqrt{1 + |\nabla u|^2}} ds + \cos \gamma \int_{\Gamma_0(\tilde{u})} ds$$

on the other hand we obtain from (2) and Cavalieri's principle

(7)
$$\int_{D(\tilde{u})} \text{div}\, Tu \, dx\, dy = \kappa \int_{D(\tilde{u})} u \, dx\, dy + \lambda a$$
$$= \kappa \int_0^a \tilde{u}(s) \, ds + \lambda a =: \nu(a)$$

52

Define $q(t) = \frac{t^2}{\sqrt{1+t^2}}$. It is an increasing function in $\mathbb{R}^+$ which is convex for $0 \leq t \leq \sqrt{2}$ and concave for $t > \sqrt{2}$. Denote by $q_0(t)$ any function which for positive t is increasing, convex and satisfies

$$(i) \quad q_0(t) \leq q(t) \quad \text{for} \quad t > 0, \quad q_0(0) = 0$$
$$(ii) \quad q_0(t)/t \to \cos\gamma \quad \text{as} \quad t \to \infty.$$

Examples
(1) $q_0(t) = \cos\gamma\, t^2/(1+t)$ \quad (This function was proposed by Talenti [8]).

$$(2) \quad q_0(t) = \begin{cases} q(t) & t \in (0, t_0) \\ \cos\gamma\, t + q(t_0) - \cos\gamma\, t_0 & t \geq t_0 \end{cases}$$

where t_0 is the positive root of $q'(t_0) = \cos\gamma$. Put

$$t_m = \begin{cases} |\nabla u| & \text{on } \Gamma(\tilde{u}) \\ m & \text{on } \Gamma_0(\tilde{u}) \end{cases}, \quad dp_m = \frac{ds}{t_m}$$

and $P_m = \oint_{\partial D(\tilde{u})} dp_m$. The latter definition makes sense if $|\nabla u| \neq 0$ on $\Gamma(\tilde{u})$. According to Sard's lemma this is true for almost all $\tilde{u} > 0$. The right-hand side of (6) can now be estimated as follows

$$(8) \qquad \nu(a) \geq \int_{\partial D(\tilde{u})} q_0(t_m)\, dp_m$$

Since q_0 is convex, Jensen's inequality applies and yields

$$(9) \qquad \nu(a) \geq P_m\, q_0(L(\partial D(\tilde{u}))/P_m)$$

This is true for all m and also if we let m tend to infinity.

$$P_m \to P = \int_{\Gamma(\tilde{u})} \frac{ds}{|\nabla u|} \quad \text{as} \quad m \to \infty.$$

From the coarea formula it follows that

$$(10) \qquad P = \frac{da}{d\tilde{u}}.$$

By the isoperimetric inequality we have $L^2(\partial D(\tilde{u})) \geq 4\pi a$. Since q_0 is monotone,

$$(11) \qquad Pq_0(L(\partial D(\tilde{u}))/P) \geq q_0(\sqrt{4\pi a}\tilde{u}'(a))/\tilde{u}'(a).$$

Finally we obtain the rearrangement inequality

$$(I) \qquad \kappa V(a) + \lambda a \geq \frac{q_0(\sqrt{4\pi a}\tilde{u}'(a))}{\tilde{u}'(a)}, \quad V(a) := \int_{D(\tilde{u})} u\, dx\, dy \quad \text{in} \quad (0, A).$$

Let us now discuss the case of equality. Consider a solution $\tilde{u}^*$ of

$$(12) \qquad \kappa V^*(a) + \lambda^* a = \frac{q_0(\sqrt{4\pi a}\tilde{u}^{*\prime}(a))}{\tilde{u}^{*\prime}(a)}, \quad V^*(a) := \int_0^a \tilde{u}^*(s)\, ds$$

After the change of variable $\pi r^2 = a$ and $u^*(r) = \tilde{u}^*(a)$, $a = \pi r^2$, (12) becomes

$$(13) \qquad \kappa u^*(r) + \lambda^* u^*(r) = \frac{1}{r}\left(\frac{q_0(u^{*\prime})r}{u^{*\prime}}\right)^{\prime} \quad \text{in} \quad (0, R).$$

If we impose the boundary conditions $u^{*\prime}(0) = 0$ and $u^{*\prime}(R) = \cos\gamma$, $\pi R^2 = A$, then u^* can be interpreted as the radial solution of

$$(14) \qquad \operatorname{div}(g(|\nabla u^*|)\nabla u^*) = \kappa u^* + \lambda^* \quad \text{in} \quad D^*, g(t) := \frac{q_0(t)}{t^2}$$

where D^* is the circle of the same area as D. The existence of such a solution is guaranteed for the examples given before. $u^{*\prime}$ is positive regardless of the sign of λ^*.

REMARK 2.1. *It can be shown that for smooth boundaries the function $\tilde{u}(a)$ is absolutely continuous.*

3 Comparison of the minimum height

Let $u(x, y)$ be the solution of (1) and let $u_0(x, y)$ be the solution of (1) without volume constraint. Besides of (1) we shall consider the following "comparison" problem

$$(15) \qquad \int_{D^*} G(|\nabla v|)\, dx\, dy + \frac{\kappa^*}{2}\int_{D^*} v^2\, dx\, dy - \cos\gamma \oint_{\partial D^*} v\, ds \to \min,$$

where $v \in K^* = \{w \in BV(D^*),\ w \geq 0,\ \int_{D^*} w^*\, dx\, dy = V^*\}$.

Here $G' = \operatorname{tg}(t) = q_0(t)/t$. The corresponding Euler equation is (14) with the boundary condition

$$g(|\nabla u^*|)\partial u^*/\partial n = \cos\gamma.$$

The solution of (15) will be denoted by u^*. Accordingly we write u_0^* for the solution of (15) with K^* replaced by $K_0^* = BV(D^*)$. In the sequel the $*$ refers to quantities related to (15).

THEOREM 3.1. *The solutions of problems (1) and (15) without volume constraints satisfy $\alpha \geq \alpha^*$.*

Proof. From (I) and (12) we infer

$$(16) \qquad \kappa[V(a) - V^*(a)] \geq \frac{q_0(\sqrt{4\pi a}\,\tilde{u}_0'(a))}{\tilde{u}_0'(a)} - \frac{q_0(\sqrt{4\pi a}\,\tilde{u}_0^{*\prime}(a))}{\tilde{u}_0^{*\prime}(a)}.$$

Integrating (14) over D^* and taking into account the boundary conditions satisfied by the minimizer of (15) we find $V_0^* = L^* \cos\gamma/\kappa$. By (4) and the isoperimetric inequality $V_0 \geq V_0^*$. Put $\delta(a) = V(a) - V^*(a)$. By the previous observation $\delta(A) \geq 0$. From (16) it then follows that $\delta(a) \geq 0$ for all a. In fact, suppose that $\delta(a)$ takes a negative minimum at a_1. Then $0 \leq \delta''(a_1) = \tilde{u}_0'(a_1) - \tilde{u}_0^{*\prime}(a_1)$. By (16) and the monotonicity of $q_0(t)/t$ (which follows immediately from the definition)

$$\kappa(V(a_1) - V^*(a_1)) \geq 0$$

which is a contradiction to our assumption. Since $V(a) \geq V^*(a)$ and $V(0) = V^*(0) = 0$ the assertion is now obvious.

REMARK 3.1. *The solution of (15) without volume constraint never coincides with the corresponding one of (1) even if we choose for q_0 the function given in the second example. This follows immediately from the fact that t_0 is always smaller than the value for which $q(t)/t = \cos\gamma$. The inequality in Theorem 3.1 is therefore never isoperimetric.*

COROLLARY 3.1. *If $V^* - V_0^* \leq V - V_0$, then the solutions of (1) and (15) satisfy $\alpha \geq \alpha^*$.*

Proof. Let $V = V_0 + hA$ and $V^* = V_0^* + h^*A$. By our assumption, $h \geq h^*$. If $h \geq -\alpha_0 := \min u_0$, then $u = u_0 + h$ is the desired solution. If $h \leq -\alpha_0$, $\bar{u} = u_0 - \alpha_0$ is a super solution and by the comparison principle [5], $\min u = 0$. Consequently $\alpha = \min(\alpha_0 + h, 0)$ and $\alpha^* = \min(\alpha_0^* + h^*, 0)$. Since $\alpha_0^* + h^* \leq \alpha_0 + h$, the proof is completed.

COROLLARY 3.2. *Assume $V \leq V^c(D)$ and $V^* - V_0^* \leq V - V_0$. Then $A_+ \geq A_+^*$.*

Proof. Suppose that the $A_+ \leq A_+^*$. $\tilde{u}$ satisfies the inequality (I) with (cf. (3)) $\lambda = \frac{\kappa}{A^+ \cos\gamma}(V_0 - V)$ and $\tilde{u}^*$ satisfies (12) with $\lambda^* = \frac{\kappa}{A_+^* \cos\gamma}(V_0^* - V^*)$. By (4) and the isoperimetric inequality $V_0 \geq V_0^*$. Thus because of our assumption $V \geq V^*$. The same arguments as for the Theorem 3.1 apply because $\lambda < \lambda^*$. Hence $V(a) \geq V^*(a)$ in $(0, A)$. By assumption $0 = \tilde{u}(a) = V'(a) < V^{*'}(a) = \tilde{u}^*(a)$ in $(A - A_+^*, A - A_+)$. Since $V(a) = V^*(a) = 0$ in $(0, A - A_+^*)$ this is impossible.

OBSERVATION 3.1. *As already noted the level lines $\Gamma(\tilde{u})$ are closed for $\tilde{u} \in (\alpha, \min_{\partial D} u)$. According to a result of Payne and Philippin [7] the function*

$$P = 2(1 - (1 + |\nabla u|^2))^{-1/2} \frac{\kappa u^2}{2}$$

takes its maximum at the points P on ∂D where $u(P) = \min_{\partial D} u(x, y)$. This together with Bernstein's result that $|\nabla u|^2$ takes its maximum on the boundary implies that

$$|\nabla u|^2 \leq ctg^2\gamma \quad in \quad D(\beta), \quad \beta = \min_{\partial D} u.$$

If $ctg\gamma \leq \sqrt{2}$, then (I) with $q_0(t)$ holds in (α, β). If this suffices to establish Finn's conjecture at least for angles close to $\pi/2$ is still an open question.

4 Additional remarks

As we have seen before the estimate $\alpha \geq \alpha^*$ provides an upper bound for V^c in terms of κ, γ, L and A. We shall construct a lower bound by means of an upper solution.

The crucial tool is the following comparison lemma due to Concus & Finn (see [5]). If $\partial D = \Gamma_0 + \Gamma_1$, $\Gamma_0 \cap \Gamma_1 = \emptyset$ and $\bar{u}$, $\underline{u}$ are two functions satisfying

$$\mathrm{div}\, T\underline{u} - \kappa\underline{u} \geq \mathrm{div}\, T\bar{u} - \kappa\bar{u} \quad in \quad D,$$

$\bar{u} \geq \underline{u}$ in Γ_0, $(n, T\bar{u}) \geq (n, T\underline{u})$ in Γ_1, then $\bar{u} \geq \underline{u}$ in D.

A suitable candidate for $\bar{u}$, already used by Finn [5], p. 113, and Gerhardt [6] is $\bar{u} = c - (R^2 - r^2)^{1/2}$ which satisfies

$$(17) \qquad \mathrm{div}\, T\bar{u} = \frac{2}{R} \quad for \quad r < R, \quad (T\bar{u}, n) = 1 \quad for \quad r = R.$$

If $c = \frac{2}{\kappa R}$, then $\bar{u}$ satisfies the comparison principle for u_0 [5]. Consequently, if r_0 is the inradius of D,

$$(18) \qquad \alpha < \frac{2}{\kappa r_0}.$$

This together with (5) implies

(19)
$$V^c \geq \frac{L \cos \gamma}{\kappa} - \frac{2A}{\kappa r_0}.$$

$\bar{u}$ can also be used to localize the set D^+ in case of a small volume V. This idea has already been applied in [4] in the case of fixed λ and will be repeated for the sake of completeness.

Let $P \in D$ be a point such that $\text{dist}(P, \partial D) = \rho$ and let B_ρ be the disk of radius ρ centered at P. Then the function $\bar{u} = \rho - (\rho^2 - r^2)^{1/2}$ satisfies

$$\text{div}\, T\bar{u} = \frac{2}{\rho} \leq \kappa \bar{u} + \frac{2}{\rho} \quad \text{in } B_\rho, \; (T\bar{u}, n) = 1 \quad \text{on } \partial B_\rho.$$

It is therefore an upper solution for the solutions u of (2) with $\lambda \geq \frac{2}{\rho}$. This construction is independent of the particular position of P. Hence $u = 0$ on $\{P \in D : \text{dist}(P, \partial D) = \rho\}$. By the comparison lemma, $u \equiv 0$ in $\{P \in D : \text{dist}(P, \partial D) > \rho\}$. Hence

(20)
$$D^+ \subset D_\rho := \{P \in D : \text{dist}(P, \partial D) < \rho\}.$$

In view of (3) we must have

(21)
$$\lambda = \frac{1}{A^+}(L \cos \gamma - \kappa V) \geq \frac{2}{\rho}.$$

Since $A^+ \leq A_\rho := \text{meas}\, D_\rho$, (21) holds for all V such that $\kappa V \leq L \cos \gamma - \frac{2A_\rho}{\rho}$. We have thus established the following result.

THEOREM 4.1. *If $V \leq (L \cos \gamma - 2A_\rho/\rho)\kappa^{-1}$, then a free boundary occurs and $D^+ \subset D_\rho$.*

Remarks

(1) Since $\lim_{\rho \to 0} A_\rho/\rho = L$, the assertion makes only sense for $\rho \geq \rho_0 > 0$.

(2) Let $\beta = \max_{0 < \rho \leq r_0}(L \cos \gamma - 2A_\rho/\rho)\kappa^{-1}$. Then $V^c \geq \beta$.

Acknowledgements This work was done when one of the authors (C. B.) was Visiting Professor of the University Complutense de Madrid. She would like to thank this University and also the European Science Foundation Program on "Mathematical Treatment of Free Boundary Problems" for their support. The research of the second author was partially supported by the DGICYT (Spain) under the project PB90/0620.

References

[1]   C. Bandle, *Isoperimetric Inequalities and their Applications*, Pitman (1980)

[2]   J.I. Diaz, J.E. Saa and U. Thiel, *Sobre la ecuacion de curvatura media prescrita y otras ecuaciones cuasilineares elipticas con soluciones anulandose localmente*, Rev. Unión Mat. Argentina 35(1990), 175–206

[3]   J.I. Diaz, *Desigualdades de tipo isoperimetrico para problemas de Plateau y capilaridad*, Rev. Acad. Canar. Cienc. 3(1991), 127–166

[4]   J.I. Diaz, *Problemas estaticos con frontera libre en Mecanica de Medios Continuos*, Rev. Real Acad. Cienc. Exactas, Fisicas y Naturales, Madrid, 83(1989),119–122

[5]   R. Finn, *Equilibrium Capillary Surfaces*, Springer (1986)

[6]   C. Gerhardt, *On the capillary problem with constant volume*, Ann. Scuola. Norm. Sup. Pisa, 2(1975), 303–320

[7]   L.E. Payne and G.A. Philippin, *Some maximum principles for nonlinear elliptic equations in divergence form with applications to capillary surfaces and to surfaces of constant mean curvature.* J. Nonlinear Analysis TMA 3(1979), 191–211

[8]   G. Talenti, *Nonlinear elliptic equations, rearrangements, of functions and Orlicz spaces*, Ann. Mat. Pura Appl. IV 120(1977), 159–184

C. B.
Mathematisches Institut
Universität Basel
Rheinsprung 21
CH–4051 BASEL (Switzerland)

J. I. D.
Departamento de Matematica
Aplicada
Universidad Complutense de
Madrid
E–28040 MADRID (Spain)

Integral Inequalities with Applications to Nonlinear Degenerate Parabolic Equations

Francisco Bernis*

Abstract

We prove some inequalities involving integrals over a real interval of products of the form $v^c |v^{(j)}|^p$, $j = 1, 2, 3$; the exponent c may be negative. Then we present applications of these inequalities to fourth order nonlinear degenerate parabolic equations, specially to the equation $u_t + (u^n u_{xxx})_x = 0$, which arises in lubrication models for thin viscous films and spreading droplets.

1 Introduction

In this note we present some inequalities involving the integrals

$$\int_J v^{b-q} |v'|^{2q} \, dx, \qquad \int_J v^b |v''|^q \, dx, \qquad \int_J v^{b+q/3} |v'''|^{2q/3} \, dx,$$

where J is the bounded open interval $(-a, a)$, $a > 0$. The exponents b and $b - q$ *may be negative* and the constants of the inequalities are *independent of the length of the interval*.

These inequalities are motivated by the study of initial–boundary value problems on the 1–D space interval J for nonlinear degenerate parabolic equations, specially for the fourth order equation

$$(1) \qquad\qquad u_t + (u^n u_{xxx})_x = 0.$$

Let us consider the boundary conditions

$$(2) \qquad \text{either (I) } u_x = u_{xxx} = 0, \quad \text{or (II) periodic boundary conditions.}$$

Equation (1) arises in lubrication models for thin viscous films and spreading droplets. In this context $u = u(x, t)$ is the height of the droplet or the thickness of the film. The lubrication approximation takes into account the smallness of the height and is a low Reynolds number limiting case of the theory of incompressible flow for Newtonian fluids. The motion is driven by surface tension, with the curvature approximated by the second derivative u_{xx}. If no slip is assumed it is obtained $n = 3$, but many authors introduce slip conditions which lead to $n = 1$ or $n = 2$ as the dominant power in the degeneracy of the equation; in the last few years the full range $0 < n \le 3$ is being considered in the literature. A guide to the physical literature on these models can be found e.g. in [2], [6] and [7]. For $n = 1$ Equation (1) also describes the flow of a thin neck of fluid of width $2u$ in a Hele–Shaw cell: see [10], [11] and [14].

A mathematical theory for Equation (1) has been developed in [5], [7], [1], [8] and [4]. Some aspects of these papers are summarized in Section 3. In [12] and [15] it is considered

---

*Departamento de Matematicas, Universidad Autonoma de Madrid, 28049 Madrid, Spain

the equation in higher dimensions. Fourth order parabolic equations with a similar type of degeneracy and/or additional lower order terms are analyzed in [9], [12], [13], [14], [15] and [18]. In addition to several variations of thin viscous flows ([9], [13], [14]), these papers include equations arising in the theory of phase separation for binary alloys (Cahn–Hilliard equation with degenerate mobility: [12], [18]) and in some plasticity models ([15]).

Another degenerate parabolic equation is (see [3], [6], [16])

$$(3) \qquad u_t + u^n u_{xxxx} = 0.$$

In [16] it is proved the existence of solution of the initial value problem for (3) with boundary conditions $u = 1$ and $u_{xx} = p$; this solution u satisfies that u and u_x are continuous for all $n > 0$. The work [6] studies (3) and other degenerate parabolic equations from the point of view of numerical simulations and matched asymptotics. In [3] are obtained some separable solutions of (3). A possible application of some integral inequalities to Equation (3) is mentioned at the end of section 3.

The inequalities presented in this note may also be useful to study the equation

$$(4) \qquad u_t + (u^n u_{xx})_{xx} = 0.$$

It may be also interesting to consider higher order generalizations of these inequalities. For example, Equation (3) suggests to study inequalities involving integrals of $v^n (v^{(4)})^2$ and the equations of order $2m + 2$ of [5] involve integrals of $v^n (v^{(2m+1)})^2$. In particular, the sixth order equation (see [17])

$$u_t - (u^n u_{xxxxx})_x = 0$$

arises in some semiconductor models.

2 Inequalities Involving up to the Second Derivative

In this section $v : \overline{J} \to \mathbf{R}$ is a function such that

$$(5) \qquad v \in C^2(\overline{J}), \quad v > 0 \text{ in } \overline{J}.$$

We consider the inequality

$$(6) \qquad \int_J v^{b-q} |v'|^{2q} \, dx \le \left| \frac{2q - 1}{b - q + 1} \right|^q \int_J v^b |v''|^q \, dx.$$

THEOREM 2.1. *Let $b, q \in \mathbf{R}$, $b \neq q - 1$, $q \ge 1$, and let v satisfy (5). Assume that either (I) $v'(-a) = v'(a) = 0$ or (II) $v(-a) = v(a)$ and $v'(-a) = v'(a)$. Then (6) holds.*

Proof. Assume first that $q > 1$. Writing $|v'|^{2q} = v'|v'|^{2q-1}\mathrm{sgn}\,v'$, integrating by parts and applying Hölder's inequality with exponents q' and q we obtain

$$\int_J v^{b-q} |v'|^{2q} = -\frac{2q - 1}{b - q + 1} \int_J v^{b-q+1} |v'|^{2q-2} v''$$

$$\le \frac{2q - 1}{|b - q + 1|} \left(\int_J v^{b-q} |v'|^{2q} \right)^{1/q'} \left(\int_J v^b |v''|^q \right)^{1/q}.$$

This implies (6). If $q = 1$ then (6) follows without applying Hölder's inequality. $\square$

The cases $q = 2$ and $q = 3$, which are the most relevant for the applications considered in this note, read

(7)
$$\int_J v^{b-2}(v')^4 \, dx \le \frac{9}{(b-1)^2} \int_J v^b(v'')^2 \, dx,$$

(8)
$$\int_J v^{b-3}(v')^6 \, dx \le \left|\frac{5}{b-2}\right|^3 \int_J v^b|v''|^3 \, dx.$$

REMARK 2.1. *We know that the constant in (7) is optimal if $-2 < b < 1$; the proof will appear elsewhere.*

In the case $q = 2$ we know an exact expression for the remainder of the inequality (6), as stated in the following theorem.

THEOREM 2.2. *Assume that v satisfies the same hypotheses as in Theorem 2.1. Let $b \in \mathbf{R}$, $b \ne -2$, and set $w = v^{(b+2)/3}$. Then*

(9)
$$\int_J v^b(v'')^2 \, dx = \frac{(1-b)^2}{9} \int_J v^{b-2}(v')^4 \, dx + \frac{9}{(b+2)^2} \int_J w(w'')^2 \, dx.$$

Proof. We start with the identity

$$\frac{3}{b+2} v^{(b+2)/6} \left(v^{(b+2)/3}\right)'' = v^{b/2}v'' + \frac{b-1}{3} v^{(b-2)/2}(v')^2.$$

Taking squares and integrating over J it follows that

$$\frac{9}{(b+2)^2} \int_J w(w'')^2 = \int_J v^b(v'')^2 + \frac{(b-1)^2}{9} \int_J v^{b-2}(v')^4 + \frac{2}{3}(b-1) \int_J v^{b-1}(v')^2 v''.$$

An integration by parts shows that the last term is equal to

$$-\frac{2}{9}(b-1)^2 \int_J v^{b-2}(v')^4$$

and this completes the proof.

Although we have assumed above that $v > 0$ in $\bar{J}$, the applications to degenerate equations require to deal with functions v having zeros. So let us drop the hypothesis $v > 0$, assume that v has a unique zero (say, at $x = 0$) and keep the other hypotheses on v of Theorem 2.1. (Assume also that $q \ge 1$). Then there appear some obstructions and critical values which we proceed to explain.

I. If $b \le -q$ the left-hand side of (6) is ∞. In fact, $v^{b-q}|v'|^{2q} = |f'|^{2q}$ with

$$f = \frac{2q}{b+q} v^{(b+q)/2q} \text{ if } b \ne -q; \quad \text{and} \quad f = \log v \text{ if } b = -q.$$

Since f is unbounded near $x = 0$ if $b \le -q$, it follows that $\int |f'|^{2q} = \infty$. (The argument is easily adapted to a function v which changes sign at $x = 0$). Conversely, let $b > -q$ and let v be a C^∞ function whose unique zero is of order m; then for m large enough both sides of (6) are finite and the inequality holds. (Split the interval and integrate by parts in $(-a, 0)$ and in $(0, a)$, see the proof of Theorem 2.1).

II. If $b \le -1/2$ and $v \ge 0$ has a zero of order 2 then both sides of (6) are ∞. This explains the crucial role of $b = -1/2$ for nonnegative functions and solutions.

III. Now consider a changing sign function v and replace v by $|v|$ in (6). It is easily checked that (6) holds if $b > q - 1$. But if $v(0) = 0$ and $v'(0) \ne 0$ then (6) does not hold if $-1 < b \le q - 1$, because the left-hand side is ∞ and the right-hand side is finite. This remark and Theorems 4.2 and 4.4 below indicate that the properties of Sections 3 and 6 do not apply to changing sign solutions.

3 Applications of the Inequalities of Section 2

Before describing the applications, we briefly summarize some aspects of the mathematical theory developed in [5], [7], [1] and [8]. Equation (1) preserves nonnegativity, i.e. $u(x,t) \geq 0$ for all time $t > 0$ if $u(x,0) \geq 0$; this is a fundamental property associated to the type of degeneracy of this equation. Notice that no maximum and comparison principles hold for fourth (and higher) order equations and that the linear equation $u_t + u_{xxxx} = 0$ does not preserve nonnegativity. The above two initial-boundary value problems for Equation (1) have a continuous weak solution for all $n > 0$ and have a strong solution if $0 < n < 3$. We refer to the above papers for the definition of weak solution; we say that a weak solution u is strong if $u(\cdot, t)$ is C^1 (in space) for almost every $t > 0$. Both weak and strong solutions are C^∞ smooth in the set $\{u > 0\} \cap \{t > 0\}$. In these notes we always deal with nonnegative strong solutions of (1).

A strong solution u is obtained in [1] and [8] as the limit of a sequence of smooth and positive functions; in the existence and regularity proof of these authors, two important properties of the solution u are

$$(10) \qquad \iint_{P_T} u^b u_{xx}^2 \, dx \, dt < \infty \quad \text{if} \quad 0 < n < 3 \quad \text{and} \quad b > \max\{-1/2, n-2\},$$

$$(11) \qquad \iint_{P_T} u^{1/2} (u^{1/2})_{xx}^2 \, dx \, dt < \infty \quad \text{if} \quad 0 < n < 3/2,$$

where $P_T = Q_T \cap \{u > 0\}$ and $Q_T = J \times (0, T)$. We proceed to explain the connection between these properties and the inequalities of Section 2. (This is an expository reflection about results included in [1] and [8]; the property (11) is obtained in [1] and used to deal with some difficult borderline cases). Assume for a moment that u is a smooth and positive solution of Equation (1) satisfying the boundary conditions (2). Multiplying the equation by $u^{1-n+b}/(1-n+b)$, $b \neq n-1$, and integrating by parts over Q_T it follows that

$$(12) \qquad M \int_J u^{2-n+b}(x, T) \, dx + F(b, u) = M \int_J u_0^{2-n+b}(x) \, dx,$$

where $M^{-1} = (1 - n + b)(2 - n + b)$, u_0 is the initial datum and

$$(13) \qquad F(b, u) = \iint_{Q_T} u^b u_{xx}^2 \, dx \, dt + \frac{b(1-b)}{3} \iint_{Q_T} u^{b-2} u_x^4 \, dx \, dt.$$

In general, u may be zero in a set of positive measure and may not be smooth in the set $\{u = 0\}$. So we require that $b > n - 2$. Then the above computation can be justified by means of an approximation process if $F(b, u) \geq 0$. (At the end of the process it is necessary to replace Q_T by $P_T = Q_T \cap \{u > 0\}$). Notice that we are interested in the smallest values of b (because u is bounded). Clearly we have

$$(14) \qquad F(b, u) \geq \iint_{Q_T} u^b u_{xx}^2 \quad \text{if} \quad 0 \leq b \leq 1.$$

The relations $b \leq 1$ and $b > n - 2$ impose the restriction $n < 3$. Furthermore, from (7) it follows that

$$(15) \qquad F(b, u) \geq \frac{2b+1}{1-b} \iint_{Q_T} u^b u_{xx}^2 \quad \text{if} \quad -1/2 < b < 0.$$

Relations (14)–(15) complete the explanation of the property (10). Finally, we consider the case $b = -1/2$. Notice that $-1/2 > n - 2$ is equivalent to $n < 3/2$. From Theorem 2.2 we obtain

$$(16) \qquad F(b, u) = 4 \iint_{Q_T} u^{1/2}(u^{1/2})^2_{xx} \quad \text{if} \quad b = -1/2.$$

This explains the property (11).

Let us point out that Properties (10)–(11) may also hold for Equation (3) with the boundary conditions (2). In fact, multiplying this equation by $u^{1-n+b}/(1+b)$ and integrating by parts in Q_T it is obtained (12) with $M^{-1} = (1+b)(2-n+b)$. It is to be checked if this "formal" computation can be justified, e.g. by means of the approximation process used by [16], or a similar one. This equation also preserves nonnegativity ([6], [16]).

4 Inequalities Involving up to the Third Derivative

In this section $v : \overline{J} \to \mathbf{R}$ is a function such that

$$(17) \qquad v \in C^3(\overline{J}), \quad v > 0 \text{ in } \overline{J}.$$

We consider the inequality

$$(18) \qquad \int_J v^b |v''|^q \, dx \le C(b, q) \int_J v^{b+(q/3)} |v'''|^{2q/3} \, dx,$$

where $C(b, q)$ is a positive constant depending only on b and q.

THEOREM 4.1. *Let v satisfy (17). Assume that either (I) $v'(-a) = v'(a) = 0$ or (II) $v(-a) = v(a)$, $v'(-a) = v'(a)$ and $v''(-a) = v''(a)$. Let $q \ge 2$. Then (18) holds if $-1/2 < b < (q-1)/2q$.*

The next theorem states that the range of the parameter b can be considerably improved in the case $q = 3$.

THEOREM 4.2. *Under the hypotheses on v of Theorem 4.1, if $-1/2 < b < 2$ there exists a constant $C(b)$ depending only on b such that*

$$(19) \qquad \int_J v^b |v''|^3 \, dx \le C(b) \int_J v^{b+1}(v''')^2 \, dx.$$

In the case $b = 0$ we have a simple formula for the constant $C(b, q)$.

THEOREM 4.3. *Under the hypotheses on v of Theorem 4.1, if $b = 0$ and $q \ge 2$ the inequality (18) holds with $C(0, q) = ((q-1)(2q-1))^{q/3}$. In particular, the following inequality holds:*

$$(20) \qquad \int_J |v''|^3 \, dx \le 10 \int_J v(v''')^2 \, dx.$$

The next theorem states that the condition $b > -1/2$ in Theorems 4.1–4.2 is sharp.

THEOREM 4.4. *Under the hypotheses on v of Theorem 4.1, let $q > 0$. Then the inequality (18) does not hold if $-1/2 - q/3 < b \le -1/2$.*

Finally we present a theorem which shows that the restrictions on the parameter b can be completely removed if we add an "auxiliary" term to the right-hand side of the inequality (18).

THEOREM 4.5. *Under the hypotheses on v of Theorem 4.1, let $q \ge 2$ and $b \in \mathbf{R}$. Then for any $\delta > 0$ there exists a constant K depending only on δ, b and q such that*

$$(21) \qquad \int_J v^b |v''|^q \, dx \le \delta \int_J v^{b+(q/3)} |v'''|^{2q/3} \, dx + K \int_J v^{b-q} |v'|^{2q} \, dx.$$

5   Proofs of the Theorems of Section 4

In this section C and C_i stand for positive constants depending only on b and q; the constants C and C_i may be different in different occurrences. We introduce the notations

$$I_1 = \int_J v^{b-q}|v'|^{2q}, \quad I_2 = \int_J v^b|v''|^q, \quad I_3 = \int_J v^{b+(q/3)}|v'''|^{2q/3}.$$

Proof of Theorem 4.1. Integrating by parts we have

$$(22) \qquad I_2 = \int v^b(|v''|^{q-1}\operatorname{sgn}v'')v'' = A + B,$$

where

$$(23) \qquad A = -(q-1)\int v^b|v''|^{q-2}v'''v', \quad B = -b\int v^{b-1}(v')^2|v''|^{q-1}\operatorname{sgn}v''.$$

We set $r = 2q/3$ and $n = b + (q/3)$. Observing that

$$\frac{1}{r} + \frac{q-2}{q} + \frac{1}{2q} = 1, \quad b = \frac{n}{r} + \frac{b(q-2)}{q} + \frac{b-q}{2q},$$

writing

$$A \le (q-1)\int v^{n/r}|v'''|\,v^{b(q-2)/q}|v''|^{q-2}\,v^{(b-q)/2q}|v'|,$$

and applying Hölder's inequality with exponents r, $q/(q-2)$ and $2q$ (with exponents $4/3$ and 4 if $q = 2$) we obtain

$$(24) \qquad A \le (q-1)I_3^{1/r}I_2^{(q-2)/2}I_1^{1/2q}.$$

Inserting (6) it follows that

$$(25) \qquad A \le (q-1)\left|\frac{2q-1}{b-q+1}\right|^{1/2}I_3^{1/r}I_2^{1-1/r} \quad \text{if} \quad b \ne q-1.$$

Next we deal with the term B in (22). By Hölder's inequality with exponents $q' = q/(q-1)$ and q we have

$$(26) \qquad B \le |b|I_2^{1-1/q}I_1^{1/q},$$

and inserting (6)

$$(27) \qquad B \le \frac{|b|(2q-1)}{|b-q+1|}I_2.$$

From (22), (25) and (27) it is clear that the inequality (18) holds if $|b|(2q-1)/|b-q+1| < 1$, which is equivalent to $-1/2 < b < (q-1)/2q$.   □

Proof of Theorem 4.2. We split the proof in three cases: $-1/2 < b < 0$, $0 \le b \le 1$ and $1 < b < 2$. The case $-1/2 < b < 0$ is included in Theorem 4.1.

Proof of the case $0 \le b \le 1$. Consider (22)-(23) with $q = 3$. Integrating B by parts we have

$$B = -b\int v^{b-1}|v''|(v')^2v'' = B_1 + B_2,$$

where

$$B_1 = \frac{b}{3}\int v^{b-1}(|v''|)'(v')^3, \quad B_2 = \frac{b(b-1)}{3}\int v^{b-2}(v')^4|v''|.$$

Since $B_2 \leq 0$ it follows that $B \leq B_1$. Notice that $(|v''|)'$ is Lipschitz continuous and $|(|v''(x)|)'| = |v'''(x)|$ for almost every $x \in J$. Writing $v^{b-1} = v^{b/2+1/2} v^{b/2-3/2}$ and applying Schwarz inequality to B_1 we obtain

$$B \leq \frac{b}{3} I_3^{1/2} I_1^{1/2},$$

and, inserting (6) with $q = 3$,

(28)
$$B \leq C I_3^{1/2} I_2^{1/2}.$$

From (22), (25) with $q = 3$ (hence, $r = 2$) and (28) it is clear that the inequality (19) holds if $0 \leq b \leq 1$.

Proof of the case $1 < b < 2$. Starting with I_1 we integrate by parts twice:

$$I_1 = \int v^{b-3}(v')^6 = -\frac{5}{b-2} \int v^{b-2}(v')^4 v'' = -\frac{5}{(b-1)(b-2)} \int (v^{b-1})' \, (v')^3 v'' = H_1 + H_2,$$

where

$$H_1 = \frac{5}{(b-1)(b-2)} \int v^{b-1}(v')^3 v''', \quad H_2 = \frac{15}{(b-1)(b-2)} \int v^{b-1}(v')^2(v'')^2$$

Notice that H_1 is very similar to B_1 above. Observing that $H_2 \leq 0$ (because $1 < b < 2$), and applying to H_1 Schwarz inequality it follows that

$$I_1 \leq H_1 \leq \frac{5}{(b-1)|b-2|} I_3^{1/2} I_1^{1/2},$$

and hence

(29)
$$I_1 \leq \left| \frac{5}{(b-1)(b-2)} \right|^2 I_3 \quad \text{if} \quad q = 3 \text{ and } 1 < b < 2.$$

Recalling that now $q = 3$ and $r = 2$, from (22), (25), (26), (29) and Young's inequality we deduce that

$$I_2 \leq C_1 I_3^{1/2} I_2^{1/2} + C_2 I_2^{2/3} I_3^{1/3} \leq \frac{1}{4} I_2 + C_3 I_3 + \frac{1}{4} I_2 + C_4 I_3.$$

Hence, the inequality (19) also holds when $1 < b < 2$. This completes the proof of Theorem 4.2. □

Proof of Theorem 4.3. Since $b = 0$ we have that $B = 0$ and, from (22) and (25),

$$I_2 \leq (q-1) \left(\frac{2q-1}{q-1} \right)^{1/2} I_3^{1/r} I_2^{1-1/r},$$

where $r = 2q/3$. This implies that Theorem 4.3 holds. □

Proof of Theorem 4.4. The function $v(x) = \delta + (a^2 - x^2)^2$, $\delta > 0$, satisfies the hypotheses. As $\delta \to 0$ the left-hand side of (18) tends to ∞ if $b \leq -1/2$, while the right-hand side remains bounded if $b > -1/2 - q/3$. □

Proof of Theorem 4.5. From (22), (24), (26) and Young's inequality it follows that

$$I_2 \leq \frac{\delta}{2} I_3 + \frac{1}{4} I_2 + K_1 I_1 + \frac{1}{4} I_2 + C_1 I_1,$$

where $K_1 = K_1(\delta, b, q)$ and $C_1 = C_1(b, q)$. This relation implies (21). □

6 Applications of the Inequalities of Section 4

The nonnegative strong solutions of Equation (1) mentioned in Section 3 satisfy

$$(30) \qquad \iint_P u^n u_{xxx}^2 \, dx \, dt \le K < \infty,$$

where $P = P_\infty$. In this section we present a method to obtain from (30) new properties of the strong solutions. In fact, weak solutions also satisfy (30), but in general they do not satisfy these new properties; hence, the method presented below cannot work for weak solutions.

In the sequel C stands for a positive constant depending only on K, $\sup_Q u$ and the parameters. Assume for the moment that u is smooth and positive. Then from Theorem 4.2 and (8) with $b = n - 1$ it follows that

$$(31) \qquad \iint_P u^{n-1} |u_{xx}|^3 \le C \quad \text{and} \quad \iint_P u^{n-4} u_x^6 \le C \quad \text{if} \quad 1/2 < n < 3.$$

Since u is bounded we have that

$$\iint_P u^{n'} u_{xxx}^2 \, dx \, dt \le C \quad \text{for all} \quad n' \ge n.$$

In the case $0 < n \le 1/2$ we apply Theorem 4.2 and (8) with $b = \delta - 1/2$, $\delta > 0$. So we obtain that for all $\delta > 0$

$$(32) \qquad \iint_P u^{\delta-1/2} |u_{xx}|^3 \le C \quad \text{and} \quad \iint_P u^{\delta-7/2} u_x^6 \le C \quad \text{if} \quad 0 < n \le 1/2.$$

Furthermore, we also have

$$(33) \qquad \left(u^{1+n/2}\right)_{xxx} \in L^2(Q) \quad \text{if} \quad 1/2 < n < 3,$$

$$(34) \qquad \left(u^{1+s}\right)_{xxx} \in L^2(Q) \quad \text{for all} \quad s > 1/4 \quad \text{if} \quad 0 < n \le 1/2,$$

where $Q = Q_\infty$. This follows from the identity

$$\frac{1}{s+1}\left(u^{s+1}\right)''' = u^s u''' + 3s u^{s-1} u' u'' + s(s-1) u^{s-2}(u')^3,$$

the inequality

$$\int \left(u^{s-1} u' u''\right)^2 \le \left(\int u^{2s-4}(u')^6\right)^{1/3} \left(\int u^{2s-1} |u''|^3\right)^{2/3},$$

and the relations (30)–(32).

In conclusion, we have sketched a way to obtain (31)–(34) from (30). A full proof requires either an appropriate approximation process or a version of Theorems 2.1 and 4.2 with weaker hypotheses on the function v. It is not evident that the main approximation process used in [1], [8] and [4] is appropriate. When $n \ge 2$ a positivity result for almost every t of [1] suggests a process providing such a proof if $2 \le n < 3$. The range $0 < n < 2$ deserves a further study.

References

[1] E. Beretta, M. Bertsch and R. Dal Passo, *Nonnegative solutions of a fourth order nonlinear degenerate parabolic equation*, Arch. Rational Mech. Anal., 129 (1995), pp. 175–200.

[2] F. Bernis, *Viscous flows, fourth order nonlinear degenerate parabolic equations and singular elliptic problems*, in Free Boundary Problems: Theory and Applications (J. I. Diaz, M. A. Herrero, A. Liñan and J. L. Vazquez, eds.), Pitman Research Notes in Mathematics 323, Longman, Harlow, 1995, pp. 40–56.

[3] F. Bernis, *On some nonlinear boundary value problems of higher order*, Nonlinear Anal., to appear.

[4] F. Bernis, *Finite speed of propagation and continuity of the interface for thin viscous flows*, Adv. Differential Equations, to appear.

[5] F. Bernis and A. Friedman, *Higher order nonlinear degenerate parabolic equations*, J. Differential Equations, 83 (1990), pp. 179–206.

[6] A. L. Bertozzi, *Symmetric singularity formation in lubrication–type equations for interface motion*, SIAM J. Applied Math., to appear.

[7] A. L. Bertozzi, M. P. Brenner, T. F. Dupont and L. P. Kadanoff, *Singularities and similarities in interface flows*, in Trends and Perspectives in Applied Mathematics (L. Sirovich, ed.), Vol. 100 of Applied Mathematical Sciences, Springer–Verlag, Berlin, 1994, pp. 155–208.

[8] A. L. Bertozzi and M. Pugh, *The lubrication approximation for thin viscous films: regularity and long time behavior of weak solutions*, Comm. Pure Appl. Math., to appear.

[9] A. L. Bertozzi and M. Pugh, *The lubrication approximation for thin viscous films: the moving contact line with a "porous media" cut off of the van der Waals interactions*, Nonlinearity, to appear.

[10] P. Constantin, T. F. Dupont, R. E. Goldstein, L. P. Kadanoff, M. J. Shelley and Su–Min Zhou, *Droplet breakup in a model of the Hele–Shaw cell*, Phys. Rev. E, 47 (1993), pp. 4169–4181.

[11] T. F. Dupont, R. E. Goldstein, L. P. Kadanoff and Su–Min Zhou, *Finite-time singularity formation in Hele–Shaw systems*, Phys. Rev. E, 47 (1993), pp. 4182–4196.

[12] C. M. Elliott and H. Garcke, *On the Cahn–Hilliard equation with degenerate mobility*, SIAM J. Math. Anal., to appear.

[13] L. Giacomelli, Preprint, Università di Firenze, 1995.

[14] R. E. Goldstein, A. I. Pesci and M. J. Shelley, *Topology transitions and singularities in viscous flows*, Physical Rev. Letters, 70 (1993), pp. 3043–3046.

[15] G. Grün, *Degenerate parabolic differential equations of fourth order and a plasticity model with nonlocal hardening*, Z. Anal. Anwendungen, to appear.

[16] K. M. Hui, *Asymptotic behaviour of solutions of the modified lubrication equation*, European J. Applied Math., to appear.

[17] J. R. King, *The isolation oxidation of silicon: the reaction-controlled case*, SIAM J. Appl. Math., 49 (1989), pp. 1064-1080.

[18] Yin Jingxue, *On the existence of nonnegative continuous solutions of the Cahn–Hilliard equation*, J. Differential Equations, 97 (1992), pp. 310–327.

Reaction - diffusion systems
modelling a class of spatially structured epidemic systems[*]

V. Capasso[†] and R.E. Wilson[‡]

Abstract

In this paper a basic introduction to important concepts of the mathematical theory of epidemics is presented, in particular with respect to infectious diseases mediated by environmental pollution.

A series of results, recently obtained by the authors in the case of a partially dissipative reaction-diffusion system with saddle point structure, are outlined. The concept of asymptotically smooth evolution operator in Banach spaces is utilized to show a bifurcation theorem for the asymptotic behaviour of the system.

1. Introduction. In a spatially homogeneous population an epidemic system consists of a nonlinear compartmental system; typical compartments include the susceptible population (S), and the infective population (I). which interact so to produce a flow (called incidence rate) from S to I.

When dealing with direct infections (measles, influenza, venereal diseases, etc.) the incidence rate is expressed in terms of a force of infection $(f.i.)(t) = [g(I(\cdot), S(\cdot))](t)$ so that, at time t the incidence rate is given by

$$(i.r.)(t) = (f.i.)(t)\, S(t) \quad .$$

The reader may recall that in the classical case [13, and following papers] the "law of mass action" had been used

$$(f.i.)(t) = k\, I(t) \quad .$$

Nowadays more general laws are used. depending upon the specific mechanism of transmission of the disease [8,14].

Structured populations are considered to include a dependence of the force of infection upon some parameter z which may describe a social group, space location, age, ... of the population.

In this case a "field of forces" of infection is introduced $(f.i.)(z; t)$ which, acting on the specific susceptible "subgroup" $s(z; t)$ provides the specific incidence rate

$$(i.r.)(z; t) \equiv (f.i.)(z; t)\, s(z; t) \quad . \quad t \in \mathbb{R}_+$$

A classical example is due to *D.G. Kendall* [12], for systems with spatial structure (see [3,5]). Let Ω denote a geographic habitat, so that $x \in \Omega$ denotes a specific location; for this case one may assume

$$(f.i.)(x; t) = \int_\Omega k(x, x')\, i(x'; t)\, dx'$$

---

[*] Work performed under the auspices of GNFM-CNR (Italy) in the framework of the ISS contract n.9203-04. Partially supportedby the MURST 40% Programme on Nonlinear Problems and Applications.

[†] Dipartimento di Matematica, Universita' di Milano, Via Saldini, 50, 20133 Milano, Italy.

[‡] Mathematical Institute, University of Oxford. 24-29 St. Giles, Oxford OX1 3LB, United Kingdom.

and consequently

$$(i.r.)(x;t) = (f.i.)(x;t)\ s(x;t) = s(x;t) \int_{\Omega} k(x,x')\,i(x';t)\,dx' \quad .$$

Here $\{i(x;t)\,;\ x \in \Omega,\ t \in \mathbb{R}_+\}$ and $\{s(x;t)\,;\ x \in \Omega,\ t \in \mathbb{R}_+\}$ denote the spatial densities of the infective population, and of the susceptible population (respectively):

$$\int_{\Omega} i(x';t)\,dx' \equiv I(t) \quad , \quad \int_{\Omega} s(x;t)\,dx \equiv S(t) \quad .$$

When dealing with indirect transmission, typical of infectious diseases transmitted via the pollution of the environment due to the infective population (typhoid fever, schistosomiasis, malaria,...), a different approach is used. The force of infection at a specific location $x \in \Omega$ may depend on the concentration of pollutant (etiological agent) available, at time $t \in \mathbb{R}_+$, in $x \in \Omega$. Thus in this case an intermediate "compartment' describing the pollutant is needed. If $\{u_1(x;t)\,;\ x \in \Omega,\ t \in \mathbb{R}_+\}$ denotes the concentration of the pollutant in the habitat Ω, then the "local" force of infection is a function $g(u_1(x;t))$; consequently the "local" incidence rate is

$$(i.r.)(t) = g(u_1(x;t))\,s(x;t) \quad .$$

On the other hand, we have to model the mechanism of production of pollutant due to the infective population. A possible model is the one proposed in [4]; if $\{u_2(x;t)\,;\ x \in \Omega,\ t \in \mathbb{R}_+\}$ denotes the concentration of the infective population in the habitat, the production term is modelled as follows,

$$g_1(u_2(\cdot\,;t))(x) = \int_{\Omega} k(x,x')\,u_2(x';t)\,dx'$$

where $k(x,x')$, $x,x' \in \Omega$ models the transfer kernel of pollutant produced by the infective population located at $x' \in \Omega$, and made available at $x \in \Omega$.

Of course the modelling of the various components of the transmission mechanism have to be embedded in a model for the whole epidemic system.

A model which includes random dispersal of the pollutant is given by

$$(1.1) \qquad \begin{cases} \dfrac{\partial}{\partial t} u_1(x;t) = d_1\,\Delta u_1(x;t) - a_{11} u_1(x;t) + \displaystyle\int_{\Omega} k(x,x')\,u_2(x';t)\,dx' \\[2ex] \dfrac{\partial}{\partial t} u_2(x;t) = \qquad\qquad\quad -a_{22} u_2(x;t) + g(u_1(x;t)) \end{cases}$$

subject to suitable boundary and initial conditions [4].

Here the susceptible population is not explicited since it is assumed that it is so large with respect to the infective part, to be considered as constant.

Further generalizations are given in [5].

System (1.1) has been widely analyzed when dealing with a force of infection g monotonically increasing and concave.

A more interesting case, which had been only partially analyzed in [7], is the one in which g is sigma-shaped, i.e. $g : \mathbb{R}_+ \to \mathbb{R}_+$ satisfies the following properties

(P1) $g \in C^2(\mathbb{R}_+)$
(P2) $g(0) = g'(0) = 0$
(P3) $g'(z) > 0$, for any $z > 0$
(P4) $\lim_{z \to +\infty} g(z) = g_\infty > 0$
(P5) $\exists \xi > 0$ s.t. $g''(z) > 0$ for $z \in (0,\xi)$ and $g''(z) < 0$ for $z \in (\xi,\infty)$.

This shape of g has been used to model an immune response that inhibits infection by small concentrations of the infectious agent; a different interpretation is given for other classes of diseases

[16]. This choice has a strong influence on the dynamical behaviour of system (1.1). In Section 2 a simplified version of system (1.1) is presented for the analysis of the steady states.

A bifurcation theorem obtained in [9] is discussed, showing the role played by the diffusion term in the dynamical behaviour of the PDE system (1.1), as opposed to the behaviour of the ODE system associated with it.

The case of pure Dirichlet boundary conditions is discussed here. Pure Neumann conditions had been analyzed in more detail in [7]; in this case it was shown that the asymptotic qualitative behaviour of the PDE system does not differ substantially from that of the associated ODE system [17].

In Section 3 the evolution problem is discussed to obtain the stability properties of the different steady states. Lack of compactness of the evolution operator of system (1.1) raises interesting problems, which are solved by referring to the concept of asymptotically smooth operators as discussed in [11].

2. Steady state analysis.

We reduce our attention to the spatial $1 - D$ case, in a bounded interval of $\mathbb{R}$.

A rescaling of system (1.1) leads to

(2.1)
$$\begin{cases} \dfrac{\partial}{\partial t} u_1(x;t) = d\, \dfrac{\partial^2}{\partial x^2} u_1(x;t) - u_1(x;t) + \alpha\, u_2(x;t) \\ \dfrac{\partial}{\partial t} u_2(x;t) = \qquad\qquad\quad -\beta u_2(x;t) + g(u_1(x;t)) \end{cases}$$

for $x \in (0,1)$.

We wish to analyze the steady states of system (2.1) subject to homogeneous Dirichlet boundary conditions. For reasons, to be clarified in Section 3, regarding the evolution system, we have confined our attention to classical solutions of the semilinear elliptic problem

(2.2)
$$\begin{cases} d\, u_1''(x) - u_1(x) + \alpha\, u_2(x) = 0 \\ g(u_1(x)) - \beta u_2(x) = 0 \end{cases}$$

for $x \in (0,1)$, subject to

(2.2')
$$u_1(0) = 0 = u_1(1) \quad .$$

By classical solution of system (2.2) we mean that, $u_1 \in C^2(\Omega; \mathbb{R}) \cap C^1(\overline{\Omega}; \mathbb{R})$, $u_2 \in C(\overline{\Omega}; \mathbb{R})$.

We may then carry out our analysis in the Banach space $X := C(\overline{\Omega}; \mathbb{R}^2) \ni \mathbf{u}$ with norm $\|\mathbf{u}\| = \|(u_1, u_2)\| = \sup_{x \in \overline{\Omega}} |u_1(x)| + \sup_{x \in \overline{\Omega}} |u_2(x)|$. X is an ordered Banach space with partial order induced by the cone

$$X_+ := \{\mathbf{u} = (u_1, u_2) \in X \ : \ u_1(x) \geq 0, \ u_2(x) \geq 0, \ x \in \overline{\Omega}\} \quad .$$

We shall say $\mathbf{u}^* \leq \mathbf{u}^{**}$ if $\mathbf{u}^{**} - \mathbf{u}^* \in X_+$; $\mathbf{u}^* < \mathbf{u}^{**}$ shall mean that $\mathbf{u}^* \leq \mathbf{u}^{**}$ and that $\mathbf{u}^* \neq \mathbf{u}^{**}$; $\mathbf{u}^* \ll \mathbf{u}^{**}$ shall mean that $u_1^{**}(x) - u_1^*(x) > 0$ and $u_2^{**}(x) - u_2^*(x) > 0$ for all $x \in \Omega$. If $\mathbf{u}^* \leq \mathbf{u}^{**}$, we may introduce the order interval

$$[\mathbf{u}^*, \mathbf{u}^{**}] = \left\{ \mathbf{u} = (u_1, u_2) \in X \ : \ \begin{matrix} u_1^*(x) \leq u_1(x) \leq u_1^{**}(x) \\ u_2^*(x) \leq u_2(x) \leq u_2^{**}(x) \end{matrix} \ , x \in \overline{\Omega} \right\} \quad .$$

[Similar notation may be adopted for comparison in $\mathbb{R}^2$].

Because of the rescaling leading to (2.2) we may assume $g_\infty = 1$ in (P4), and $d, \alpha, \beta > 0$.

We reduce system (2.1) to

(2.3)
$$d\, u_1''(x) - f_\gamma(u_1(x)) = 0 \quad , \quad x \in (0,1) \quad ,$$

by introducing $\gamma := \dfrac{\alpha}{\beta}$ and

$$(2.4) \qquad\qquad f_\gamma(z) := z - \gamma\, g(z) \quad, \quad z \in \mathbb{R}_+ \ .$$

By ignoring boundary conditions (for the present), equation (2.3) is equivalent to the ODE system

$$(2.5) \qquad\qquad \begin{cases} u'(x) = v(x) \\ v'(x) = f_\gamma(u(x))\,/\,d \end{cases}$$

for $x \in \mathbb{R}$.

System (2.5) is a classical Hamiltonian system with total energy [1]

$$(2.6) \qquad\qquad H(u,v) := \frac{v^2}{2} - \frac{1}{d} F_\gamma(u) \quad, \quad (u,v) \in \mathbb{R}^2$$

where

$$(2.7) \qquad\qquad F_\gamma(z) := \int_0^z f_\gamma(s)\, ds \quad, \quad z \in \mathbb{R}_+$$

The orbits of system (2.5) in the (u,v) phase plane must lie on the level curves of the total energy $H(u,v) = const$; i.e. the general integral of (2.5) is given by

$$\frac{v^2}{2} - \frac{1}{d} F_\gamma(u) = \frac{c}{d} \quad, \quad c \in \mathbb{R}$$

The critical points of (2.5) are given by

$$(2.8) \qquad\qquad \begin{cases} v = 0 \\ f_\gamma(u) = 0 \end{cases}$$

Due to the sigma shape of g, there exists a value $\gamma_{crit} > 0$ such that

- for $\gamma < \gamma_{crit}$, $f_\gamma(z) > 0$ for any $z > 0$;
- for $\gamma = \gamma_{crit}$ $f_\gamma(z) \geq 0$, and there exists a unique $z^c > 0$ such that $f_\gamma(z^c) = 0$ and $f'_\gamma(z^c) = 0$;
- for $\gamma > \gamma_{crit}$, f_γ admits exactly two distinct zeros z^- and z^+ such that $0 < z^- < z^+$, $f'_\gamma(z^-) < 0$ and $f'_\gamma(z^+) > 0$. As functions of γ, z^- and z^+ are (strictly) monotone decreasing and monotone increasing, respectively.

The critical points of (2.5) are then $(0,0), (0, z^-(\gamma)), (0, z^+(\gamma))$.

F_γ has local minima at 0 and $z^+(\gamma)$; and a local maximum at $z^-(\gamma)$. Hence both $(0,0)$ and $(0, z^+(\gamma))$ are saddle points; and $(0, z^-(\gamma))$ is a centre [1,9].

In [7] it was proven that system (2.2), when subject to homogeneous Dirichlet boundary conditions, may not have nontrivial equilibria for

$$\gamma < \gamma_{crit}\,(1 + d\,\lambda_1) \ .$$

where $\lambda_1 > 0$ is the first eigenvalue of the elliptic problem

$$\begin{cases} \Delta\phi + \lambda\phi = 0 \\ \phi(0) = \phi(1) = 0 \end{cases} \ .$$

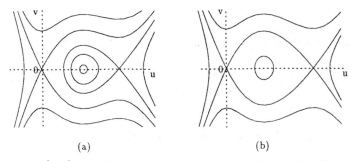

(a)   (b)

Fig. 2.1. (u, v) phase plane topologies for (a) $\gamma_{crit} < \gamma < \gamma_D$, and for (b) $\gamma = \gamma_D$.

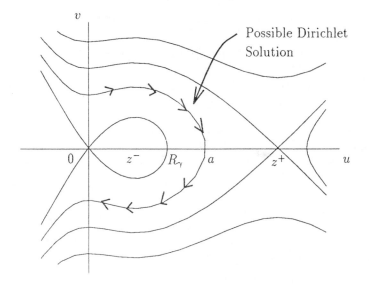

Fig. 2.2. Phase plane topology for $\gamma > \gamma_D$.

More has been obtained in [9]. Let

$$\mathcal{F}(\gamma) := F_\gamma(z^+(\gamma)) = \int_0^{z^+(\gamma)} f_\gamma(s)\, ds \quad .$$

In [9] it has been shown that $\mathcal{F}$ admits a unique zero at some $\gamma_D > \gamma_{crit}$ and

(i) for $\gamma_{crit} < \gamma < \gamma_D$, $\mathcal{F}(\gamma) > 0$
(ii) $\mathcal{F}(\gamma_D) = 0$
(iii) for $\gamma > \gamma_D$, $\mathcal{F}(\gamma) < 0$.

Figures 2.1 (a),(b) and 2.2 show the phase plane topologies for these three cases. It is clear that the only case which allows solutions of (2.3) satisfying homogeneous Dirichlet boundary conditions is $\gamma > \gamma_D$, as illustrated in Fig.2.2 . By using "time" map arguments, the solutions we look for, must be such that

(2.9)
$$1 = 2 \int_0^a \frac{\sqrt{\frac{d}{2}}}{F_\gamma(u) - F_\gamma(a)} du$$

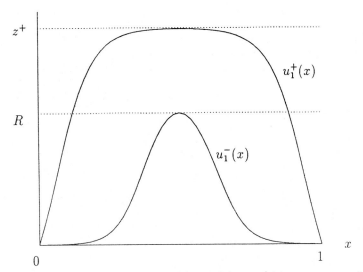

Fig.2.3. *Shooting method computations of the steady states* $u_1^-(x)$ *and* $u_1^+(x)$, *with* $g(u) = u^2/(1 + u^2)$. $\gamma = 2.3 > \gamma_D \simeq 2.17$, $d = 0.0025 < d_{crit}$. *To four decimal places,* $a^+ = 1.7105$, $z^+ = 1.7179$. $a^- = 1.0420$. $R = 1.0420$ *(but note* $a^- > R$).

where a is the intersection of the relevant trajectory with the u-axis in Fig.2.2 . We must take $a \in (R_\gamma, z^+(\gamma))$, where R_γ is the least positive zero of F_γ.

By using now the diffusion coefficient d as a second bifurcation parameter, the following theorem was proven in [9], via an analysis of (2.9).

Theorem 2.1. *Consider system (2.2) subject to homogeneous Dirichlet boundary conditions.*

1. *If* $0 \leq \gamma \leq \gamma_D$, *then the system admits only the trivial solution* $u_1(x) = u_2(x) = 0$, *for* $x \in [0, 1]$.
2. *If* $\gamma > \gamma_D$, *then a critical value* $d_{crit}(\gamma) > 0$ *exists such that*
   (a) *for* $d > d_{crit}(\gamma)$, $\mathbf{u} = 0$ *is the only solution;*
   (b) *for* $d = d_{crit}(\gamma)$ *there is exactly one nontrivial solution* $\mathbf{u}^c$ *such that* $(u_1^c(x), u_2^c(x)) \gg 0$;
   (c) *for* $0 < d < d_{crit}(\gamma)$, *there are exactly two nontrivial solutions* $\mathbf{u}^- = (u_1^-(x), u_2^-(x))$ *and* $\mathbf{u}^+ = (u_1^+(x), u_2^+(x))$ *such that* $0 \ll \mathbf{u}^- \ll \mathbf{u}^+$.

It can be shown that d_{crit} is increasing with γ, so that as γ is increased, we do not require diffusion to be as small for the existence of nontrivial solutions to (2.2).

Theorem 2.1 is a two-parameter threshold result, requiring both $\gamma > \gamma_D$ and $d > d_{crit}$ for the existence of nontrivial solutions. Because d_{crit} depends on γ in a complicated way, it is difficult to identify a single threshold parameter.

Hence suppose that $\gamma > \gamma_D$, and $d < d_{crit}(\gamma)$ (Theorem 2.1, case 2(c)), so that problem (2.2) with homogeneous Dirichlet boundary conditions admits exactly two nontrivial solutions $\mathbf{u}^-$ and $\mathbf{u}^+$. If a^- and a^+ denote the amplitudes (*max*) of $u_1^-(x)$ and $u_1^+(x)$ respectively, then we have

$$R_\gamma < a^- < a^+ < z^+(\gamma) \quad .$$

Also, by simple comparison arguments, $u_1^-(x) < u_1^+(x)$ for all $x \in (0, 1)$; consequently $u_2^-(x) < u_2^+(x)$ for all $x \in (0, 1)$, since the second equation of system (2.2) is monotone. Hence $\mathbf{u}^- \ll \mathbf{u}^+$, as claimed in Theorem 2.1.

We may say more concerning the structure of $\mathbf{u}^-$ and $\mathbf{u}^+$. First note that $\mathbf{u}^-$ and $\mathbf{u}^+$ are symmetric about $x = 1/2$, by the $v \to -v$ symmetry in the phase plane. Now, in the singular limit

$d \uparrow 0$, it can then be shown that $a^- \downarrow R_\gamma$ and $a^+ \uparrow z^+(\gamma)$. In this way, the phase plane trajectory segment $(u_1^-, u_1^{-\prime})$ approaches the homoclinic orbit attached to $(0,0)$; that for $(u_1^+, u_1^{+\prime})$ approaches the heteroclinic separatrices attached to $(z^+(\gamma), 0)$.

In the first case, the largest contribution to the time map comes from the neighbourhood of the equilibrium point $(0,0)$, so that $u_1^-(x), u_1^{-\prime}(x) \simeq 0$ through most of $(0,1)$: there is a 'fast' trip around the homoclinic loop; this corresponds to a 'hump' of width $O(\sqrt{d})$ centred at $x = 1/2$, whose height is bounded below by R_γ, no matter how small d is.

In the second case, the most significant contribution to the time map comes from the neighbourhood of $(z^+(\gamma), 0)$, so that $u_1^+(x) \simeq z^+(\gamma)$, $u_1^{+\prime}(x) \simeq 0$ throughout most of the domain. There are boundary layers of width $O(\sqrt{d})$ connecting the solution to the homogeneous Dirichlet data at $x = 0.1$.

In Figure 2.3, some sample plots of $u_1^-(x)$ and $u_1^+(x)$ are given. The infective human concentration, u_2^- and u_2^+, have similar structure to their bacterial counterparts.

3. The evolution problem.

Because of the assumptions made on g, system (2.1) is a quasimonotone (increasing) reaction-diffusion system (see [5] p.230ff ; this term means that $\dfrac{dz_i}{dt}$ is nondecreasing in z_j for $i \neq j$).

Under the above assumptions it can be shown (see [6]) that a unique solution exists for problem (2.1), subject to homogeneous Dirichlet boundary conditions. for any sufficiently regular initial conditions. Moreover, due to the assumptions made on g, any solution may be extended to all $t \in [0, \infty)$. Furthermore, under sufficient regularity assumptions on the data, it may be shown that solutions are classical solutions: that is

$$u_1 \in C^{2,1}(\Omega \times (0, +\infty); \mathbb{R}) \cap C^{1,0}(\overline{\Omega} \times (0, +\infty); \mathbb{R}),$$
$$u_2 \in C^{0,1}(\overline{\Omega} \times (0, +\infty); \mathbb{R}) \quad .$$

Hence, for any sufficiently smooth initial condition $\mathbf{u}_0 \in X_+$, we may obtain a unique global solution $\{\mathbf{u}(t), t \in \mathbb{R}_+\}$ in X. This identifies an evolution operator $\{\mathcal{U}(t). \in \mathbb{R}_+\}$ on X_+ which satisfies the following properties:

1. $\mathcal{U}(0) = I$.
2. $\mathcal{U}(t)\mathcal{U}(s) = \mathcal{U}(t + s)$ for s . $t \in \mathbb{R}_+$.
3. $\mathcal{U}(t)0 = 0$, $t \in \mathbb{R}_+$.
4. For any $t \geq 0$, the map $\mathbf{u}_0 \in X_+ \to \mathcal{U}(t)\mathbf{u}_0 \in X$ is continuous. uniformly for $t \in [t_1, t_2] \subset \mathbb{R}_+$.
5. For any $\mathbf{u}_0 \in X_+$, the map $t \in \mathbb{R}_+ \to \mathcal{U}(t)\mathbf{u}_0$ is continuous.

Due to the quasimonotonicity of system (2.1), we may use comparison theorems (see [5]) to state that

6. $\mathbf{u}_0$, $\overline{\mathbf{u}}_0 \in X_+$, $\mathbf{u}_0 \leq \overline{\mathbf{u}}_0 \Rightarrow \mathcal{U}(t)\mathbf{u}_0 \leq \mathcal{U}(t)\overline{\mathbf{u}}_0$, $t > 0$.

These properties imply the strong properties of $\mathcal{U}(t)$:

7. $\mathbf{u}_0 \in X_+ \Rightarrow \mathcal{U}(t)\mathbf{u}_0 \in X_+$. $t \in \mathbb{R}_+$.
8. $\mathbf{u}_0 \in X_+, \mathbf{u}_0 \neq 0 \Rightarrow \mathcal{U}(t)\mathbf{u}_0 \gg 0$, $t > 0$.
9. $\mathbf{u}_0$, $\overline{\mathbf{u}}_0 \in X_+$, $\mathbf{u}_0 \leq \overline{\mathbf{u}}_0$. $\mathbf{u}_0 \neq \overline{\mathbf{u}}_0$, $\Rightarrow \mathcal{U}(t)\mathbf{u}_0 \ll \mathcal{U}(t)\overline{\mathbf{u}}_0$, $t > 0$.

As far as the asymptotic behaviour of system (2.1). the most interesting case is $\gamma > \gamma_D$. and $0 < d < d_{crit}(\gamma)$ so that the conclusions of Theorem 2.1. case 2(c), hold.

Due to the monotonicity properties of the evolution operator $\{\mathcal{U}(t), t \in \mathbb{R}_+\}$ of system (2.1), we may claim that if $\mathbf{u}(t)$ is a solution to system (2.1) with initial data $\mathbf{u}_0$, then (i) $0 \leq \mathbf{u}_0 \leq \mathbf{u}^-$ implies that $0 \leq \mathbf{u}(t) \leq \mathbf{u}^-$ for all $t \geq 0$; on the other hand, (ii) $\mathbf{u}^- \leq \mathbf{u}_0 \leq \mathbf{u}^+$ implies that $\mathbf{u}^- \leq \mathbf{u}(t) \leq \mathbf{u}^+$ for all $t \geq 0$. In this way solutions whose initial data are 'sandwiched' by a pair of steady states remain 'sandwiched' for all time. We also know from [7] that $(0,0)$ is locally asymptotically stable.

As an extension of the analysis carried out in [7] for the ODE system associated with (2.1) we conjecture that $\mathbf{u}^- = (u_1^-(x), u_2^-(x))$. $x \in (0,1)$, is a saddle point for system (2.2) while $\mathbf{0}$ and $\mathbf{u}^+ = (u_1^+(x), u_2^+(x))$, $x \in (0,1)$, are stable attractors. To this end, let us observe that

Lemma 3.1. *Under the assumptions of Theorem 2.1., case 2(c), the system*

$$(3.1) \quad \begin{cases} d\, w_1''(x) - w_1'(x) + (\alpha - \epsilon)\, w_2(x) = 0 \\ \qquad\qquad g(w_1(x)) - \beta\, w_2(x) = 0 \end{cases}$$

for $x \in \Omega$, and subject to homogeneous Dirichlet boundary conditions, admits two nontrivial classical solutions for suitable choice of $\epsilon > 0$. Each of them is a lower solution of Problem (2.2).

Lemma 3.2. *Under the assumptions of Theorem 2.1, case 2(c), the system*

$$(3.2) \quad \begin{cases} -z_1 + \alpha z_2 = 0 \\ -z_2 + g(z_1) = 0 \end{cases}$$

admits two nontrivial solutions. Each of them is a spatially homogeneous upper solution for system (2.2). subject to homogeneous Dirichlet boundary conditions.

Let us denote by $\underline{\mathbf{u}} = \{(\underline{u}_1(x), \underline{u}_2(x))$. $x \in \Omega\}$ the smallest of the nontrivial solutions of problem (3.1). and by $\overline{\mathbf{u}} = \{(\overline{u}_1(x), \overline{u}_2(x))$, $x \in \Omega\}$ the largest of the nontrivial solutions of problem (3.2).

By use of comparison theorems we may claim that the largest nontrivial solution $\mathbf{u}^+$ of system (2.2) satisfies

$$\underline{\mathbf{u}} \le \mathbf{u}^+ \le \overline{\mathbf{u}} \quad .$$

Further, we may claim that $\mathcal{U}(t)\,\underline{\mathbf{u}}$ is monotone nondecreasing in $t \in \mathbb{R}_+$, while $\mathcal{U}(t)\,\overline{\mathbf{u}}$ is monotone nonincreasing in $t \in \mathbb{R}_+$. and by further use of comparison theorems

$$\mathcal{U}(t)\,\underline{\mathbf{u}} \le \mathbf{u}^+ \le \mathcal{U}(t)\,\overline{\mathbf{u}} \quad .$$

for $t \ge 0$. We have thus built a family of *contracting rectangles* containing the equilibrium point $\mathbf{u}^+$ in the function space X.

We cannot directly apply the convergence theorem of contracting rectangles in Banach spaces [5] because the evolution operator $\mathcal{U}(t)$ is not compact. This is because the second equation in system (2.1) lacks diffusion.

A careful analysis of such a situation has been carried out more recently by Hale [11] (see also [15]. and [10]). making use of the concept of asymptotically smooth operator. In order to apply their results we need to require that g is sufficiently smooth and the initial conditions have sufficient regularity.

The relevant remark is the following. System (2.1) can be rewritten as an abstract evolution problem in X:

$$(3.3) \quad \frac{d}{dt}\underline{u}(t) = A\,\underline{u}(t) + F(\underline{u}(t)) \quad , t > 0$$

subject to a suitable initial condition

$$(3.3o) \quad \underline{u}^0 \in X \quad .$$

Here

$$A = \begin{pmatrix} d\Delta - a_{11} & 0 \\ 0 & -a_{22} \end{pmatrix} \quad , \quad F(\underline{u}) = \begin{pmatrix} a_{12}u_2 \\ g(u_1) \end{pmatrix} \quad .$$

The solution of (3.3), (3.3o) satisfies

$$\mathcal{U}(t)\,\underline{u}^0 := \underline{u}(t) = e^{At}\,\underline{u}^0 + \int_0^t e^{A(t-s)}\,F[\mathcal{U}(s)]\,ds$$
$$=: S(t)\,\underline{u}^0 + T(t)\,(\underline{u}^0) \quad .$$

It can be shown [2] that, for g sufficiently smooth and $\underline{u}^0$ sufficiently regular, we are in the conditions of Lemma 3.2.3 in [11], so that $\{\mathcal{U}(t)\ ,\ t \geq 0\}$ is asymptotically smooth in X.

We have already proved that for a sufficiently smooth initial condition $\underline{u}^0 \in X$, the trajectory $\gamma^+(\underline{u}^0)$ is bounded in X, so that we may apply also Lemma 3.2.1 in [11] as a basis for the LaSalle Invariance Principle for the family of contracting rectangles (Theorem B32 in [5]).

Theorem 3.3. *If g is sufficiently smooth then the non trivial equilibrium u^+ is GAS with respect to sufficiently smooth initial conditions $u^0 \in X_+$.*

Numerical simulations carried out in [9] confirm this result, for a large selection of initial conditions.

4. Conclusions. In previous papers [4-7] it was shown that the concavity of the force of infection induces the uniqueness of a nontrivial steady state for system (2.1), together with its global asymptotic stability (above a parameter threshold). Thus any, even small, outbreak of an epidemic would tend to an endemic state. This situation had been considered rather unrealistic from an epidemiological point of view. In this paper the choice of a sigma shaped force of infection induces a double parameter bifurcation, to three steady states two of which are nontrivial. The central steady state exhibits a saddle point behaviour. The spatial structure of this state strongly depends upon the value of the diffusion coefficient d; for small values of d it is very localized with a spatial width of the order of $\sqrt{d}$. Thus a small diffusion coefficient is responsible for the spatial localization of the outbreak of a "small" epidemic which will tend to extinction; on the other hand large outbreaks, above the saddle solution, will tend to a pratically spatially homogeneous endemic state, with boundary layers at the extremes of the interval $[0,1]$, with width again of the order of $\sqrt{d}$.

REFERENCE

[1] Arnold. V.I. : *Ordinary Differential Equations.* MIT press. Cambridge Mass., 1973.

[2] Barbu, V., Capasso, V. : *On the compactness of trajectories for partially dissipative reaction-diffusion systems.* In preparation.

[3] Capasso, V. : *Global solution for a diffusive nonlinear deterministic epidemic model.* SIAM J. Appl. Math. **35** (1978), 274–284.

[4] Capasso, V. : *Asymptotic stability for an integro-differential reaction- diffusion system.* J. Math. Anal. Appl. **103** (1984), 575–588.

[5] Capasso, V. :*Mathematical Structures of Epidemic Systems.* Lecture Notes Biomath. Vol. 97 Springer-Verlag, Heidelberg, 1993.

[6] Capasso, V., Kunisch K. : *A reaction-diffusion system arising in modelling man-environment diseases.* Quart. Appl. Math. **46** (1988), 431–450.

[7] Capasso, V., Maddalena, L. : *Saddle point behavior for a reaction-diffusion system. Application to a class of epidemic models.* Math. Comp. Simulation **24** (1982), 540–547.

[8] Capasso, V., Serio, G. :*A generalization of the Kermack-McKendrick deterministic epidemic model.* Math. Biosci. **42** (1978), 41–61.

[9] Capasso, V., Wilson, R.E.: *Analysis of a reaction-diffusion system modelling man-environment-man epidemics.* Quaderno n.19/1995, Dipartimento di Matematica, Universita' di Milano, Italy.

[10] Fabrie, P., Galusinski, C. :*Exponential attractors for a partially dissipative reaction system.* Rapport Interne n. 95006 (1995), Math. Appl. Bordeaux, France.

[11] Hale. J.K.:*Asymptotic Behaviour of Dissipative Systems*. Math. Surveys and Monographs n.25 AMS. Providence (R.I.), 1988.

[12] Kendall. D. G. : *Mathematical models of the spread of infection*. In : Mathematics and Computer Science in Biology and Medicine. H.M.S.O., London, 1965, pp. 213–225.

[13] Kermack. W.O., McKendrick, A.G. : *Contributions to the mathematical theory of epidemics*. Part I. Proc. Roy. Soc., A, **115** (1927), 700–721.

[14] Liu, W.M., Hethcote, H.M., Levin, S.A. : *Dynamical behavior of epidemiological models with nonlinear incidence rates*. J. Math. Biol. **25** (1987), 359–380.

[15] Marion. M. : *Inertial manifolds associated to partly dissipative reaction-diffusion systems*. J. Math. Anal. Appl. **143**, (1989), 295–326.

[16] Nåsell. I.. Hirsch, W.M. : *The transmission dynamics of schistosomiasis*. Comm. Pure and Appl. Math. **26** (1973), 395–453.

[17] Smoller. J. : *Shock Waves and Reaction-Diffusion Equations*. Springer-Verlag, Berlin, Heidelberg. 1993.

Quenching for Coupled Degenerate Parabolic Equations

C. Y. Chan* K. K. Nip[†]

Abstract

Quenching phenomena for a first initial-boundary value problem involving a coupled system of degenerate parabolic equations are studied. Existence of a unique solution is also investigated.

1 Introduction

Let $\Omega_{h,\sigma} = (h,a) \times (0,\sigma)$, and $\partial\Omega_{h,\sigma}$ be its parabolic boundary $([h,a] \times \{0\}) \cup (\{h,a\} \times (0,\sigma))$. For any set S, we denote its closure by $\overline{S}$. For convenience, we denote $\Omega_{0,T}$ and $\partial\Omega_{0,T}$, respectively, by Ω and $\partial\Omega$, where $T \leq \infty$. Let us denote a column vector as the transpose of a row vector $(\varsigma_1, \varsigma_2)$ by $(\varsigma_1, \varsigma_2)^T$. Also let $\mathbf{w}$ and $\mathbf{g(w)}$ denote, respectively, the vectors $(w_1, w_2)^T$ and $(g_1(w_1, w_2), g_2(w_1, w_2))^T$. Chan and Fung [1] studied the following problem:

$$(1) \qquad \mathbf{w}_{xx} - \mathbf{w}_t = -\mathbf{g(w)} \text{ in } \Omega, \ \mathbf{w} = \mathbf{0} \text{ on } \partial\Omega,$$

where

(i) $\mathbf{g(0)} > \mathbf{0}$;

(ii) there exists a positive constant column vector $\mathbf{q} = (q_1, q_2)^T$ with $q_1 \geq q_2$ such that $\mathbf{g(w)}$ is of class C^2 in $(-\infty, q_1) \times (-\infty, q_2)$, and $\lim_{w_2 \to q_2} g_1(\mathbf{w}) = \infty$;

(iii) to any ϵ with $0 < \epsilon < q_2$, there exists a constant C such that

$$g_2(\mathbf{w}) - g_1(\mathbf{w}) \geq C(w_2 - w_1)$$

for $\mathbf{w} \leq \mathbf{q} - \epsilon$ where ϵ is the column vector $(\epsilon, \epsilon)^T$;

(iv) the first and second partial derivatives of $\mathbf{g(w)}$ are nonnegative, and the first partial derivatives are bounded for $\mathbf{w} \leq \mathbf{q} - \epsilon$.

A solution $\mathbf{w}$ of the problem (1) is said to quench if

$$\max\{w_2(x,t) : 0 \leq x \leq a\} \to q_2^- \text{ as } t \to T^-$$

for some finite time T; a^* is called the critical length of the problem (1) if $\mathbf{w}$ exists globally for $a < a^*$ and $\mathbf{w}$ quenches for $a > a^*$. The condition (iii) is used to show $w_1 \leq w_2$. That is,

$$w_1 \leq w_2 \leq q_2 \leq q_1 \text{ in } \Omega.$$

---

*Department of Mathematics, University of Southwestern Louisiana, Lafayette, LA 70504-1010
[†]Department of Mathematics, Southern University, Baton Rouge, LA 70813

Under this condition, they showed existence of a unique critical length. Without using $w_1 \leq w_2$ and $q_1 \geq q_2$, they showed, under certain conditions, the blow-up of $w_{1_t}(a/2, t)$ and $\mathbf{w}_t(a/2, t)$ when $t \to T^-$.

Let

$$L = \frac{\partial^2}{\partial x^2} - x^q \frac{\partial}{\partial t},$$

where q is a real constant. Chan and Kong [3] studied quenching phenomena for problems of the type,

$$Lw = -g(w) \text{ in } \Omega, \; w = 0 \text{ on } \partial\Omega.$$

Our main purpose here is to extend the results of Chan and Fung, and Chan and Kong to the following degenerate first initial-boundary value problem:

$$(2) \qquad\qquad L\mathbf{u} = -\mathbf{f}(\mathbf{u}) \text{ in } \Omega, \; \mathbf{u} = \mathbf{0} \text{ on } \partial\Omega,$$

where

(a) $\mathbf{f}(\mathbf{0}) > \mathbf{0}$;

(b) $f_1(\mathbf{u}) \in C^2\left((-\infty, c_1] \times (-\infty, c_2)\right)$ and $f_2(\mathbf{u}) \in C^2\left((-\infty, c_1) \times (-\infty, c_2]\right)$ for some positive constants c_1 and c_2;

(c) all the first partial derivatives of $\mathbf{f}$ are positive;

(d) $\int_0^{c_2} f_1(c_1, u_2)du_2 + \int_0^{c_1} f_2(u_1, c_2)du_1 < \infty$, and the matrix

$$\begin{bmatrix} f_{i,1,1} & f_{i,1,2} \\ f_{i,2,1} & f_{i,2,2} \end{bmatrix}$$

is positive semidefinite with the subscript $, k$ denoting the partial derivative with respect to the kth argument. Unlike Chan and Fung, we do not assume the condition (iii). Therefore, we can adopt a more general definition of quenching: For $i \neq j$, when $f_i(\mathbf{u}) \to \infty$ as $u_j \to c_j^-$, a solution $\mathbf{u}$ is said to quench if there exists a finite time T such that

$$\max\{u_j(x, t) : 0 \leq x \leq a\} \to c_j^- \text{ as } t \to T^-.$$

Let $\mathbf{c} = (c_1, c_2)^T$ and $\boldsymbol{\infty} = (\infty, \infty)^T$. When $\mathbf{f}(\mathbf{u}) \to \boldsymbol{\infty}$ as $\mathbf{u} \to \mathbf{c}^-$, then a solution $\mathbf{u}$ is said to quench if there exists a finite time T such that either

$$\max\{u_1(x, t) : 0 \leq x \leq a\} \to c_1^- \text{ as } t \to T^-$$

or

$$\max\{u_2(x, t) : 0 \leq x \leq a\} \to c_2^- \text{ as } t \to T^-.$$

In section 2, we prove a maximum principle for our problem and establish existence of a unique classical solution before its quenching time. In section 3, we show existence of the critical length, and the blow-up of u_{j_t} if $u_i \to c_i^-$ as $t \to T$ for $i \neq j$.

2 Existence and Uniqueness

Let

$$\mathbf{M} = \begin{bmatrix} m_{11} & m_{12} \\ m_{21} & m_{22} \end{bmatrix},$$

where $m_{ij}(x, t), i, j = 1, 2$, are bounded in Ω with m_{12} and m_{21} being nonnegative.

LEMMA 2.1. *If*

$$L\mathbf{z} + \mathbf{Mz} > \mathbf{0} \text{ in } \Omega, \mathbf{z} < \mathbf{0} \text{ on } \partial\Omega,$$

then $\mathbf{z} < \mathbf{0}$ *in* Ω.

Proof. Let us assume that $z_i \geq 0$ somewhere in Ω for some i. Let $\bar{t}$ be the least upper bound of t at which $\mathbf{z} < \mathbf{0}$ in Ω. By continuity, $\mathbf{z}(x,\bar{t}) \leq \mathbf{0}$ in $(0,a)$, which we denote by D. Without loss of generality, let us assume $z_1(\bar{x},\bar{t}) = 0$ and $z_2(\bar{x},\bar{t}) \leq 0$ for some $\bar{x}$ in D. It follows from $z_1(x,\bar{t}) \leq 0$ in D, and $z_1(\bar{x},\bar{t}) = 0$ that $z_{1_{xx}}(\bar{x},\bar{t}) \leq 0$. Since $z_1(\bar{x},t) < 0$ for $0 < t < \bar{t}$, and $z_1(\bar{x},\bar{t}) = 0$, we have $z_{1_t}(\bar{x},\bar{t}) \geq 0$. Thus, $Lz_1 \leq 0$ at the point $(\bar{x},\bar{t})$. It follows from $z_1(\bar{x},\bar{t}) = 0$, $m_{12} \geq 0$, and $z_2(\bar{x},\bar{t}) \leq 0$ that at $(\bar{x},\bar{t})$,

$$Lz_1 + m_{11}z_1 + m_{12}z_2 \leq 0,$$

which gives a contradiction. $\square$

The following result gives a maximum principle (c.f. Protter and Weinberger [9, pp. 188–190]).

THEOREM 2.1. *If*

$$L\mathbf{z} + M\mathbf{z} \geq \mathbf{0} \text{ in } \Omega, \mathbf{z} \leq \mathbf{0} \text{ on } \partial\Omega,$$

then $\mathbf{z} \leq \mathbf{0}$ *in* Ω*. Moreover, if* $z_i = 0$ *at an interior point* (x_0,t_0)*, then* $z_i \equiv 0$ *for* $t \leq t_0$*.*

Proof. Since m_{ij} are bounded in Ω, there exists a positive constant α such that

$$\alpha > \sum_{j=1}^{2} m_{ij} \text{ in } \Omega \text{ for } i = 1,2.$$

For some positive constants k_1 and k_2, let $\mathbf{w}(x,t) = \mathbf{z}(x,t) - k_2\mathbf{l}(x)$, where $\mathbf{l}(x)$ is the column vector $(l(x),l(x))^T$ with $l(x)$ satisfying

$$l'' < -\alpha l \text{ in } D, l(0) = k_1 = l(a).$$

We note that $l > 0$ in D. In Ω,

$$
\begin{aligned}
L\mathbf{w} + M\mathbf{w} &> L\mathbf{z} + k_2\alpha\mathbf{l} + M\mathbf{z} - k_2 M\mathbf{l} \\
&> L\mathbf{z} + M\mathbf{z} \\
&\geq \mathbf{0}.
\end{aligned}
$$

Since $\mathbf{z} \leq \mathbf{0}$ on $\partial\Omega$ and $l \geq k_1 > 0$ on $\overline{D}$, we have $\mathbf{w} < \mathbf{0}$ on $\partial\Omega$. By Lemma 2.1 , $\mathbf{w} < \mathbf{0}$ in Ω, and hence, $\mathbf{z} < k_2\mathbf{l}$ in Ω. Since k_2 is an arbitrary constant, we have $\mathbf{z} \leq \mathbf{0}$ in Ω.

It follows from $m_{12} \geq 0$, $m_{21} \geq 0$, and $z_i \leq 0$ in Ω that $Lz_i + m_{ii}z_i \geq 0$ in Ω. If $z_i = 0$ at some interior point (x_0,t_0), then $z_i \equiv 0$ for $t \leq t_0$ by the strong maximum principle (cf. Friedman [5, p. 39]). $\square$

From $\mathbf{f}(\mathbf{0}) > \mathbf{0}$ and Theorem 2.1, $\mathbf{0}$ is a lower solution of the problem (2). Hence, $\mathbf{u} > \mathbf{0}$ in Ω. Let

$$H(v) = f_1\left(c_1, \frac{c_2}{c_1}v\right) + \frac{c_1}{c_2}f_2(v,c_2).$$

Our next result gives an upper solution.

LEMMA 2.2. *There exist positive constants* $t_0 (\leq T)$ *and* $\bar{c}_1 (\in (0,c_1))$ *such that the problem (2) has an upper solution* $\mathbf{h} = (h, c_2h/c_1)^T$*, where* $h \in C^{2,1}\left(\overline{\Omega}_{0,t_0}\right)$*,* $h \in (0,\bar{c}_1]$*, and* h *depends only on* f_1*,* f_2*,* a *and* q*.*

Proof. By Theorem 2 of Chan and Kong [3], there exist positive constants $t_0 (\leq T)$ and $\bar{c}_1 (\in (0,c_1))$ such that the problem,

$$Lw = -H(w) \text{ in } \Omega, w = 0 \text{ on } \partial\Omega,$$

has an upper solution $h \in C^{2,1}\left(\overline{\Omega}_{0,t_0}\right)$, $h \in (0, \bar{c}_1]$, and h depends only on H, a and q. In Ω_{0,t_0},

$$
\begin{aligned}
Lh &\leq -H(h) \\
&< -f_1\left(h, \frac{c_2}{c_1}h\right);
\end{aligned}
$$

$$
\begin{aligned}
L\left(\frac{c_2}{c_1}h\right) &\leq -\frac{c_2}{c_1}H(h) \\
&< -f_2(h, c_2) \\
&\leq -f_2\left(h, \frac{c_2}{c_1}h\right).
\end{aligned}
$$

That is, $L\mathbf{h} \leq -\mathbf{f}(\mathbf{h})$ in Ω_{0,t_0}. The lemma then follows. □

We now establish an existence theorem for our coupled system by modifying the method of Floater [4], who considered a blow-up problem for a scalar equation.

THEOREM 2.2. *The problem (2) has a unique classical solution in* Ω_{0,t_0}.

Proof. Let ϵ be a positive constant, and $\mathbf{u}^{(\epsilon)} = (u_1^{(\epsilon)}, u_2^{(\epsilon)})^T$. We consider the problem,

(3) $\qquad x^{-q}L\mathbf{u}^{(\epsilon)} = -x^{-q}\mathbf{f}\left(\mathbf{u}^{(\epsilon)}\right)$ in Ω_{ϵ,t_0}, $\mathbf{u}^{(\epsilon)} = \mathbf{0}$ on $\partial\Omega_{\epsilon,t_0}$.

By Lemma 2.2, $\mathbf{h}$ is an upper solution. Since $\mathbf{0}$ is a lower solution, it follows from Theorem 4.4.6 of Ladde, Lakshmikantham and Vatsala [7, p. 160] that the problem (3) has a solution $\mathbf{u}^{(\epsilon)} \in C^{2+\beta,1+\beta/2}\left([\epsilon, a] \times [0, t_0)\right)$. Let $\bar{c}_2 = (c_2/c_1)\bar{c}_1$, and $\bar{\mathbf{c}} = (\bar{c}_1, \bar{c}_2)^T$. Since $0 < h \leq \bar{c}_1 < c_1$, we have

$$0 < \mathbf{u}^{(\epsilon)} < \mathbf{h} \leq \bar{\mathbf{c}} < \mathbf{c}.$$

Let ϵ_1 and ϵ_2 be some positive constants such that $\epsilon_1 < \epsilon_2 < a$, and $\mathbf{u}^{(\epsilon_1)}$ and $\mathbf{u}^{(\epsilon_2)}$ be solutions of the problem (3) corresponding to $\epsilon = \epsilon_1$ and $\epsilon = \epsilon_2$ respectively. Let $\mathbf{w} = \mathbf{u}^{(\epsilon_1)} - \mathbf{u}^{(\epsilon_2)}$, and

$$
\mathbf{F}(\boldsymbol{\rho}_1, \boldsymbol{\rho}_2) = \left[\begin{array}{cc} f_{1,1}(\boldsymbol{\rho}_1) & f_{1,2}(\boldsymbol{\rho}_1) \\ f_{2,1}(\boldsymbol{\rho}_2) & f_{2,2}(\boldsymbol{\rho}_2) \end{array}\right].
$$

Then,

$$L\mathbf{w} + \mathbf{F}(\boldsymbol{\xi}_1, \boldsymbol{\xi}_2)\mathbf{w} = \mathbf{0} \text{ in } \Omega_{\epsilon_2,t_0},$$

where $\boldsymbol{\xi}_i, i = 1, 2$, lie between $\mathbf{u}^{(\epsilon_1)}$ and $\mathbf{u}^{(\epsilon_2)}$. Since $\mathbf{u}^{(\epsilon_1)} = \mathbf{0} = \mathbf{u}^{(\epsilon_2)}$ on $([\epsilon_2, a] \times \{0\}) \cup (\{a\} \times (0, t_0))$, $\mathbf{u}^{(\epsilon_1)} > \mathbf{0}$ on $\{\epsilon_2\} \times (0, t_0)$, and $\mathbf{u}^{(\epsilon_2)} = \mathbf{0}$ on $\{\epsilon_2\} \times (0, t_0)$, we have $\mathbf{w} \geq \mathbf{0}$ on $\partial\Omega_{\epsilon_2,t_0}$. From Theorem 2.1, $\mathbf{u}^{(\epsilon_1)} > \mathbf{u}^{(\epsilon_2)}$ in Ω_{ϵ_2,t_0}. Let $\mathbf{u} = \lim_{\epsilon \to 0} \mathbf{u}^{(\epsilon)}$.

For $(x_1, t_1) \in \Omega_{0,t_0}$, let $b_1, b_2, b_3, d_1, d_2, d_3$ and t_2 be positive constants such that $0 < b_1 < b_2 < b_3 < x_1 < d_3 < d_2 < d_1 < a$, and $0 < t_1 < t_2 < t_0$. For $k = 1, 2, 3$, let $Q_k = (b_k, d_k) \times (0, t_2)$. Then,

$$(x_1, t_1) \in Q_3 \subset Q_2 \subset Q_1 \subset \Omega_{0,t_0}.$$

Let us consider those ϵ with $\epsilon < b_1$. Since $C^{2+\beta,1+\beta/2}(Q_1) \subset W_p^{2,1}(Q_1)$ for any $p > 0$ and $\mathbf{u}^{(\epsilon)} \in C^{2+\beta,1+\beta/2}(Q_1)$, we have $\mathbf{u}^{(\epsilon)} \in W_p^{2,1}(Q_1)$. By (4.10.12) of Ladyženskaja, Solonnikov and Ural'ceva [8, p. 355],

$$\left\|u_i^{(\epsilon)}\right\|_{W_p^{2,1}(Q_2)} \leq N_{i,1}\left\|x^{-q}f_i(\mathbf{u}^{(\epsilon)})\right\|_{L^p(Q_1)} + N_{i,2}\left\|u_i^{(\epsilon)}\right\|_{L^s(Q_1)}$$

for some positive constants $N_{i,1}$, $N_{i,2}$ and s with $1 \leq s < p$. Let us choose $p > 2/(1-\beta)$. By an embedding theorem of Ladyženskaja, Solonnikov and Ural'ceva [8, pp. 61 and 80], we can embed $W_p^{2,1}(Q_2)$ into $C^{\beta,\beta/2}(Q_2)$, and

$$\left\| u_i^{(\epsilon)} \right\|_{C^{\beta,\beta/2}(Q_2)} \leq K \left\| u_i^{(\epsilon)} \right\|_{W_p^{2,1}(Q_2)}$$

for some positive constant K. Thus,

$$
\begin{aligned}
\left\| u_i^{(\epsilon)} \right\|_{C^{\beta,\beta/2}(Q_2)} &\leq K \left(N_{i,1} \left\| x^{-q} f_i(\mathbf{u}^{(\epsilon)}) \right\|_{L^p(Q_1)} + N_{i,2} \left\| u_i^{(\epsilon)} \right\|_{L^s(Q_1)} \right) \\
&\leq K \left(N_{i,1} \left\| b_1^{-q} f_i(\overline{\mathbf{c}}) \right\|_{L^p(Q_1)} + N_{i,2} \left\| \overline{c}_i \right\|_{L^s(Q_1)} \right),
\end{aligned}
$$

which is a constant independent of ϵ. It then follows from Theorem 4.10.1 of Ladyženskaja, Solonnikov and Ural'ceva [8, pp. 351–352] that

$$(4) \qquad \left\| u_i^{(\epsilon)} \right\|_{C^{2+\beta,1+\beta/2}(Q_3)} \leq N_{i,3} \left\| x^{-q} f_i(\mathbf{u}^{(\epsilon)}) \right\|_{C^{\beta,\beta/2}(Q_2)} + N_{i,4} \max_{Q_2} \left| u_i^{(\epsilon)} \right|$$

for some positive constants $N_{i,3}$ and $N_{i,4}$. To estimate the first term on the right-hand side, we note that
(i)

$$\max_{Q_2} \left| x^{-q} f_i(\mathbf{u}^{(\epsilon)}) \right| \leq b_2^{-q} f_i(\overline{\mathbf{c}});$$

(ii) for $x, \bar{x} \in (b_2, d_2)$ with $x \neq \bar{x}$,

$$
\begin{aligned}
&\left| x^{-q} f_i\left(\mathbf{u}^{(\epsilon)}(x,t) \right) - \bar{x}^{-q} f_i\left(\mathbf{u}^{(\epsilon)}(\bar{x},t) \right) \right| \\
&\leq \left| x^{-q} \left[f_i\left(\mathbf{u}^{(\epsilon)}(x,t) \right) - f_i\left(\mathbf{u}^{(\epsilon)}(\bar{x},t) \right) \right] \right| + \left| (x^{-q} - \bar{x}^{-q}) f_i\left(\mathbf{u}^{(\epsilon)}(\bar{x},t) \right) \right| \\
&= x^{-q} \left| f_{i,1}(\boldsymbol{\xi}_i) \left(u_1^{(\epsilon)}(x,t) - u_1^{(\epsilon)}(\bar{x},t) \right) + f_{i,2}(\boldsymbol{\xi}_i) \left(u_2^{(\epsilon)}(x,t) - u_2^{(\epsilon)}(\bar{x},t) \right) \right| \\
&\quad + \left| x^{-q} - \bar{x}^{-q} \right| \left| f_i\left(\mathbf{u}^{(\epsilon)}(\bar{x},t) \right) \right|,
\end{aligned}
$$

where $\boldsymbol{\xi}_i$ lies between $\mathbf{u}^{(\epsilon)}(x,t)$ and $\mathbf{u}^{(\epsilon)}(\bar{x},t)$. Let

$$N_{i,5} = \max \left\{ \sup_{Q_2} f_{i,1}, \sup_{Q_2} f_{i,2} \right\}.$$

Then,

$$
\begin{aligned}
&\left\langle x^{-q} f_i(\mathbf{u}^{(\epsilon)}) \right\rangle_{x,Q_2}^{\beta} \\
&\equiv \sup_{Q_2} \frac{\left| x^{-q} f_i\left(\mathbf{u}^{(\epsilon)}(x,t) \right) - \bar{x}^{-q} f_i\left(\mathbf{u}^{(\epsilon)}(\bar{x},t) \right) \right|}{|x - \bar{x}|^{\beta}} \\
&\leq b_2^{-q} N_{i,5} \sup_{Q_2} \left(\frac{\left| u_1^{(\epsilon)}(x,t) - u_1^{(\epsilon)}(\bar{x},t) \right| + \left| u_2^{(\epsilon)}(x,t) - u_2^{(\epsilon)}(\bar{x},t) \right|}{|x - \bar{x}|^{\beta}} \right) \\
&\quad + f_i(\overline{\mathbf{c}}) \sup_{Q_2} \frac{\left| x^{-q} - \bar{x}^{-q} \right|}{|x - \bar{x}|^{\beta}} \\
&= b_2^{-q} N_{i,5} \left(\left\langle u_1^{(\epsilon)} \right\rangle_{x,Q_2}^{\beta} + \left\langle u_2^{(\epsilon)} \right\rangle_{x,Q_2}^{\beta} \right) + f_i(\overline{\mathbf{c}}) \left\langle x^{-q} \right\rangle_{x,Q_2}^{\beta};
\end{aligned}
$$

(iii) for $t, \bar{t} \in (0, t_2)$ with $t \neq \bar{t}$,

$$\left| x^{-q} f_i\left(\mathbf{u}^{(\epsilon)}(x,t)\right) - x^{-q} f_i\left(\mathbf{u}^{(\epsilon)}(x,\bar{t})\right) \right|$$
$$\leq \quad x^{-q} \left| f_{i,1}(\boldsymbol{\xi}_i)\left(u_1^{(\epsilon)}(x,t) - u_1^{(\epsilon)}(x,\bar{t})\right) + f_{i,2}(\boldsymbol{\xi}_i)\left(u_2^{(\epsilon)}(x,t) - u_2^{(\epsilon)}(x,\bar{t})\right) \right|,$$

where $\boldsymbol{\xi}_i$ lies between $\mathbf{u}^{(\epsilon)}(x,t)$ and $\mathbf{u}^{(\epsilon)}(x,\bar{t})$. Then,

$$\left\langle x^{-q} f_i(\mathbf{u}^{(\epsilon)}) \right\rangle_{t,Q_2}^{\beta/2}$$
$$\equiv \quad \sup_{Q_2} \frac{\left| x^{-q} f_i\left(\mathbf{u}^{(\epsilon)}(x,t)\right) - x^{-q} f_i\left(\mathbf{u}^{(\epsilon)}(x,\bar{t})\right) \right|}{|t - \bar{t}|^{\beta/2}}$$
$$\leq \quad b_2^{-q} N_{i,5} \sup_{Q_2} \left(\frac{\left| u_1^{(\epsilon)}(x,t) - u_1^{(\epsilon)}(x,\bar{t}) \right| + \left| u_2^{(\epsilon)}(x,t) - u_2^{(\epsilon)}(x,\bar{t}) \right|}{|t - \bar{t}|^{\beta/2}} \right)$$
$$= \quad b_2^{-q} N_{i,5} \left(\left\langle u_1^{(\epsilon)} \right\rangle_{t,Q_2}^{\beta/2} + \left\langle u_2^{(\epsilon)} \right\rangle_{t,Q_2}^{\beta/2} \right).$$

By (i), (ii) and (iii),

$$\left\| x^{-q} f_i(\mathbf{u}^{(\epsilon)}) \right\|_{C^{\beta,\beta/2}(Q_2)}$$
$$\equiv \quad \max_{Q_2} \left| x^{-q} f_i(\mathbf{u}^{(\epsilon)}) \right| + \left\langle x^{-q} f_i(\mathbf{u}^{(\epsilon)}) \right\rangle_{x,Q_2}^{\beta} + \left\langle x^{-q} f_i(\mathbf{u}^{(\epsilon)}) \right\rangle_{t,Q_2}^{\beta/2}$$
$$\leq \quad b_2^{-q} f_i(\overline{\mathbf{c}}) + b_2^{-q} N_{i,5} \left(\left\langle u_1^{(\epsilon)} \right\rangle_{x,Q_2}^{\beta} + \left\langle u_2^{(\epsilon)} \right\rangle_{x,Q_2}^{\beta} \right) + f_i(\overline{\mathbf{c}}) \left\langle x^{-q} \right\rangle_{x,Q_2}^{\beta}$$
$$\quad + b_2^{-q} N_{i,5} \left(\left\langle u_1^{(\epsilon)} \right\rangle_{t,Q_2}^{\beta/2} + \left\langle u_2^{(\epsilon)} \right\rangle_{t,Q_2}^{\beta/2} \right)$$
$$\leq \quad f_i(\overline{\mathbf{c}}) \left(b_2^{-q} + \left\langle x^{-q} \right\rangle_{x,Q_2}^{\beta} \right) + b_2^{-q} N_{i,5} \left(\left\| u_1^{(\epsilon)} \right\|_{C^{\beta,\beta/2}(Q_2)} + \left\| u_2^{(\epsilon)} \right\|_{C^{\beta,\beta/2}(Q_2)} \right).$$

It then follows from (4) that

$$\left\| u_i^{(\epsilon)} \right\|_{C^{2+\beta,1+\beta/2}(Q_3)} \leq \quad N_{i,3} f_i(\overline{\mathbf{c}}) \left(b_2^{-q} + \left\langle x^{-q} \right\rangle_{x,Q_2}^{\beta} \right)$$
$$\quad + b_2^{-q} N_{i,3} N_{i,5} \left(\left\| u_1^{(\epsilon)} \right\|_{C^{\beta,\beta/2}(Q_2)} + \left\| u_2^{(\epsilon)} \right\|_{C^{\beta,\beta/2}(Q_2)} \right)$$
$$\quad + N_{i,4} \max_{Q_2} \left| u_i^{(\epsilon)} \right|,$$

which is a constant independent of ϵ.

For (x,t) and $(\bar{x}, \bar{t})$ in Q_3, we have

$$\left| u_i^{(\epsilon)}(x,t) - u_i^{(\epsilon)}(\bar{x},\bar{t}) \right|$$
$$\leq \quad \left| u_i^{(\epsilon)}(x,t) - u_i^{(\epsilon)}(\bar{x},t) \right| + \left| u_i^{(\epsilon)}(\bar{x},t) - u_i^{(\epsilon)}(\bar{x},\bar{t}) \right|$$
$$= \quad |x - \bar{x}|^{\beta} \frac{\left| u_i^{(\epsilon)}(x,t) - u_i^{(\epsilon)}(\bar{x},t) \right|}{|x - \bar{x}|^{\beta}} + |t - \bar{t}|^{\beta/2} \frac{\left| u_i^{(\epsilon)}(\bar{x},t) - u_i^{(\epsilon)}(\bar{x},\bar{t}) \right|}{|t - \bar{t}|^{\beta/2}}$$
$$\leq \quad |x - \bar{x}|^{\beta} \left\langle u_i^{(\epsilon)} \right\rangle_{x,Q_3}^{\beta} + |t - \bar{t}|^{\beta/2} \left\langle u_i^{(\epsilon)} \right\rangle_{t,Q_3}^{\beta/2}$$
$$\leq \quad \left(|x - \bar{x}|^{\beta} + |t - \bar{t}|^{\beta/2} \right) \left\| u_i^{(\epsilon)} \right\|_{C^{\beta,\beta/2}(Q_3)}.$$

That is, $u_i^{(\epsilon)}$ is equicontinuous in Q_3. Similarly, we can show that $u_{it}^{(\epsilon)}$, $u_{ix}^{(\epsilon)}$ and $u_{ixx}^{(\epsilon)}$ are equicontinuous in Q_3. By the Ascoli-Arzela Theorem, $\mathbf{u} \in C^{2+\beta,1+\beta/2}(Q_3)$ and the partial derivatives of $\mathbf{u}$ are the limits of the corresponding converging subsequences of the partial derivatives of $\mathbf{u}^{(\epsilon)}$. Hence, $\mathbf{u}$ is $C^{2,1}$ at (x_1, t_1). Since $\mathbf{u} = \lim_{\epsilon \to 0} \mathbf{u}^{(\epsilon)}$ and $\mathbf{u}^{(\epsilon)} \in C([\epsilon, a] \times [0, t_0))$, it follows that $\mathbf{u}$ is continuous on $\partial \Omega_{0, t_0}$.

By Theorem 2.1, $\mathbf{u}$ is unique. □

Let T be the supremum over t_0 for which there is a unique solution in $(0, t_0)$. Then, $\mathbf{u}$ is the unique classical solution in Ω.

3 Quenching Phenomena

We now give a criterion for quenching to occur.

THEOREM 3.1. *If* $f_i(\mathbf{u}) \to \infty$ *as* $u_j \to c_j^-$ *for* $i \neq j$, *and*

$$a > 4\sqrt{\frac{2c_i}{f_i(0)}},$$

then $\mathbf{u}$ *quenches in a finite time. If* $\mathbf{f}(\mathbf{u}) \to \infty$ *as* $\mathbf{u} \to \mathbf{c}^-$, *and*

$$a > \min_{k=1,2} \left\{ 4\sqrt{\frac{2c_k}{f_k(0)}} \right\},$$

then $\mathbf{u}$ *quenches in a finite time.*

Proof. We prove this by contradiction. Without loss of generality, let us suppose $u_2 < c_2$ in $\overline{\Omega}_{0,\infty}$ since the proof in the other case $u_1 < c_1$ in $\overline{\Omega}_{0,\infty}$ is similar.

Case 1: $q \leq 0$.

Let

$$\tau = \frac{c_1 a^{q+2}}{2^q \left(a^2 f_1(0) - 32c_1 \right)}.$$

Since $a > 4\sqrt{2c_1/f_1(0)}$, it follows that $a^2 f_1(0) - 32c_1 > 0$, and hence, $\tau > 0$. Let

$$\gamma = \frac{2^{q+4} f_1(0)}{2^{q+5}\tau + a^{q+2}},$$

$\varphi(x, t) = \gamma(x - a/2)(a - x)t$, and $w = u_1 - \varphi$. In $\Omega_{a/2, \tau}$,

$$
\begin{aligned}
Lw &= -f_1(\mathbf{u}) + 2\gamma t + \gamma x^q \left(x - \frac{a}{2} \right)(a - x) \\
&\leq -f_1(0) + \gamma \left[2\tau + \left(\frac{a}{2} \right)^q \frac{a^2}{16} \right] \\
&= 0.
\end{aligned}
$$

Since $\varphi = 0$ on $\partial \Omega_{a/2,\tau}$ and $u_1 \geq 0$ on $\partial \Omega_{a/2,\tau}$, we have $w \geq 0$ on $\partial \Omega_{a/2,\tau}$. By the weak maximum principle, $w \geq 0$ in $\Omega_{a/2,\tau}$. Thus,

$$
\begin{aligned}
u_1 \left(\frac{3a}{4}, \tau \right) &\geq \varphi \left(\frac{3a}{4}, \tau \right) \\
&= c_1.
\end{aligned}
$$

This gives a contradiction.

Case 2: $q > 0$.

Let

$$\bar{\tau} = \frac{c_1 a^{q+2}}{a^2 f_1(0) - 32c_1},$$

$$\bar{\gamma} = \frac{16 f_1(0)}{32\bar{\tau} + a^{q+2}},$$

$\bar{\varphi}(x,t) = \bar{\gamma}(x - a/2)(a - x)t$, and $\bar{w} = u_1 - \bar{\varphi}$. An argument similar to that in Case 1 gives $\bar{w} \geq 0$ in $\Omega_{a/2,\bar{\tau}}$. Thus,

$$u_1\left(\frac{3a}{4}, \bar{\tau}\right) \geq \bar{\varphi}\left(\frac{3a}{4}, \bar{\tau}\right)$$

$$= c_1.$$

This gives a contradiction.

Let $c_i/f_i(0) = \min_{k=1,2} \{c_k/f_k(0)\}$. If $\mathbf{f}(\mathbf{u}) \to \infty$ as $\mathbf{u} \to \mathbf{c}^-$, and $a > 4\sqrt{2c_i/f_i(0)}$, then the proof follows by assuming $u_j < c_j$ for $i \neq j$ (instead of $u_2 < c_2$) in $\overline{\Omega}_{0,\infty}$. $\square$

To establish existence of the critical length a^*, we need the following three lemmas. Let

$$G(v) = \int_0^v H(u)du.$$

By using the condition (d), $G(c_1) < \infty$. A proof similar to that of Lemma 5 of Chan and Ke [2] gives the following result.

LEMMA 3.1. *There exists a unique positive solution v satisfying*

(5) $$v'' = -H(v) \text{ in } (0, b_s^*), \quad v(0) = 0, \quad v(b_s^*) = c_1,$$

where

$$b_s^* = \frac{1}{\sqrt{2}} \int_0^{c_1} \frac{ds}{\sqrt{G(c_1) - G(s)}}.$$

LEMMA 3.2. *(a) If $a \leq b_s^*$, then $\mathbf{u} < \mathbf{c}$ in Ω.*
(b) If $b_s^* < a \leq 2b_s^*$, then $\mathbf{u} < \mathbf{c}$ in Ω.
(c) If $a > 2b_s^*$, then $\mathbf{u} < \mathbf{c}$ in $([0, b_s^*] \cup [a - b_s^*, a]) \times (0, T)$.

Proof. (a) Let v be the unique solution of the problem (5), $\mathbf{v} = (v, (c_2/c_1)v)^T$, and $\mathbf{w} = \mathbf{u} - \mathbf{v}$. Then,

$$Lw_1 > -f_1(\mathbf{u}) + f_1\left(c_1, \frac{c_2}{c_1}v\right)$$

$$> -f_1(c_1, u_2) + f_1\left(c_1, \frac{c_2}{c_1}v\right)$$

$$= -f_{1,2}(c_1, \eta)\left(u_2 - \frac{c_2}{c_1}v\right);$$

$$Lw_2 > -f_2(\mathbf{u}) + f_2(v, c_2)$$

$$> -f_2(u_1, c_2) + f_2(v, c_2)$$

$$= -f_{2,1}(\xi, c_2)(u_1 - v),$$

where ξ lies between u_1 and v, and η lies between u_2 and $(c_2/c_1)v$. Hence,

(6) $$L\mathbf{w} + \begin{bmatrix} 0 & f_{1,2}(c_1, \eta) \\ f_{2,1}(\xi, c_2) & 0 \end{bmatrix} \mathbf{w} > \mathbf{0} \text{ in } \Omega.$$

Since $\mathbf{v} \geq \mathbf{0}$ on $\partial\Omega$ and $\mathbf{u} = \mathbf{0}$ on $\partial\Omega$, $\mathbf{w} \leq \mathbf{0}$ on $\partial\Omega$. It follows from Theorem 2.1 that $\mathbf{w} \leq \mathbf{0}$ in Ω. That is, $\mathbf{u} \leq \mathbf{v} < \mathbf{c}$ in Ω.

(b) Let $\omega = (0, b_s^*) \times (0, T)$, $\partial\omega$ be its parabolic boundary, and $\mathbf{w} = \mathbf{u} - \mathbf{v}$. Since $\mathbf{u} = \mathbf{0}$ and $\mathbf{v} \geq \mathbf{0}$ on $([0, b_s^*] \times \{0\}) \cup (\{0\} \times (0, T))$, we have $\mathbf{w} \leq \mathbf{0}$ there; furthermore, $\mathbf{w}(b_s^*, t) \leq \mathbf{0}$. By Theorem 2.1, $\mathbf{w} \leq \mathbf{0}$ in ω. That is, $\mathbf{u} \leq \mathbf{v} < \mathbf{c}$ in ω.

Let $\tilde{\mathbf{w}}(x, t) = \mathbf{u}(x, t) - \mathbf{v}(a - x)$. From (6),

$$L\tilde{\mathbf{w}} + \begin{bmatrix} 0 & f_{1,2}(c_1, \eta) \\ f_{2,1}(\xi, c_2) & 0 \end{bmatrix} \tilde{\mathbf{w}} > \mathbf{0} \text{ in } \Omega_{a-b_s^*, T},$$

where ξ lies between $u_1(x, t)$ and $v(a - x)$, and η lies between $u_2(x, t)$ and $(c_2/c_1)v(a - x)$. Since $\mathbf{u}(x, t) = \mathbf{0}$ and $\mathbf{v}(a - x) \geq \mathbf{0}$ on $([a - b_s^*, a] \times \{0\}) \cup (\{a\} \times (0, T))$, we have $\tilde{\mathbf{w}} \leq \mathbf{0}$ there; furthermore, $\tilde{\mathbf{w}}(a - b_s^*, t) \leq \mathbf{0}$. By Theorem 2.1, $\tilde{\mathbf{w}} \leq \mathbf{0}$ in $\Omega_{a-b_s^*, T}$. That is, $\mathbf{u}(x, t) \leq \mathbf{v}(a - x) < \mathbf{c}$ in $\Omega_{a-b_s^*, T}$.

Since $b_s^* < a \leq 2b_s^*$, $\omega \cup \Omega_{a-b_s^*, T} = \Omega$. Thus, $\mathbf{u} < \mathbf{c}$ in Ω.

(c) It follows from part (b) that $\mathbf{u} < \mathbf{c}$ in $\omega \cup \Omega_{a-b_s^*, T}$. Since $a > 2b_s^*$, $\omega \cup \Omega_{a-b_s^*, T} = ([0, b_s^*] \cup [a - b_s^*, a]) \times (0, T)$. Thus, $\mathbf{u} < \mathbf{c}$ in $([0, b_s^*] \cup [a - b_s^*, a]) \times (0, T)$. $\square$

Let $\mathbf{u}(x, t; a)$ be the solution of the problem (2) with Ω denoted by Ω_a.

LEMMA 3.3. *If $a_1 < a_2$, then $\mathbf{u}(x, t; a_1) < \mathbf{u}(x, t; a_2)$ in Ω_{a_1}.*

Proof. Let $\mathbf{w} = \mathbf{u}(x, t; a_1) - \mathbf{u}(x, t; a_2)$. Then,

$$L\mathbf{w} + \mathbf{F}(\xi_1, \xi_2)\mathbf{w} = \mathbf{0} \text{ in } \Omega_{a_1},$$

where ξ_i, $i = 1, 2$, lie between $\mathbf{u}(x, t; a_1)$ and $\mathbf{u}(x, t; a_2)$. Since $\mathbf{u}(x, t; a_1) = \mathbf{0}$ on $\partial\Omega_{a_1}$ and $\mathbf{u}(x, t; a_2) \geq \mathbf{0}$ on $\partial\Omega_{a_1}$, we have $\mathbf{w} \leq \mathbf{0}$ on $\partial\Omega_{a_1}$. By Theorem 2.1, $\mathbf{w} < \mathbf{0}$ in Ω_{a_1}, and hence the lemma is proved. $\square$

THEOREM 3.2. *The critical length $a^* = \sup\{a : \mathbf{u}(x, t; a) \text{ exists globally}\}$.*

Proof. It follows from Lemma 3.2 that $\mathbf{u}$ is bounded away from $\mathbf{c}$ for $a \leq 2b_s^*$. By Lemma 3.3, $\mathbf{u}(x, t; a_1) < \mathbf{u}(x, t; a_2)$ in Ω_{a_1} for $a_1 < a_2$. In the case $f_i(\mathbf{u}) \to \infty$ as $u_j \to c_j^-$ for $i \neq j$, and $a > 4\sqrt{2c_i/f_i(0)}$, we have from Theorem 3.1 that $\mathbf{u}$ quenches in a finite time. If $\mathbf{f}(\mathbf{u}) \to \infty$ as $\mathbf{u} \to \mathbf{c}^-$, and $a > \min_{k=1,2}\left\{4\sqrt{2c_k/f_k(0)}\right\}$, then $\mathbf{u}$ also quenches in a finite time. Thus, the theorem is proved. $\square$

We show that $\mathbf{u}$ is a strictly increasing function of t.

LEMMA 3.4. *In Ω, $\mathbf{u}_t > \mathbf{0}$.*

Proof. Let $\mathbf{w} = \mathbf{u}(x, t + \delta) - \mathbf{u}(x, t)$ for some $\delta > 0$. Then,

$$L\mathbf{w} + \mathbf{F}(\xi_1, \xi_2)\mathbf{w} = \mathbf{0} \text{ in } \Omega,$$

where ξ_i, $i = 1, 2$, lie between $\mathbf{u}(x, t)$ and $\mathbf{u}(x, t + \delta)$. Since $\mathbf{w}(0, t) = \mathbf{0} = \mathbf{w}(a, t)$, and $\mathbf{w}(x, 0) = \mathbf{u}(x, \delta) - \mathbf{u}(x, 0) > \mathbf{0}$, it follows from Theorem 2.1 that $\mathbf{w} > \mathbf{0}$ in Ω, and hence the lemma is proved. $\square$

Using Lemma 3.2(c), we can modify the argument of Friedman and McLeod [6] for blow-up problems and the proof of Theorem 4 of Chan and Kong [3] to prove the following result.

THEOREM 3.3. *If $f_i(\mathbf{u}) \to \infty$ as $u_j \to c_j^-$ for $i \neq j$, and $u_j \to c_j^-$ as $t \to T^-$, then $u_{i_t} \to \infty$ as $t \to T^-$. If $\mathbf{f}(\mathbf{u}) \to \infty$ as $\mathbf{u} \to \mathbf{c}^-$, and $\mathbf{u} \to \mathbf{c}^-$ as $t \to T^-$, then $\mathbf{u}_t \to \infty$ as $t \to T^-$.*

References

[1] C. Y. Chan and D. T. Fung, *Quenching for coupled semilinear reaction-diffusion problems*, Nonlinear Anal., 21 (1993), pp. 143–152.

[2] C. Y. Chan and L. Ke, *Beyond quenching for singular reaction-diffusion problems*, Math. Methods Appl. Sci., 17 (1994), pp. 1–9.

[3] C. Y. Chan and P. C. Kong, *Quenching for degenerate semilinear parabolic equations*, Appl. Anal., 54 (1994), pp. 17–25.

[4] M. S. Floater, *Blow-up at the boundary for degenerate semilinear parabolic equations*, Arch. Rat. Mech. Anal., 114 (1991), pp. 57–77.

[5] A. Friedman, *Partial Differential Equations of Parabolic Type*, Prentice-Hall, Inc., Englewood Cliffs, NJ, 1964.

[6] A. Friedman and B. McLeod, *Blow-up of positive solutions of semilinear heat equations*, Indiana Univ. Math. J., 34 (1985), pp. 425–447.

[7] G. S. Ladde, V. Lakshmikantham, and A. S. Vatsala, *Monotone Iterative Techniques for Nonlinear Differential Equations*, Pitman Advanced Publishing Program, Boston, MA, 1985.

[8] O. A. Ladyženskaja, V. A. Solonnikov, and N. N. Ural'ceva, *Linear and Quasilinear Equations of Parabolic Type*, Transl. Math. Monographs, Vol. 23, American Mathematical Society, Providence, RI, 1968.

[9] M. H. Protter, and H. F. Weinberger, *Maximum Principles in Differential Equations*, Prentice-Hall, Inc., Englewood Cliffs, NJ, 1967.

Numerical Mountain Pass Methods for Nonlinear Boundary Value Problems

by

Y.S. Choi[1] and P.J. McKenna[2]
Department of Mathematics
University of Connecticut
Storrs, Connecticut 06268

Abstract

In the last two decades, mini-max methods have led to considerable progress in the understanding of the existence of multiple solutions of boundary value problems. We survey how the mountain pass theorem can be used to construct a robust numerical algorithm that can be applied to a wide variety of nonlinear problems.

1 Introduction.

In the nineteen sixties and seventies, the numerical solution of the problem

$$\Delta u + f(x, u) = 0 \qquad \text{in } \Omega \qquad (1.1)$$

with Dirichlet boundary conditions $u|_{\partial\Omega} = 0$, where Ω is a nice bounded region in R^2 was treated by a number of authors, including Bers, Douglas, Parter, Ciarlet, Schultz and Varga, usually under assumptions which guaranteed the existence of a unique solution of (1). A typical restriction was that $f_u(x, u) < \lambda_1$, the first eigenvalue of the Laplacian with Dirichlet boundary conditions.

This type of assumption guaranteed existence and uniqueness of the solution and allowed the proof of this by either a contraction mapping theorem (thereby creating a constructive iterative scheme) or a monotone iterative scheme involving starting at an upper solution.

None of these approaches gave much insight into how to numerically find solutions of boundary value problems when there were multiple solutions of a nonobvious type.

---

[1] Research partially supported by NSF grant DMS-9208636.

[2] Research partially supported by NSF grant DMS-9102632.

The purpose of this paper is to outline recent advances in this area, where methods of differential inequalities and maximum principles, variational methods, and numerical methods are combined to give new ways of finding multiple solutions of boundary value problems.

The typical situation will be that there is an "obvious" solution and some additional information that there are more solutions existing. The other key ingredient is that the boundary value problem is equivalent to finding critical points of a functional on a well-defined Hilbert space.

One then uses a numerical variation of the mountain pass theorem to find these other solutions. We use the words "obvious solution" rather loosely. We mean that the solution is readily calculable by explicit calculation or that its existence is known by the existence of upper and lower solutions, so that it can be readily calculated by an iterative scheme.

First a few words about the mountain pass theorem.

Since its introduction in 1973, by Ambrosetti and Rabinowitz the Mountain Pass Theorem has been used extensively as a tool for proving the existence of critical points of nonlinear functionals. Typically, these functionals are such that their critical points are solutions of semilinear partial differential equations or periodic solutions of nonlinear Hamiltonian systems.

The usual ingredients to apply the mountain-pass theorem are:

1. a functional $I(u)$ defined on a Banach space B

2. an already known critical point e_1 which is a *local minimum*

3. the existence of another point e_2 such that $I(e_1) > I(e_2)$

The theorem then states that if one considers the totality of all paths joining e_1 and e_2 and looks at the infimum of the maxima of the functional I along those paths, one obtains a critical value.

For example, consider the semilinear elliptic equation

$$\Delta u + u^3 = 0 \qquad \text{in } \Omega \qquad (1.2)$$

with Dirichlet boundary conditions $u|_{\partial\Omega} = 0$, where Ω is a nice bounded region in R^2.

Solutions of this equation correspond to critical points of the functional

$$I(u) = \int_\Omega (\frac{1}{2}|\nabla u|^2 - u^4/4)dx \qquad (1.3)$$

defined on the Hilbert space $H_0^1(\Omega)$.

It is clear that $u \equiv 0$ is a solution of the equation. It becomes clear, after some some thought, that in fact, this solution corresponds to a local minimum of the functional I. It is also clear that if we pick another positive function V in the Banach space, then $I(cV) \to -\infty$ as $c \to +\infty$. Thus we have the ingredients of the mountain pass theorem, and we can conclude the existence of a separate solution v, with the property that $I(v) > 0$.

This idea, although highly nonconstructive, can be used to numerically solve the equation.

On a finite-dimensional approximating subspace, one takes a piecewise linear path from the local minimum to some point whose altitude $I(e_2)$ is lower than that of the minimum. One then finds the maximum of I along that path. One then alters or deforms the path by pushing the point at which the local maximum is located in the direction of *descent*.

(How exactly one chooses an appropriate direction for the descent direction will depend on the problem. However, in most cases, when easily computable, this direction is located by finding the Fréchet derivative of I at that point.)

One repeats this step, stopping only when the critical point is reached, after which no further lowering of the local maximum is possible. In this way, we are sure of avoiding the "obvious" solution, namely the local minimum. In the next sections, we show how these ideas can be used on a variety of nonlinear boundary value problems.

2 The semilinear elliptic case

We wil look at results for two different semilinear problems, with two different types of nonlinearity $f(x, u)$.

2.1 A piecewise linear nonlinearity.

For our first example, we consider the boundary value problem

$$\Delta u + bu^+ = \sin \pi x \sin \pi y \tag{2.1}$$

on the region $\Omega = (0, 1) \times (0, 1) \in R^2$, with Dirichlet boundary conditions.

It is easy to verify, using the techniques of that if $2\pi^2 < b < 5\pi^2$, then the equation has exactly two solutions, namely the "obvious" ones, $u_1(x, y) = 1/(b - 2\pi^2) \sin \pi x \sin \pi y$ and $u_2(x, y) = -1/(2\pi^2) \sin \pi x \sin \pi y$. It is noted that u_1 is always positive and u_2 is always negative.

It is also easy to verify that for all b, the function u_2 is a local minimum of the associated functional

$$I(u) = \int_\Omega (\frac{1}{2}(|\nabla u|^2 - b(u^+)^2) + u \sin \pi x \sin \pi y) dx dy . \tag{2.2}$$

The fact that this functional satisfies the (PS) condition is easily verified.

As soon as b increases beyond the second eigenvalue $\lambda = 5\pi^2$, additional solutions appear.

One of the interesting features of this example is that it has one exact solution which can be obtained as a saddle point, namely u_1, and in addition has four more solutions which are almost exact, in the sense that we can use standard ordinary differential equation methods to solve the two-point boundary value problem for $Y(y)$, and this gives us the solutions of the form $u(x, y) = \sin \pi x \, Y(y)$ and its rotations. Thus, we have an excellent means of checking the results of the algorithm. The solutions found are indeed of high accuracy.

Many different solutions can be found by beginning the search in a different invariant subspace. Thus , the algorithm was actually successful in finding *eight* different solutions to this problem.

2.2   Connections with Upper and Lower Solutions

In the last example, the two points where we begin and end the path were easily calculable. Now we turn to a case where the the local minimum is not so obvious.

We will discuss the example

$$\Delta u + \lambda e^u = 0 \; in \; \Omega \; , \tag{2.3}$$

subject to the Dirichlet boundary conditions

$$u|_{\partial\Omega} = 0, \tag{2.4}$$

where Ω is the unit square in R^2. It is known that there exists a $\lambda^* > 0$ such that when $\lambda > \lambda^*$ there is no solution, while for $0 < \lambda < \lambda^*$ two solutions exist. From our numerical calculation, $\lambda^* > 5.5$. We will take $\lambda = 5.5$ in the equation (2.3) in all subsequent discussion.

As a reminder, we define a function $\bar{u} \in C^2(\bar{\Omega})$ as an upper solution for the equations (2.3) and (2.4) if

$$\Delta \bar{u} + \lambda e^{\bar{u}} \leq 0 \; in \; \bar{\Omega} \tag{2.5}$$

and

$$\bar{u}|_{\partial\Omega} \geq 0 \; . \tag{2.6}$$

An analogous definition for $\underline{u}$ as an lower solution holds when the inequality signs are reversed.

For $\lambda = 5.5 < \lambda^*$, a solution of the equation (2.3) can be shown to exist by using the technique of upper and lower solutions.

The standard methods of differential inequalities say that this solution can be obtained as an asymptotic limit, as $t \to \infty$, of the solution of the corresponding nonlinear parabolic equation starting with initial data being the lower solution $\underline{u}$. In our case, $\underline{u} = 0$.

It is a well-known result that if you find a solution by the method of upper and lower solutions, that solution is a local minimum of the associated variational problem, for example [2].

It is then easy to find a point in the space where the functional I is lower than this local minimum, and the second solution can be found using the algorithm.

3   When the obvious solution is not a local minimum

In the previous section, the solutions found by the variational method had Morse index one (at least generically). However, this is rarely the case in the direct variational formulation for Hamiltonian systems and semilinear wave equation.

One way has been found to deal with this difficulty, at least in some special cases. In this section, we describe how one can sometimes employ a dual variational formulation which will enable us to study a new functional in which we have the two key ingredients for the 'mountain pass theorem', namely an "obvious" local minimum, and a point lower than this minimum. This technique has already been used in existence and multiplicity results for Hamiltonian system by Rabinowitz, Clark and Ekeland, and others.

The first problem of this type, [3] on which the mountain pass algorithm was tested was on periodic solutions of the equation

$$u_{tt} - u_{xx} + g(u) = f(x, t)$$

The following model problem was studied, with with γ small.

$$u_{tt} - u_{xx} + au^+ - bu^- = -\sin(\pi x)(1 - \gamma \sin(2\pi t)) \tag{3.1}$$

subject to the boundary conditions

$$u(0, t) = u(1, t) = 0 \ , \ \ u(x, t+1) = u(x, t) \ . \tag{3.2}$$

Let $\Omega \equiv [0, 1] \times [0, 1]$. Define the linear operator A acting on a subspace of the Hilbert space $L^2(\Omega)$ such that $Au \equiv u_{tt} - u_{xx}$.

The operator A can be sh9own to have an orthonormal basis of eigenfunctions, with eigenvalues λ_i going to $\pm\infty$. There is an "obvious" solution of (1), namely

$$u(x, t) = (-\frac{1}{\pi^2 + b} + \frac{\gamma}{-3\pi^2 + b} \sin(2\pi t)) \sin(\pi x) \tag{3.3}$$

We now form the duality formulation of the semilinear wave equation subject to the symmetric boundary conditions so as to make this solution a local minimum. The conditions $a > b \geq 0$ and γ being sufficiently small are assumed to be satisfied.

Let β be a constant satisfying the conditions:

$$a > b > \beta > -\pi^2 \text{ and for any integer } i, \ a, b, \beta \neq -\lambda_i \text{ , and } -\lambda_i \notin [\beta, b] \text{ .} \tag{3.4}$$

Define:

$$\begin{aligned}
Au &\equiv u_{tt} - u_{xx} \text{ ,} \\
B_\beta(u) &\equiv (a - \beta)u^+ - (b - \beta)u^- \text{ ,} \\
f(x, t) &\equiv -\sin(\pi x)(1 - \gamma \sin(2\pi t)) \text{ .}
\end{aligned} \tag{3.5}$$

The equation (3.1) can be recast as

$$(A + \beta)u + B_\beta(u) = f \text{ .} \tag{3.6}$$

For any sufficiently smooth $v \in \bar{S}$, we can find unique $u \in \bar{S}$ such that

$$-(A + \beta)u = v \text{ .} \tag{3.7}$$

Since $a > b > \beta$, the operator $B_\beta : R \to R$ is a strictly monotone increasing function and hence is invertible. The nonlinear wave equation (3.6) then becomes:

$$(A + \beta)^{-1}v + B_\beta^{-1}(v + f) = 0 \text{ .} \tag{3.8}$$

The solution of this equation can be obtained as the critical points of the functional $I : \bar{S} \to R$ defined by:

$$I(v) \equiv \int_\Omega \frac{1}{2}(A + \beta)^{-1}v \cdot v + J_\beta^*(v + f) \tag{3.9}$$

Now, we have an exact "obvious" solution

$$v_{min} = \sin(\pi x)[\frac{\pi^2 + \beta}{\pi^2 + b} + \frac{(3\pi^2 - \beta)\gamma}{-3\pi^2 + b} \sin(2\pi t)] \text{ .} \tag{3.10}$$

and one can verify that this critical point is known to be a strict local minimum in $\bar{S}$, hence explaining the subscript in v_{min}.

It is shown by explicit calculation that for certain choices of the constants a, b one can explicitly calculate a point lower than the local minimum and use the mountain pass algorithm on the functional I.

We summarize the results by saying that when we were able to make the right choices of a, b so that we could choose e_1, e_2 explicitly, the algorithm worked well and gave accurate and checkable solutions.

4 Two suspension bridge problems

It can be objected that some of the problems treated in the last section were of mathematical interest only. In this section we treat two applications which arise naturally in applications, concerning large oscillations in suspension bridges.

4.1 Standing waves in suspension bridges

Recall the problem. The model for the suspension bridge is given by

$$u_{tt} + u_{xxxx} + bu^+ = 1 + \epsilon h(x,t) \quad \text{in } [0,\pi] \times R$$

$$u(0,t) = u(\pi,t) = u_{xx}(0,t) = u_{xx}(\pi,t) = 0 \tag{4.1}$$

$$u \text{ is } \pi\text{-periodic in } t \text{ and even in } x$$

If we let $h(x,t) = \sin x \sin 2t$. This yields a near equilibrium solution given by

$$u(x,t) = y(x) + \frac{\epsilon}{b-3} \sin x \sin 2t. \tag{4.2}$$

where $y(x)$ satisfies the ordinary differential equation $y^{(4)} + by = 1$, with $y(0) = y''(0) = y(\pi) = y''(\pi) = 0$. Given this obvious "near-equilibrium" solution, how do we find the other solutions whose existence has been proved in the literature?

The treatment has some similarities with the vibrating string equation. Let $\Omega \equiv [0,\pi] \times [0,\pi]$. Define the linear operator L acting on a subspace of the Hilbert space $L^2(\Omega)$ such that $Lu = u_{tt} + u_{xxxx}$. The eigenvalue problem $Lu = \lambda u$ is known to have eigenfunctions given by $\phi_{mn} = \sin 2mt \sin nx$, $\cos 2mt \sin nx$, $(m, n = 1, 2, 3, \ldots)$ and $\phi_{0n} = \sin nx$ for $m = 0$, $n \geq 1$. Corresponding eigenvalues are given by $\lambda_{mn} = n^4 - 4m^2$. Suitably normalized they become $\phi_{mn} = \frac{2}{\pi} \sin 2mt \sin nx$ for $m > 0$ and $\phi_{0n} = \frac{\sqrt{2}}{\pi} \sin nx$ for $m = 0$.

As before, to ensure invertibility and compactness, we restrict to symmetric subspaces of $L^2(\Omega)$.

Recast equation (4.1) as

$$-(L + \beta)u + B_\beta(u) = -f.$$

where

$$B_\beta(u) = (\beta - b)u^+ - \beta u^-$$

For $\beta > b$, the operator $B_\beta : R \to R$ is a strictly monotone increasing function and hence is invertible. We end up with the equivalent formulation that our desired solutions can be found as critical points of the functional $I : \overline{S} \to R$ defined by:

$$I(v) \equiv \int_{\Omega} \frac{1}{2}(L + \beta)^{-1} v \cdot v + J_{\beta}^*(-v - f) \tag{4.3}$$

One now verifies that the obvious solution, for appropriate choice of β is a local minimum. The other difficulty is the choice of the second point e_2 which is lower that the local minimum. This is not given explicitly, but is found numerically by starting at random points away from the local minimum, and then searching until a lower point is found.

This method worked well on this problem and gave results which were in some restricted cases, checkable against solutions of ordinary differential equations, and thus could be shown to be highly accurate.

4.2 Travelling waves and homoclinic orbits

In looking for travelling wave solutions of the suspension bridge equation, one puts $u(x, t) = z(x - ct) + 1$ and gets the equation

$$z^{(4)} + c^2 z'' + (z + 1)^+ - 1 = 0 \tag{4.4}$$

Reasonable solutions as candidates for travelling waves are homroclinic orbits of the above equation. Since the linearisation of this equation near zero has stable and unstable submanifolds of dimension two, these can be difficult to calculate. The mountain pass algorithm can be applied to the functional

$$I(z) = \int_R \left(\frac{|z''|^2}{2} - \frac{c^2|z'|^2}{2} \right) dx + \int_{z \le -1} \left(|z| - \frac{1}{2} \right) dx + \int_{z > -1} \left(\frac{|z|^2}{2} \right) dx \tag{4.5}$$

defined on H^2, and its Fréchet derivative

$$I'(z)\phi = \int_R \left(z''\phi'' - c^2 z'\phi' \right) dx - \int_{z \le -1} \phi \, dx + \int_{z > -1} z\phi \, dx \tag{4.6}$$

for $z, \phi \in H^2$. Then the critical points of I correspond to the weak solutions of (4.4).

It turns out that zero is a local minimum, one can find a point where the functional is negative and one can use the theory to prove the existence of critical points, and the algorithm can be used to calculate them, replacing the real line with a sufficiently large interval.

As a bonus, when the nonlinearity $(z + 1)^+ - 1$ is replaced with an analytic nonlinearity $e^{y-1} - 1$, these solutions have extremely interesting interaction properties, [1], holding their shape like solitons when colliding. The reason for this is not clear.

5 Numerical solutions of Monge-Ampere equations

Monge-Ampére equations are extremely nonlinear and have many interesting technical difficulties. They are of the form

$$\begin{cases} (det(\nabla^2 u))^{\frac{1}{n}} = g(x, u, \nabla u) & \text{in } \Omega \\ u = 0 & \text{on } \partial\Omega \end{cases} \tag{5.1}$$

where Ω is a bounded convex domain in R^n, $n \geq 2$.

Solutions of the equation correspond to critical points of the functional

$$J(u) = \frac{-1}{n+1} \int_\Omega u(det\nabla^2 u) - \int_\Omega G(x, -u)dx$$

where $G(x, z) = \int_0^z g(x, s)ds$.

A fact that will be used later is that along the flow

$$\begin{cases} u_t = \log(det(\nabla^2 u)) - \log g(x, -u) \\ u(x, 0) \text{ be a given convex function with } u \mid_{\partial\Omega} (x, t) = 0. \end{cases} \tag{5.2}$$

J is non-increasing. The concepts of upper and lower solutions exists here and it appears (experimentally) that if one has a solution between an upper and lower solution, then one has a local minimum.

Thus one has the two key ingredients for the mountain pass algorithm, namely the existence of a local minimum, and a way to find a direction to deform points so that the functional decreases in that direction.

The algorithm has been used to solve the model problem $M(u) = 1 + u^4$, in the case of radial symmetry, and good results have been obtained. Also, monotonicity results on iterative schemes beginning with an upper solution have been established in [5]

6 Future work in this area

Probably, the most important problems are error estimates and rates of convergence. Presumably, one should be able, assuming one is in the neighbourhood of a nondegenerate critical point of the right Morse index, to estimate the rate of convergence of this algorithm. This has not been done.

With interval arithmetic, it may be possible to establish existence of solutions to the original equation, based on these type of calculations. This remains to be done.

It is not clear how to form a dual variational problem for systems of ordinary differential equations. We expect further interesting work here.

For reasons of space, we have omitted a complete bibliography. Complete references to section 2 can be found in [2]. Complete references to section 3 can be found in [3]. Complete references to section 4.1 can be found in [4]. Complete references to section 4.2 can be found in [1]. Complete references to section 5 can be found in [5]. Copies of these papers are available from the second author, as well as a video-tape showing numerical results on the soliton-like interaction properties of the waves mentioned in 4.2.

References

[1] Y. Chen and P.J. McKenna, Numerical variational methods for approximationg travelling waves in a nonlinearly suspended beam, submitted.

[2] Y. S. Choi and P. J. McKenna. *A mountain pass method for the numerical solutions of semilinear elliptic problems.* Nonlinear Analysis, Theory, Methods and Applications, Vol. 20 No. 4, 417-437, 1993.

[3] Y. S. Choi, P. J. McKenna and M. Romano. *A mountain pass method for the numerical solution of semilinear wave equations.* Numer. Math., 64:487-509, 1993.

[4] L. Humphreys, *Numerical Mountain Pass Solutions of a Suspension Bridge Equation*, submitted.

[5] C. Wang, *Numerical and Theoretical Results of the Monge-Ampére Equation,* Ph.D. Thesis, University of Connecticut, 1995.

Spectral Theory of the Magnetic Far Field Operator in an Orthotropic Medium[*]

David Colton[†]   Claudia Erbe[†]

1   Introduction

The inverse scattering problem for electromagnetic waves has attracted increased attention in recent years and the mathematical foundation of the area has now been developed to a reasonable degree ([1]). Of central importance in this development is the realization that the inverse scattering problem is both nonlinear and improperly posed. Due to this fact, the numerical implementation of inversion algorithms for determining the index of refraction from scattering data has until now been restricted to two-dimensional problems since the complexity of solving genuinely three-dimensional problems is at this time daunting ([1]). However, in a recent paper ([3]) Colton and Kress have suggested an alternative direction in inverse scattering theory than that persued to date where now one only tries to obtain lower and upper bounds for a few relevant features of the scattering object rather than attempting a complete reconstruction. In practice this is sometimes all that is needed and, if this is so, the need to solve a large nonlinear optimization problem can be avoided. We note with satisfaction that the spirit of this approach for studying nonlinear problems is the same as that employed by Ivar Stakgold in various papers over the years and hence this theme seems to us particularly appropriate for a contribution to Ivar's seventieth birthday volume.

The results of Colton and Kress mentioned in the previous paragraph were for the case of isotropic media. However, in many cases of practical importance the assumption of an isotropic medium is unwarranted. One such example is in the use of microwaves for purposes of medical imaging since nerves and organs such as the heart and liver are strongly anisotropic. Of particular importance in this case is to be able to detect changes in the structure of such tissues (e.g. due to a tumor) in a relatively simple manner. Another example is the non-destructive testing of aircraft windshields and canopies. In particular, it is hypothesized that the decline in impact toughness of the windshield is due to the fact that ultraviolet radiation causes the polycarbonate thermoplastic from which the windshield is made to change its anisotropic structure. The specific problem is whether this change in structure can be determined from electromagnetic scattering experiments in a quick and simple manner.

In both of the above examples the problem is to detect changes in an anisotropic medium from measurements of scattered electromagnetic waves. For the case of an isotropic medium, this problem was investigated by Colton and Kress in [3] where it was shown that the location of the eigenvalues of the electric far field operator determine a lower bound for

---

[*]This research was supported in part by grants from the Air Force Office of Scientific Research and the Deutsche Forschungsgemeinschaft.
[†]Department of Mathematical Sciences, University of Delaware, Newark, Delaware 19716

a quantity of physical interest.

In this paper we will extend this investigation to include the case of anisotropic media. For the sake of simplicity, we will restrict our attention to the case of orthotropic media since in this case we can reduce the problem to a scalar problem in two dimensions. Our aim is to establish three basic spectral properties of the (compact) magnetic far field operator F in terms of the refractive index N (which, due to anisotropy, is now a matrix):

1. If N is real and symmetric then F is normal and its eigenvalues lie on a circle C in the complex plane.

2. If N is complex and satisfies a particular coercivity condition then there exist an infinite number of eigenvalues and they all lie inside C.

3. If N is complex and satisfies the coercivity condition then the modulus of the largest eigenvalue yields a lower bound for a quantity involving the norm of $I - N$ and the coercivity coefficient γ.

2   Maxwell's Equations in an Orthotropic Medium

We consider time-harmonic electromagnetic waves propagating in an inhomogeneous anisotropic medium. In this case, if ω is the frequency, ε_0 is the permittivity and μ_0 the permeability of free space, then Maxwell's equations become

$$\text{(1)} \qquad\qquad\qquad \text{curl } E - ik H = 0$$

$$\text{(2)} \qquad\qquad\qquad \text{curl } H + ik \mathcal{N} E = 0$$

where E is the electric field, H the magnetic field; the wave number $k > 0$ is defined by $k^2 = \varepsilon_0 \mu_0 \omega^2$ and the refractive index $\mathcal{N} = \mathcal{N}(x)$ is assumed to be a matrix given by

$$\text{(3)} \qquad\qquad\qquad \mathcal{N}(x) = \frac{1}{\varepsilon_0} \left(\varepsilon(x) + i \frac{\sigma(x)}{\omega} \right).$$

We consider the scattering of electromagnetic waves by an inhomogeneous medium characterized by the refractive index $\mathcal{N} = \mathcal{N}(x)$. In particular let E^i, H^i be a solution of Maxwell's equation for a homogeneous medium, i.e. $\mathcal{N} = I$, in the whole space $\mathbb{R}^3$. We then look for a solution $E, H \in C^1(\mathbb{R}^3)$ of (1), (2) in $\mathbb{R}^3$ such that $E = E^i + E^s$, $H = H^i + H^s$ where the incident field E^i, H^i is as given above and E^s, H^s denotes the scattered field satisfying the Silver-Müller radiation condition.

$$\text{(4)} \qquad\qquad\qquad \lim_{|x| \to \infty} (H^s \times x - |x| E^s) = 0$$

uniformly in all directions $\hat{x} = \frac{x}{|x|}$. We will now restrict our attention to a special case of anisotropy, i.e. orthotropy, where the refractive index $\mathcal{N}$ is represented by a matrix of the form

$$\text{(5)} \qquad\qquad\qquad \mathcal{N}(x) = \begin{pmatrix} n_{11}(x) & n_{12}(x) & 0 \\ n_{21}(x) & n_{22}(x) & 0 \\ 0 & 0 & n_{33}(x) \end{pmatrix}.$$

We further assume that the scattering object is an infinite cylinder with axis in the $\hat{e}_z$ direction and $\mathcal{N}$ is independent of the z coordinate. Then if E^i is polarized parallel to the $\hat{e}_z$ direction it can be shown that the total field $E = (0, 0, E_3)$ satisfies

(6)
$$\left(\Delta + k^2 n_{33}(x)\right) E_3 = 0.$$

For this problem Colton and Kress have gained information about the refractive index n_{33} by studying the spectral properties of the electric far field operator ([3]). The case when the magnetic field is polarized parallel to the $\hat{e}_z$ direction leads in a similar fashion to the equation

(7)
$$\left(\nabla \cdot N(x)\nabla + k^2\right) H_3 = 0$$

where

(8)
$$N = \frac{1}{n_{11}n_{22} - n_{12}n_{21}} \begin{pmatrix} n_{11} & n_{21} \\ n_{12} & n_{22} \end{pmatrix}.$$

In this case the scattering problem is to find a solution H_3 of (7) such that $H_3 = H_3^i + H_3^s$ where H_3^i is an incident plane wave propagating in the direction d and H_3^s is the scattered wave, i.e. H_3^s satisfies the *Sommerfeld radiation condition*

(9)
$$\lim_{r \to \infty} \sqrt{r} \left(\frac{\partial}{\partial r} H_3^s - ik H_3^s \right) = 0$$

uniformly for all directions where $r = |x|$, $x \in \mathbb{R}^2$.

3 Spectral Theory of the Magnetic Far Field Operator

Motivated by the above discussion, we consider the scattering problem

(10)
$$\nabla \cdot (N\nabla u) + k^2 u = 0 \quad \text{in } \mathbb{R}^2$$
$$u = u^i + u^s$$
$$\lim_{r \to \infty} \sqrt{r} \left(\frac{\partial}{\partial r} u^s - ik u^s \right) = 0$$

where $u^i = e^{ikx \cdot d}$ and $N \in C^1(\mathbb{R}^2)$ is a 2×2 matrix such that $N = I$ for $x \in \mathbb{R}^2 \backslash \overline{D}$ where D is a bounded domain with smooth boundary ∂D. From Green's theorem we can obtain the equation

(11)
$$u^s(x) = - \int\int_{\mathbb{R}^2} \Phi \nabla \cdot ((N - I)\nabla u)\, dy,$$

where

(12)
$$\Phi(x, y) = \frac{i}{4} H_0^{(1)}(k|x - y|), \quad x \neq y$$

and $H_0^{(1)}$ denotes a Hankel function of the first kind of order zero. From (11) we have that

(13)
$$u^s(x) = \frac{e^{ikr}}{\sqrt{r}} u_\infty(\hat{x}; d) + O\left(\frac{1}{r^{\frac{3}{2}}}\right)$$

where the *far field pattern* u_∞ is given by

(14)
$$u_\infty(\hat{x}; d) = -\frac{1}{\sqrt{8\pi k}} e^{i\frac{\pi}{4}} \int\int_{\mathbb{R}^2} e^{-ik\hat{x} \cdot y} \nabla \cdot ((N - I)\nabla u)\, dy.$$

In order to proceed we will need the following reciprocity principle which, using the fact that N is symmetric, can be proved in exactly the same way as in the isotropic case (c.f. [1], p. 216).

Reciprocity Principle: If N is symmetric we have that

$$u_\infty(\hat{x}; d) = u_\infty(-d; -\hat{x}) \text{ for } \hat{x}, d \in \Omega = \{x \in \mathbf{R}^2 \mid |x| = 1\}.$$

Remark: As discussed in Ishimaru ([5]), electromagnetic fields in an isotropic media do not in general obey the reciprocity principle unless conditions of symmetry are imposed.

We are now in a position to investigate the spectral properties of the *magnetic far field operator* $F : L^2(\Omega) \to L^2(\Omega)$ defined by

$$(15) \qquad (Fg)(\hat{x}) := \int_\Omega u_\infty(\hat{x}; d)g(d)ds(d).$$

We note that F is compact and $(Fg)(\hat{x})$ is the far field pattern corresponding to problem (10) with the incident field $e^{ikx \cdot d}$ replaced by the *Herglotz wave function*

$$(16) \qquad v_g^i(x) = \int_\Omega e^{ikx \cdot d}g(d)ds(d)$$

with kernel $g \in L^2(\Omega)$.

LEMMA 3.1. *Suppose there exists an eigenvalue λ of F with eigenfunction g and let v_g be the solution of (10) with the incoming wave $e^{idx \cdot d}$ replaced by the Herglotz wave function v_g^i. Then*

$$(17) \qquad \int\int_D \text{Im} \left(\nabla v_g \cdot \overline{N \nabla v_g} \right) dy = \left(\sqrt{8\pi k}\, \text{Im} \left(e^{-i\frac{\pi}{4}} \lambda \right) - k|\lambda|^2 \right) \|g\|^2$$

where $\| \cdot \|$ is the norm on $L^2(\Omega)$.

Proof. Let $v_g = v_g^i + v_g^s$ be as stated in the theorem. Then using (10), Green's second indentity and the identities (c.f. [3])

$$(18) \qquad \int_{\partial D} \left(v^s \frac{\partial \overline{w}^s}{\partial n} - \overline{w}^s \frac{\partial v^s}{\partial n} \right) ds = -2ik \int_\Omega v_\infty \overline{w}_\infty ds$$

and

$$(19) \qquad \int_{\partial D} \left(v^s \frac{\partial \overline{w}_h^i}{\partial n} - \overline{w}_h^i \frac{\partial v^s}{\partial n} \right) ds = \sqrt{8\pi k}\, e^{-i\frac{\pi}{4}} \int_\Omega \overline{h}(d)v_\infty(d)ds(d)$$

we have

$$2i \int\int_D \text{Im} \left(\nabla v_g \cdot \overline{N \nabla v_g} \right) dy = 2i \int\int_D \text{Im} \left(v_g \nabla \cdot \left(\overline{N \nabla v_g} \right) + \nabla v_g \cdot \overline{N \nabla v_g} \right) dy$$

$$= \int\int_D \left(v_g \nabla \cdot \left(\overline{N \nabla v_g} \right) + \nabla v_g \cdot \overline{N \nabla v_g} - \overline{v}_g \nabla \cdot (N \nabla v_g) - \overline{\nabla v}_g \cdot N \nabla v_g \right) dy$$

$$= \int_{\partial D} \left(v_g \left(\overline{N \nabla v_g} \right) \cdot n - \overline{v}_g \left(N \nabla v_g \right) \cdot n \right) ds$$

$$= \int_{\partial D} \left(v_g \frac{\partial \overline{v}_g}{\partial n} - \overline{v}_g \frac{\partial v_g}{\partial n} \right) ds$$

$$= \int_{\partial D} \left(v_g^s \frac{\partial \overline{v_g^s}}{\partial n} - \overline{v}_g^s \frac{\partial v_g^s}{\partial n} \right) ds + \int_{\partial D} \left(v_g^s \frac{\partial \overline{v_g^i}}{\partial n} - \overline{v}_g^i \frac{\partial v_g^s}{\partial n} \right) ds$$

$$+ \int_{\partial D} \left(v_g^i \frac{\partial \overline{v_g^s}}{\partial n} - \overline{v}_g^s \frac{\partial v_g^i}{\partial n} \right) ds + \int_{\partial D} \left(v_g^i \frac{\partial \overline{v_g^i}}{\partial n} - \overline{v}_g^i \frac{\partial v_g^i}{\partial n} \right) ds$$

$$= -2ik \int_\Omega |v_{g,\infty}^s|^2 ds + \sqrt{8\pi k} e^{-i\frac{\pi}{4}} \int_\Omega \overline{g} v_{g,\infty}^s ds - \sqrt{8\pi k} e^{i\frac{\pi}{4}} \int_\Omega g \overline{v_{g,\infty}^s} ds$$

$$= -2ik \|Fg\|^2 + \sqrt{8\pi k} e^{-i\frac{\pi}{4}} (Fg, g) - \sqrt{8\pi k} e^{i\frac{\pi}{4}} (g, Fg)$$

where $(\cdot, \cdot)$ denotes the inner product and $\| \cdot \|$ the norm on $L^2(\Omega)$. Setting $Fg = \lambda g$, the lemma follows. $\square$

THEOREM 3.1. *If N is real and symmetric then the far field operator F is normal (and hence has eigenvalues) and its eigenvalues lie on the circle*

$$\sqrt{8\pi k} \, \mathrm{Im} \left(e^{-i\frac{\pi}{4}} \lambda \right) - k |\lambda|^2 = 0.$$

Proof. Let v_g^i and v_h^i be the Herglotz wave functions with kernels $g, h \in L^2(\Omega)$ respectively and let v_g, v_h be the solutions of the scattering problem (10) with the incoming wave $e^{ikx \cdot d}$ replaced by v_g^i and v_h^i respectively. Then we have

$$2i \int\int_D \mathrm{Im} \left(\nabla v_g \cdot \overline{N \nabla v_h} \right) dy = 2i \int\int_D \mathrm{Im} \left(\nabla \cdot \left(v_g \overline{N \nabla v_h} \right) - v_g \nabla \cdot \left(\overline{N \nabla v_h} \right) \right) dy$$

$$= \int\int_D \left(\nabla \cdot \left(v_g \overline{N \nabla v_h} \right) - \nabla \cdot (\overline{v}_g N \nabla v_h) - v_g \nabla \cdot \left(\overline{N \nabla v_h} \right) + \overline{v}_g \nabla \cdot (N \nabla v_h) \right) dy$$

$$= \int_{\partial D} \left(\left(v_g \overline{N \nabla v_h} \right) \cdot n - (\overline{v}_g N \nabla v_h) \cdot n \right) ds + k^2 \int\int_D (v_g \overline{v}_h - \overline{v}_g v_h) dy$$

$$= \int_{\partial D} \left(v_g \frac{\partial \overline{v}_h}{\partial n} - \overline{v}_g \frac{\partial v_h}{\partial n} \right) ds + k^2 \int\int_D (v_g \overline{v}_h - \overline{v}_g v_h) dy.$$

Assuming N to be real and symmetric, we have

$$\int\int_D \mathrm{Im} \left(\nabla v_g \cdot \overline{N \nabla v_h} \right) dy = 0.$$

Hence

$$\int\limits_{\partial D} \left(v_g \frac{\partial \overline{v_h}}{\partial n} - \overline{v_g} \frac{\partial v_h}{\partial n} \right) ds = k^2 \int\limits_{D}\int \left(\overline{v_g} v_h - v_g \overline{v_h} \right) dy$$

for all $g, h \in L^2(\Omega)$. In particular, interchanging g and h and taking conjugates yields

$$\int\limits_{\partial D} \left(\overline{v_h} \frac{\partial v_g}{\partial n} - v_h \frac{\partial \overline{v_g}}{\partial n} \right) ds = k^2 \int\limits_{D}\int \left(v_h \overline{v_g} - \overline{v_h} v_g \right) dy = \int\limits_{\partial D} \left(v_g \frac{\partial \overline{v_h}}{\partial n} - \overline{v_g} \frac{\partial v_h}{\partial n} \right) ds.$$

From this we conclude that

$$\int\limits_{\partial D} \left(\overline{v_h} \frac{\partial v_g}{\partial n} - v_g \frac{\partial \overline{v_h}}{\partial n} \right) ds = \int\limits_{\partial D} \left(v_h \frac{\partial \overline{v_g}}{\partial n} - \overline{v_g} \frac{\partial v_h}{\partial n} \right) ds,$$

i.e. the above integral is real for any $g, h \in L^2(\Omega)$. Replacing h by ih, we can conclude that

$$\int\limits_{\partial D} \left(v_h \frac{\partial \overline{v_g}}{\partial n} - \overline{v_g} \frac{\partial v_h}{\partial n} \right) ds = 0$$

for all $g, h \in L^2(\Omega)$. Interchanging g and h and using the identities (18) and (19) now shows that (c.f. [2])

$$(20) \qquad 0 = \int\limits_{\partial D} \left(v_g \frac{\partial \overline{v_h}}{\partial n} - \overline{v_h} \frac{\partial v_g}{\partial n} \right) ds$$

$$= -2ik(Fg, Fh) + \sqrt{8\pi k} e^{-i\frac{\pi}{4}} (Fg, h) - \sqrt{8\pi k} e^{i\frac{\pi}{4}} (g, Fh).$$

To show the normality of F, it suffices to prove the normality of $\mathcal{F} := e^{-i\frac{\pi}{4}} F$. Using the identity (20) and the reciprocity principle this can be done exactly as in [2]. Considering the equality (17) and noting that if N is real and symmetric the integral on the left hand side vanishes, we see that if $Fg = \lambda g$ then $\sqrt{8\pi k} \operatorname{Im} \left(e^{-i\frac{\pi}{4}} \lambda \right) - k|\lambda|^2 = 0$. The proof of the theorem is now complete. $\qquad \square$

Definition The complex matrix N is *coercive* if

$$(21) \qquad \operatorname{Im} \left(a \cdot \overline{N a} \right) \geq \gamma(x) |a|^2$$

for every $a \in \mathbf{R}^2$ where $\gamma(x) > 0$ and $x \in D$.

THEOREM 3.2. *If N is complex and coercive then the far field operator F has an infinite number of eigenvalues which lie inside the circle*

$$\sqrt{8\pi k} \operatorname{Im} \left(e^{-i\frac{\pi}{4}} \lambda \right) - k|\lambda|^2 = 0.$$

Proof. Instead of F we can consider the operator $\mathcal{F} := e^{-i\frac{\pi}{4}} F$. From [4], p. 118–119, we can conclude that $\mathcal{F}$ is a trace class operator. From the proof of the previous lemma we have that

$$2i \int\limits_{D}\int \operatorname{Im} \left(\nabla v_g \cdot \overline{N \nabla v_g} \right) dy = -2ik\|Fg\|^2 + \sqrt{8\pi k} (Fg, g) - \sqrt{8\pi k} (g, Fg)$$

for any $g \in L^2(\Omega)$. Thus, using the coercivity of N,

$$\text{Im}(\mathcal{F}g, g) = \frac{1}{2i}((\mathcal{F}g, g) - (g, \mathcal{F}g))$$

$$= \frac{1}{\sqrt{8\pi k}} \left(\int \int_D \text{Im}\left(\nabla v_g \cdot \overline{N \nabla v_g}\right) dy + k^2 \|\mathcal{F}g\|^2 \right) \geq 0$$

for every $g \in L^2(\Omega)$. In particular, if $\text{Im}(\mathcal{F}g, g) = 0$ then $\nabla v_g = 0$ for $x \in D$. Since v_g is a solution of $\nabla \cdot (N \nabla v_g) + k^2 v_g = 0$ we conclude $v_g = 0$ for $x \in D$ and hence by the unique continuation principle (c.f. [1]) $v_g = 0$ for $x \in \mathbb{R}^2$. Since $v_g = v_g^i + v_g^s$ we now have that v_g^i is an entire solution of the Helmholtz equation satisfying the radiation condition and hence from Green's theorem and Rellich's lemma $v_g^i = 0$ for all $x \in \mathbb{R}^2$. But this implies $g = 0$ (c.f. [1], p. 55).

The above considerations show that 1) $\text{Im}(\mathcal{F}g, g) \geq 0$ and 2) $\lambda = 0$ is not an eigenvalue of $\mathcal{F}$. Hence, from Lidski's theorem (c.f. [6], p. 149), we can conclude that $\mathcal{F}$ has an infinite number of eigenvalues. From the previous lemma and coercivity we have that the eigenvalues of $F = e^{i\frac{\pi}{4}} \mathcal{F}$ all lie inside the circle $\sqrt{8\pi k} \, \text{Im}\left(e^{-i\frac{\pi}{4}} \lambda\right) - k|\lambda|^2 = 0$. □

THEOREM 3.3. *If N is complex and coercive then*

$$4|\lambda|^2 \left(k \int \int_D \frac{\|I - N\|^2}{\gamma} dy \right)^{-1} \leq \sqrt{8\pi k} \, \text{Im}\left(e^{-i\frac{\pi}{4}} \lambda\right) - k|\lambda|^2$$

for any eigenvalue λ of the far field operator F where $\|\cdot\|$ denotes the matrix norm (defined as the sum of the absolute values of the entries of the matrix).

Proof. If g is the eigenfunction of F corresponding to the eigenvalue λ we have from (14) that

$$\lambda g(\hat{x}) = (Fg)(\hat{x}) = -\frac{1}{\sqrt{8\pi k}} e^{i\frac{\pi}{4}} \int \int_D e^{-ik\hat{x}\cdot y} \nabla \cdot ((N - I)\nabla v_g) dy$$

where v_g as is previously defined. Hence

(22)
$$\lambda \|g\|^2 = -\frac{1}{\sqrt{8\pi k}} e^{i\frac{\pi}{4}} \int \int_D \overline{v_g^i} \nabla \cdot ((N - I)\nabla v_g) dy$$

$$= \frac{1}{\sqrt{8\pi i}} e^{i\frac{\pi}{4}} \int \int_D \nabla \overline{v_g^i} \cdot (I - N)\nabla v_g dy$$

where v_g^i is the Herglotz wave function with kernel g.

It is easily verified that v_g^i satisfies the estimate $|\nabla v_g^i|^2 \leq 2\pi k^2 \|g\|^2$ for every $y \in D$. Hence by the Schwarz inequality applied to (22) we now have that

$$|\lambda|^2 \|g\|^4 = \frac{1}{8\pi k} |\int \int_D \nabla \overline{v_g^i} \cdot (I - N)\nabla v_g dy|^2$$

$$\leq \frac{1}{8\pi k} \int \int_D \frac{|\nabla v_g^i|^2 \cdot |(I-N)\nabla v_g|^2}{\text{Im}\left(\nabla v_g \cdot \overline{N\nabla v_g}\right)} dy \cdot \int \int_D \text{Im}\left(\nabla v_g \cdot \overline{N\nabla v_g}\right) dy$$

$$\leq \frac{k}{4}\|g\|^2 \int \int_D \frac{|(I-N)\nabla v_g|^2}{\text{Im}\left(\nabla v_g \cdot \overline{N\nabla v_g}\right)} dy \cdot \int \int_D \text{Im}\left(\nabla v_g \cdot \overline{N\nabla v_g}\right) dy$$

and the theorem follows from the coercivity condition and the previous lemma. □

Remark : Let r be the radius of the smallest circle with center on $\text{Im}\lambda = -\text{Re}\,\lambda, \text{Im}\lambda \geq 0$, and passing through the origin which contains all the eigenvalues of F. Then the above theorem shows that a knowledge of r provides a lower bound for

$$(23) \qquad k \int \int_D \frac{\|I-N\|^2}{\gamma} dy.$$

References

[1] D. Colton and R. Kress, *Inverse Acoustic and Electromagnetic Scattering Theory*, Springer-Verlag, Berlin, 1992.

[2] D. Colton and R. Kress, Eigenvalues of the far field operator and inverse scattering theory, *SIAM J. Math. Anal.* 26, No. 3, (1995), pp. 601-615.

[3] D. Colton and R. Kress, Eigenvalues of the far field operator for the Helmholtz equation in an absorbing medium, *SIAM J. Applied Math.*, to appear.

[4] I.C. Gohberg and M.G. Krein, *Introduction to the Theory of Linear Nonselfadjoint Operators*, American Mathematical Society, Providence, Rhode Island, 1969.

[5] A. Ishimaru, *Electromagnetic Wave Propagation. Radiation and Scattering*, Prentice Hall, Englewood Cliffs, New Jersey, 1991.

[6] J.R. Ringrose, *Compact Non-Self Adjoint Operators*, Van Nostrand Reinhold, London, 1971.

CONDUCTION-CONVECTION WITH CHANGE OF PHASE

E. DiBenedetto
M. O'Leary

ABSTRACT. We discuss the conduction-convection system with change of phase, and review recent results. We present a generalization of the existence theorem proven in [13].

§1. Introduction.

The purpose of this note is to review the results on conduction convection systems and to give some generalizations of the results in [20].

We begin by considering the physics of the problem. Suppose we have a bounded region in $\mathbf{R}^3$ in which we have water and ice, and assume that melting and freezing takes place solely when the temperature $u(x,t)$ is zero. In the ice the temperature is determined by the heat equation, while in the water we have convective motions driven by temperature induced density variations that mix the fluid in an attempt to bring the system to equilibrium. Such motion in a gravitational field is called *free convection*. The dynamic state of the liquid is determined by the velocity $\mathbf{v}$ and the pressure p, and we assume the fluid evolves via free convection [28].

Although we have described a water/ice system, this model applies to other systems of physical interest, for example in solidification of alloys in the continuous casting of metals, and in welding [9,10,34].

To make our ideas precise, let Ω be a fixed bounded domain in $\mathbf{R}^3$ with smooth boundary $\partial\Omega$, and for $0 < T < \infty$, set $\Omega_T = \Omega \times (0, T]$. Let u denote the temperature, so at any time t, the set $[u(\cdot, t) < 0]$ is ice, while $[u(\cdot, t) > 0]$ is water. We do not identify a phase for the set $[u(\cdot, t) = 0]$; it may be ice or water, or a mixture of both. We remark that our results will allow this set to be not only nonempty, but to have positive measure.

To model this system, we begin by assuming that $\Omega_T = [u > 0] \cup [u < 0] \cup \Gamma$ where Γ is a smooth surface with a well defined normal. We model the dynamic state

Partially supported by NSF grant DMS-9104088

of the liquid with the Navier–Stokes equations and the Boussinesq approximation [31,34]

$$
(1.1) \quad
\begin{cases}
\mathbf{v}_t - \triangle\mathbf{v} + (\mathbf{v}\cdot\nabla)\mathbf{v} + \nabla p = \mathbf{f}(u) & \text{in } [u > 0], \\
\mathbf{f}(u) \equiv -g\rho(1 - \alpha_0 u)_+\mathbf{k}, \\
\operatorname{div}\mathbf{v} = 0,
\end{cases}
$$

where α_0 is the dilation parameter, g is the gravitational constant, and $\mathbf{k}$ is the unit vector ascending along the vertical. Also, we have normalized the kinematic viscosity to 1.

Heat flow in the liquid set is governed by the equation

$$
(1.2) \quad u_t - k_1\triangle u + \lambda(\mathbf{v}\cdot\nabla u) = 0,
$$

where k_1 is the thermal conductivity of water, and $\lambda > 0$ is Grashoff's number. For a discussion on the physics of these equations, see [1,28].

In the ice, $\mathbf{v} \equiv 0$, and the temperature satisfies

$$
(1.3) \quad u_t - k_2\triangle u = 0,
$$

where k_2 is the thermal conductivity of ice.

This system is complemented with the following initial and boundary conditions:

$$
(1.4) \quad
\begin{cases}
u(x,0) = u_0(x) & \text{in } \Omega, \\
u(x,t)|_{\partial\Omega\times(0,T]} = g(x,t).
\end{cases}
$$

We assume the fluid adheres to the solid boundary, so

$$
(1.5) \quad
\begin{cases}
\mathbf{v}(x,0) = \mathbf{v}_0(x) & \text{in } \Omega, \\
\mathbf{v}|_{\partial[u\geq0]} = 0.
\end{cases}
$$

On the solid-liquid interface Γ, we prescribe the Stefan condition

$$
(1.6) \quad
\begin{cases}
u(x,t)|_\Gamma = 0, \\
(k_1\nabla u^+ - k_2\nabla u^-)\cdot\mathbf{N_x} = LN_t,
\end{cases}
$$

where $\mathbf{N}$ is the normal to Γ with components $\mathbf{N} = (\mathbf{N_x}, N_t)$ and L is the latent heat [1,30]. This relationship measures, roughly speaking, the amount of heat used in the melting process. Curvature effects and dendritic phenomena are ignored.

To obtain a weak formulation of this problem, take (1.2), multiply by $\phi \in C_0^\infty(\Omega_T)$, and integrate over $[u > 0]$. Using the boundary conditions, we obtain

$$
(1.7) \quad \iint_{[u>0]} \{-u\phi_t - k_1 u\triangle\phi + \lambda u\mathbf{v}\cdot\nabla\phi\}\,dx\,dt = k_1\int_\Gamma \phi\nabla u\cdot\mathbf{N_x}\,d\sigma(x,t),
$$

where $d\sigma$ is surface measure on Γ. A similar calculation in $[u < 0]$ gives.

$$(1.8) \qquad \iint_{[u<0]} \{-u\phi_t - k_2 u \triangle\phi\} \, dx\, dt = -k_2 \int_\Gamma \phi \nabla u \cdot \mathbf{N_x} \, d\sigma(x,t).$$

Define

$$K(u) = \begin{cases} k_1 u & \text{if } u \geq 0, \\ k_2 u & \text{if } u < 0. \end{cases}$$

Since $\mathbf{v} = 0$ in the set $[u < 0]$, we can add (1.7) and (1.8) to obtain

$$\iint_{\Omega_T} -u\phi_t - K(u)\triangle\phi + \lambda u \mathbf{v} \cdot \nabla\phi\, dx\, dt = \int_\Gamma \phi(k_1 \nabla u^+ - k_2 \nabla u^-) \cdot \mathbf{N_x}\, d\sigma.$$

Define the graph

$$H(s) = \begin{cases} L & \text{if } s > 0, \\ [0, L] & \text{if } s = 0, \\ 0 & \text{if } s < 0. \end{cases}$$

Then, using (1.6) and the fact that Γ is a surface,

$$\iint_{\Omega_T} H(u)\phi_t\, dx\, dt = \iint_{[u>0]} L\phi_t\, dx\, dt = \int_\Gamma \phi N_t\, d\sigma$$

$$= \int_\Gamma \phi(k_1 \nabla u^+ - k_2 \nabla u^-) \cdot \mathbf{N_x}\, d\sigma.$$

So, if we set $\beta(s) = s + H(s)$, then for all $\phi \in C_0^\infty(\Omega_T)$ we have

$$(1.9) \qquad \iint_{\Omega_T} -\beta(u)\phi_t - K(u)\triangle\phi + \lambda u\mathbf{v} \cdot \nabla\phi = 0.$$

Thus, in the sense of distributions

$$\frac{\partial}{\partial t}\beta(u) - \triangle K(u) + \lambda\mathbf{v} \cdot \nabla u = 0.$$

With this in mind, we will establish the solvability of the conduction-convection system:

$$(1.10) \qquad \begin{cases} u \in C(0, T; L_2(\Omega_T)) \cap L_2(0, T; W_2^1(\Omega_T)), \\ \dfrac{\partial}{\partial t}\beta(u) - \triangle K(u) + \lambda\mathbf{v} \cdot \nabla u = 0 \quad \text{weakly in } \Omega_T, \\ u|_{t=0} = u_0 \in C^\alpha(\Omega) \cap L_\infty(\Omega), \\ u|_{\partial\Omega \times (0,T)} = g \in C^{2+\alpha}(\partial\Omega \times [0, T]); \end{cases}$$

$$(1.11) \qquad \begin{cases} \mathbf{v} \in [L_\infty(0, T; L_2(\Omega)) \cap L_2(0, T; \overset{\circ}{W}{}_2^1(\Omega))]^3, \\ \text{div } \mathbf{v} = 0 \quad \text{weakly in } \Omega_T, \end{cases}$$

$$(1.12) \quad \begin{cases} \mathbf{v}_t - \triangle \mathbf{v} + (\mathbf{v} \cdot \nabla)\mathbf{v} + \nabla p = \mathbf{f}(u) & \text{weakly in } [u > 0], \\ \mathbf{v} = 0 & \text{a.e. in } [u < 0], \\ \mathbf{v}|_{t=0} = \mathbf{v}_0. \end{cases}$$

With the given space specifications, the notion of a solution to (1.10) is standard (see for example [8,12,21,23]). We interpret the symbol $(x,t) \to \beta(u(x,t))$ to be a selection $(x,t) \to w(x,t) \subset \beta(u(x,t))$ out of the graph $\beta(u(x,t))$. For this reason, to the initial data in (1.10) we must add the selection $w_0(x) \subset \beta(u_0(x))$. A common device consists in prescribing $u_0(x) \neq 0$ a.e in Ω so that $\beta(u_0(x))$ is unambiguously defined almost everywhere.

The interpretation of the Navier-Stokes equations requires some analysis of the topological structure of $[u > 0]$. If such a set were open for example, we could define them weakly by testing them against vector fields $\psi \in C_0^\infty([u > 0])$ that are solenoidal in the space variables. This analysis is the primary difficulty of this problem. Various aspects of the generalized conduction-convection system have been considered in [11,13,14,19,34] while the classical problem has been studied in [2,3,4,15,16,17] for short times. In the two-dimensional case, i.e. $\Omega \subset \mathbf{R}^2$, an existence theorem has been proven in [12]. Its main feature is that $\Omega \subset \mathbf{R}^2$ implies $\mathbf{v} \in [L_4(\Omega_T)]^2$, and this is sufficient to ensure continuity of u. To see that this is plausible, consider weak solutions of

$$w_t - \triangle w = \mathbf{v} \cdot \nabla w \quad \text{in } \Omega_T$$

where w is in the same space as u, and $\Omega \subset \mathbf{R}^N$. Then w is continuous if $\mathbf{v} \in L_q(\Omega_T)$ for some $q > N + 2$, while w may not be continuous if $q < N + 2$; see [27]. The results in [12] were obtained by using both the limit integrability $\mathbf{v} \in L_4(\Omega_T)$ and the extra information div $\mathbf{v} = 0$.

It does not appear to be possible to use this idea in the three dimensional case. Indeed, for $N = 3$, Sobolev embedding merely gives

$$L_\infty(0,T; L_2(\Omega)) \cap L_2(0,T; \overset{\circ}{W}_2^1(\Omega)) \hookrightarrow L_{10/3}(\Omega_T)$$

which is insufficient integrability to use the ideas of [12]. However, in [20,33], we were able to derive topological information about the set $[u > 0]$ independently of the regularity of the temperature.

In [20], the flow was driven by the Stokes system, while here we use the full Navier-Stokes equations. This note describes the work in [20], and presents the necessary modifications for that result to hold in this case. Our result is that the system (1.10)-(1.12) is solvable, in a way to be made precise, everywhere in Ω_T. Moreover, u is continuous in Ω_T except for a closed set $\mathcal{E}_0$ of Hausdorff dimension at most 5/3 and whose Hausdorff measure is arbitrarily small.

§2. Statement of Results.

To fix our notation, let B_ρ be the ball in of radius ρ about the origin in $\mathbf{R}^3$, and let $Q_\rho \equiv B_\rho \times (-\rho^2, 0)$ be the parabolic cylinder with "vertex" at the origin. A congruent cylinder with "vertex" at (x, t) will be denoted by $[(x, t) + Q_\rho]$.

Given a set $\mathcal{E}_0$, we define its k-dimensional parabolic measure $\mathcal{P}_k(\mathcal{E}_0)$ for $k > 0$ by

$$\mathcal{P}_k(\mathcal{E}_0) \equiv \liminf_{\delta \downarrow 0} \left\{ \sum_{i=0}^{\infty} \rho_i{}^k \,\middle|\, \mathcal{E}_0 \subseteq \bigcup_{i=0}^{\infty} [(x_i, t_i) + Q_{\rho_i}]; \rho_i < \delta \right\},$$

while its parabolic dimension is $\dim_\mathcal{P}(\mathcal{E}_0) \equiv \inf\{k | \mathcal{P}_k(\mathcal{E}_0) > 0\}$. Hausdorff measure $\mathcal{H}_k(\mathcal{E}_0)$, and dimension $\dim_\mathcal{H}(\mathcal{E}_0)$ are defined analogously, where we replace parabolic cylinders by arbitrary closed sets of diameter less than δ. In particular, it follows that $\mathcal{H}_k(\mathcal{E}_0) \leq \mathcal{P}_k(\mathcal{E}_0)$, and $\dim_\mathcal{H}(\mathcal{E}_0) \leq \dim_\mathcal{P}(\mathcal{E}_0)$. See also [22,32].

Our main result is

Theorem. *For any $T > 0$ there exists a temperature distribution $u : \Omega_T \longrightarrow \mathbf{R}$, a velocity field $\mathbf{v} : \Omega_T \longrightarrow \mathbf{R}^3$, and a set $\mathcal{E}_0$ so that (1.10) and (1.11) hold,*

$$(2.1) \qquad \mathcal{P}_{5/3}(\mathcal{E}_0) = 0, \qquad\qquad \dim_\mathcal{P}(\mathcal{E}_0) < 5/3,$$

and

$$(2.2) \qquad u \in C(\Omega_T \backslash \mathcal{E}_0), \qquad [u > 0] \cap (\Omega_T \backslash \mathcal{E}_0) \quad \textit{is an open set.}$$

Moreover

$$(2.3) \qquad \iint_{[u>0] \cap (\Omega_T \backslash \mathcal{E}_0)} [-\mathbf{v} \cdot \psi_t + \nabla \mathbf{v} : \nabla \psi + (\mathbf{v} \cdot \nabla \mathbf{v}) \cdot \psi - \mathbf{f}(u) \cdot \psi] \, dx dt = 0,$$

for all vector fields ψ satisfying

$$(2.4) \qquad \begin{cases} \psi \in [\overset{\circ}{W}{}_2^1(0, T; L_2(\Omega) \cap L_2(0, T; \overset{\circ}{W}{}_2^1(\Omega)))]^3, \\ \operatorname{supp} \psi \subseteq [u > 0] \cap (\Omega_T \backslash \mathcal{E}_0), \\ \operatorname{div} \psi(\cdot, t) = 0 \quad \textit{weakly in } \Omega, \end{cases}$$

where $\nabla \mathbf{v} : \nabla \psi = \sum_{i,j=1}^3 \frac{\partial v_i}{\partial x_j} \frac{\partial \psi_i}{\partial x_j}$.
The initial data $\mathbf{v}_0$ is taken in the sense

$$(2.5) \qquad \lim_{t \downarrow 0} \int_\Omega \mathbf{v}(x, t) \phi(x) dx = \int_\Omega \mathbf{v}_0(x) \phi(x) dx$$

for all smooth solenoidal vector fields ϕ with $\operatorname{supp} \phi \subset [u_0 > 0]$.
Finally $\mathbf{v} = 0$ a.e. on the set $[u < 0]$.

Remark. In [13], where the flow is governed by the Stokes equations, we were able to conclude that $\mathcal{E}_0 \subseteq \partial[u > 0]$. This does not appear possible in this case, because solutions of the Navier-Stokes equations may not be regular, even in the set $[u > 0]$.

§3. Proof of the existence theorem.

Although our problem is non-variational, we use a technique of quasi-variational inequalities.

For $\epsilon > 0$, consider the family of penalized problems:

(3.1)
$$\begin{cases} u_\epsilon \in C(0, T; L_2(\Omega_T)) \cap L_2(0, T; W_2^1(\Omega_T)), \\ \dfrac{\partial}{\partial t} \beta_\epsilon(u_\epsilon) - \triangle K_\epsilon(u_\epsilon) + \lambda \mathbf{v}_\epsilon^* \cdot \nabla B_\epsilon(u_\epsilon) = 0 \quad \text{weakly in } \Omega_T, \\ u_\epsilon|_{t=0} = u_0 \in C^\alpha(\Omega) \cap L_\infty(\Omega), \\ u_\epsilon|_{\partial\Omega \times (0,T)} = g \in C^{2+\alpha}(\partial\Omega \times [0, T]); \end{cases}$$

(3.2)
$$\begin{cases} \mathbf{v}_\epsilon \in [L_\infty(0, T; L_2(\Omega)) \cap L_2(0, T; \overset{\circ}{W}_2^1(\Omega))]^3, \\ \operatorname{div} \mathbf{v}_\epsilon = 0 \quad \text{weakly in } \Omega_T, \\ \mathbf{v}_\epsilon|_{t=0} = \mathbf{v}_0, \\ \dfrac{\partial \mathbf{v}_\epsilon}{\partial t} - \triangle \mathbf{v}_\epsilon + (\mathbf{v}_\epsilon \cdot \nabla)\mathbf{v}_\epsilon + \dfrac{1}{\epsilon}\alpha_\epsilon(u_\epsilon^-)\mathbf{v}_\epsilon + \nabla p = \mathbf{f}(u_\epsilon) \text{ weakly in } \Omega_T. \end{cases}$$

Here β_ϵ and K_ϵ are smooth approximations of β and K, $\mathbf{v}_\epsilon^*$ is a smooth mollification of $\mathbf{v}$ preserving its solenoidal nature,

$$B_\epsilon(s) = \begin{cases} s & \text{if } s < \frac{1}{\epsilon}, \\ \frac{1}{\epsilon} & \text{if } s \geq \frac{1}{\epsilon}, \end{cases}$$

and $\alpha_\epsilon(s)$ is a smooth mollification of $B_\epsilon(s)$. This approximation scheme allows us to view the equations in (3.2) over the whole of Ω_T, but they are 'penalized' in the set $[u < 0]$.

Proposition 1. *For any $\epsilon > 0$ the system (3.1)-(3.2) has a solution; further, there is a constant γ depending on the data and Ω but independent of ϵ so that*

(3.3) $\qquad \|u_\epsilon\|_{\infty,\Omega_T} \leq \gamma,$

(3.4) $\qquad \|\nabla u_\epsilon\|_{2,\Omega_T} \leq \gamma,$

(3.5) $\qquad \displaystyle\int_0^{T-h} h^{-1}\|u_\epsilon(t+h) - u_\epsilon(t)\|_{2,\Omega}^2\, dt \leq \gamma, \, 0 < h < T,$

(3.6) $\qquad \displaystyle\operatorname*{ess\,sup}_{0 \leq t \leq T} \|\mathbf{v}_\epsilon(\cdot, t)\|_{2,\Omega} + \|\nabla \mathbf{v}_\epsilon\|_{2,\Omega_T} \leq \gamma,$

(3.7) $\qquad \dfrac{1}{\epsilon} \displaystyle\iint_{\Omega_T} \alpha_\epsilon(u_\epsilon^-)|\mathbf{v}_\epsilon|^2\, dx\, dt \leq \gamma.$

Sketch of Proof. We proceed by Galerkin approximation; for each approximating stage N we choose $\tilde{u}_\epsilon^N$ and the solve the approximating system of (3.2) with $u_\epsilon = \tilde{u}_\epsilon^N$ to obtain $\mathbf{v}_\epsilon^N$ which serves as input in the approximation of (3.1) to obtain u_ϵ^N. We show this map has a fixed point by Leray-Schauder, and pass to the limit in N to obtain the solutions $u_\epsilon, \mathbf{v}_\epsilon$. The estimates (3.3)-(3.7) are consequences of standard theory [26, 27]. $\square$

As a consequence, we may choose subnets, relabelled ϵ so that

(3.8) $\qquad u_\epsilon, K_\epsilon(u_\epsilon) \longrightarrow u, K(u)$ strongly in $L_2(\Omega_T)$,

(3.9) $\qquad \nabla u_\epsilon, \nabla K_\epsilon(u_\epsilon) \longrightarrow \nabla u, \nabla K(u)$ weakly in $L_2(\Omega_T)$,

(3.10) $\qquad \beta_\epsilon(u_\epsilon) \longrightarrow w \subset \beta(u)$ weakly in $L_2(\Omega_T)$,

(3.11) $\qquad \mathbf{v}_\epsilon, \mathbf{v}_\epsilon^*, \nabla \mathbf{v}_\epsilon \longrightarrow \mathbf{v}, \nabla \mathbf{v}$ weakly in $L_2(\Omega_T)$,

(3.12) $\qquad \operatorname{div} \mathbf{v} = 0$ weakly in $L_2(\Omega_T)$;

further, for all h sufficiently small

(3.13) $\qquad \displaystyle\int_0^{T-h} h^{-1} \|u_\epsilon(\cdot, t+h) - u_\epsilon(\cdot, t)\|_{2,\Omega}^2 \, dt \leq \gamma$

so the data on the parabolic boundary of Ω_T are preserved in the limit, weakly as $t \downarrow 0$, and in the sense of traces on $\partial\Omega \times (0, T)$.

Now the above estimates are sufficient to allow us to pass to the limit as $\epsilon \downarrow 0$ in the temperature equations (3.1), so that u solves (1.10).

For the velocity, pass to the limit in (3.7) to see that, by weak lower semicontinuity,

$$\iint_{\Omega_T} u^- |\mathbf{v}|^2 = 0,$$

so that $\mathbf{v} = 0$ a.e. on $[u < 0]$. Before we can continue, however, we need more information on the structure of the set $[u > 0]$.

§4. The singular set.

Let $V(Q_\rho)$ be the space

$$V(Q_\rho) \equiv L_\infty(-\rho^2, 0; L_2(B_\rho)) \cap L_2(-\rho^2, 0; \overset{\circ}{W}_2^1(B_\rho)),$$

with norm

$$\|\phi\|_{V(Q_\rho)} \equiv \operatorname*{ess\,sup}_{-\rho^2 \leq \tau \leq 0} \|\phi(\cdot, \tau)\|_{L_2(B_\rho)} + \|\nabla\phi\|_{L_2(Q_\rho)}.$$

Define the set $\mathcal{E}_0$ by

(4.1) $\qquad \mathcal{E}_0 \equiv \left\{ (x_0, t_0) \,\middle|\, \limsup_{\rho \downarrow 0} \rho^{-1/2} \|\mathbf{v}\|_{10/3;[(x_0,t_0)+Q_\rho]} = \infty \right\}.$

The standard covering arguments in [20] show that $\dim_P \mathcal{E}_0 < 5/3$ and $\mathcal{P}_{5/3}(\mathcal{E}_0) = 0$.

Given a point $(x_0, t_0) \in \Omega_T \backslash \mathcal{E}_0$, we define $\rho_0 = \min\{\text{dist}(x_0, \partial\Omega); \sqrt{t_0}\}$ and

$$\mathcal{C}_0 = \sup_{0 < \rho < \rho_0} \rho^{-1/2} \|\mathbf{v}\|_{10/3;[(x_0,t_0)+Q_\rho]}.$$

This choice of $\mathcal{E}_0$ and $\mathcal{C}_0$ allows us to prove the following technical estimates for cylinders $[(x_0, t_0) + Q_\rho]$ where $(x_0, t_0) \in \Omega_T \backslash \mathcal{E}_0$, and

$$\iint_{Q_\rho} u \, dx = \frac{1}{\text{meas } Q_\rho} \iint_{Q_\rho} u \, dx.$$

Lemma 1. *Let $0 < \sigma < 1$, and let ζ be a cutoff function defined in $[(x_0, t_0) + Q_\rho]$ that equals 1 on $[(x_0, t_0) + Q_{\sigma\rho}]$. Then there exists a constant γ that depends on the data and $\mathcal{C}_0$ so that for all η, $\inf u < \eta < \sup u$, we have*

(4.2)
$$\|(u - \eta)_\pm \zeta\|_{V(Q_\rho)}^2 \le$$
$$\le \gamma(\mathcal{C}_0) \frac{\rho^3 \|(u - \eta)_\pm\|_{\infty;Q_\rho}^\theta}{(1 - \sigma)^2} \left(\iint_{Q_\rho} \chi[(u - \eta)_\pm > 0] dx dt \right)^{\frac{7}{10}},$$

where $\theta = 1 + \frac{1}{2}(1 \mp \text{sign } \eta)$.

Lemma 2. *Let $\eta > 0$, $0 < \nu < \|(u - \eta)_+\|_{\infty,Q_\rho}$, and*

(4.3)
$$\Psi = \ln^+ \left\{ \frac{\|(u - \eta)_+\|_{\infty,Q_\rho}}{\|(u - \eta)_+\|_{\infty,Q_\rho} - (u - \eta)_+ + \nu} \right\}.$$

Let $\xi(x)$ be a smooth cutoff function in B_ρ that equals 1 in $B_{\sigma\rho}$. Then there is a constant γ depending only on the data and $\mathcal{C}_0$ so that for all $-\rho^2 < t_0 < t \le 0$

(4.4)
$$\int_{B_\rho \times \{t\}} \Psi^2 \xi^2 \, dx \le \int_{B_\rho \times \{t_0\}} \Psi^2 \xi^2 \, dx + \frac{\gamma\rho^3}{(1 - \sigma^2)} \|\Psi\|_{\infty,Q_\rho}.$$

Proofs of these estimates can be found in [20]; the importance of these estimates is that they are exactly what is required in [18] to prove the continuity of u at the point (x_0, t_0). Thus we know that $u \in C(\Omega_T \backslash \mathcal{E}_0)$, and $[u > 0] \cap (\Omega_T \backslash \mathcal{E}_0)$ is an open set.

Consider the set

$$\mathcal{U}^+ \equiv \{(x, t) \in \Omega_T | u > \eta \text{ a.e. in } [(x, t) + Q_\rho] \text{ for some } \eta, \rho > 0\}.$$

We know $\mathcal{U}^+$ is open and nonempty because $(\Omega_T \backslash \mathcal{E}_0) \cap [u > 0] \subseteq \mathcal{U}^+$ We want to show that $u_\epsilon^- = 0$ in this set for sufficiently small ϵ, so that we can pass to the limit as $\epsilon \downarrow 0$ in (3.2). We have the following:

Proposition 2. *Let* $[(x_0, t_0) + Q_\rho] \subset \Omega_T$, *and let* $\eta > 0$ *be fixed. For every* $\sigma \in (0,1)$ *there exists a number* $\delta \in (0,1)$ *depending on the data,* ρ, η, $\mathcal{E}_0$, *and* σ *so that if*

$$(4.5) \qquad \text{meas}\{(x,t) \in [(x_0, t_0) + Q_\rho] \mid u_\epsilon < \eta\} < \delta \, \text{meas} \, Q_\rho,$$

then

$$(4.6) \qquad u_\epsilon(x,t) \geq \frac{1}{2}\eta \qquad a.e. \ (x,t) \in [(x_0, t_0) + Q_{\sigma\rho}].$$

This result is a consequence of the following technical lemma:

Lemma 3. *There exist a constant* γ *depending on the data and independent of* ϵ, *so that for all* $\eta > 0$

$$(4.7) \qquad \begin{aligned} \|(u_\epsilon - \eta)_-\zeta\|^2_{V(Q_\rho)} &\leq \frac{\gamma}{(1-\sigma)^2\rho^2}\left\{\|(u_\epsilon - \eta)_-\|^2_{2,Q_\rho} + \|(u_\epsilon - \eta)_-\|_{1,Q_\rho}\right\} \\ &+ \gamma\rho^3 \frac{\|(u_\epsilon - \eta)_-\|^2_{\infty,Q_\rho}}{(1-\sigma)} \frac{\|\mathbf{v}_\epsilon\|_{10/3;Q_\rho}}{\sqrt{\rho}}\left\{\iint_{Q_\rho}\chi\left[u_\epsilon < \eta\right]\delta\right\}^{\frac{7}{10}}. \end{aligned}$$

This estimate allows us to write a recursion relation for

$$(4.8) \qquad Y_n = \iint_{Q_{\rho n}}\chi[u_\epsilon < \eta_n]dxdt,$$

where $\rho_n = \sigma\rho + (1-\sigma)\rho/2^n$, $\eta_n = \eta(1 + 1/2^n)$, of the form

$$(4.9) \qquad Y_{n+1} \leq C_{\eta,\sigma}4^n \frac{\|\mathbf{v}_\epsilon\|_{10/3;Q_\rho}}{\sqrt{\rho}}Y_n^{1+\frac{1}{10}},$$

so $Y_n \to 0$ by fast geometric convergence ([27], p. 95) provided Y_0 is sufficiently small. Since this is implied by (4.5), proposition 2 follows.

§5 The Navier-Stokes equations within $\mathcal{U}^+$.

A consequence of Proposition 2 is the following:

Lemma 4. *Let* $\mathcal{K}$ *be a compact subset of* $\mathcal{U}^+$, *and let* η *be a fixed positive number. Then there exists* ϵ_0 *depending only on* $\mathcal{K}$, *and* η *so that*

$$u_\epsilon(x,t) \geq \eta$$

for all $(x,t) \in \mathcal{K}$, *and for all* $\epsilon < \epsilon_0$.

This finally allows us to pass to the limit in the Navier-Stokes system (3.2) in the liquid set. If we were working with the Stokes equations, as in [20], we could test (3.2) against smooth solenoidal functions supported in $\mathcal{U}^+$ and pass to the limit. Then the weak convergence of $\mathbf{v}_\epsilon$ established in (3.11) suffices to show that $\mathbf{v}$ also satisfies the Stokes system. For the Navier Stokes equations, however, we also need to show that the nonlinear term $(\mathbf{v}_\epsilon \cdot \nabla)\mathbf{v}_\epsilon$ converges to the corresponding nonlinear term for the limiting velocity. This is the purpose of the next lemma.

Lemma 5. *There is a subsequence ϵ_ℓ so that if $[(x_0,t_0)+Q_{2\rho}] \subseteq \mathcal{U}^+$ then $\mathbf{v}_{\epsilon_\ell} \longrightarrow \mathbf{v}$ strongly in $L_2[(x_0,t_0) + Q_\rho]$*

Proof. The proof is a modification of a technique found in [26] to prove strong convergence of the Galerkin approximations of solutions to the Navier-Stokes equations. By translation, we may assume that $(x_0,t_0) \equiv (0,0)$. Let $\{\mathbf{c}_k(x)\}$ be the eigenfunctions of the Stokes operator on B_ρ, that form a basis for the space $J(\Omega) = \{\mathbf{v} \in \overset{o}{W}{}^1_2(\Omega) : \text{div } \mathbf{v} = 0\}$ and are orthonormal in $L_2(B_\rho)$. Choose ϵ_0 as in Lemma 4, so that for all $\epsilon < \epsilon_0$ we have weakly in Q_ρ

(5.1) $$\frac{\partial}{\partial t}\mathbf{v}_\epsilon - \triangle\mathbf{v}_\epsilon + (\mathbf{v}_\epsilon\cdot\nabla)\mathbf{v}_\epsilon + \nabla p = \mathbf{f}(u_\epsilon),$$

(5.2) $$\text{div } \mathbf{v}_\epsilon = 0.$$

For any subsequence ϵ_j, there is a set $\Sigma \subset (-\rho^2, 0)$ of measure zero so that if $t_1, t_2 \notin \Sigma$, then

(5.3)
$$\int_{t_1}^{t_2}\int_{B_\rho} \left(\nabla\mathbf{v}_\epsilon{:}\nabla\mathbf{c}_k + (\mathbf{v}_\epsilon\cdot\nabla)\mathbf{v}_\epsilon\cdot\mathbf{c}_k - \mathbf{f}(u_\epsilon)\cdot\mathbf{c}_k\right)dx\,dt =$$
$$-\int_{B_\rho}\left(\mathbf{v}_\epsilon(x,t_2)\cdot\mathbf{c}_k(x) - \mathbf{v}_\epsilon(x,t_1)\cdot\mathbf{c}_k(x)\right)dx$$

for all j and k. We can define

(5.4) $$\psi_{\epsilon,k}(t) \equiv \int_{B_\rho}\mathbf{v}_\epsilon(x,t)\cdot\mathbf{c}_k(x)dx$$

for almost every $-\rho^2 < t < 0$, and can use (5.3) extend the definition to all t, so that $\psi_{\epsilon,k}$ is equicontinuous in t, uniformly in ϵ for each fixed k. We can use a standard diagonal process to select a subsequence ϵ_ℓ so that $\psi_{\epsilon_\ell,k}$ converges to some function ψ_k as $\ell \to \infty$ uniformly in t for each fixed k. Indeed, for all $\delta > 0$, there exists $\epsilon_\ell = \epsilon_\ell(\delta, k)$ so that for all $\epsilon_m > \epsilon_\ell$

$$\sup_{-\rho^2 \leq t \leq 0} |\psi_{\epsilon_\ell,k}(t) - \psi_{\epsilon_m,k}(t)| < \delta$$

By a vector version of Friedrich's Lemma ([27], p. 72), $\mathbf{v}_{\epsilon_\ell}$ converges strongly in $L_2(Q_\rho)$. Indeed, given $\eta > 0$ there exists a number N_η and a constant C independent of ϵ so that:

$$\int_{-\rho^2}^{0}\int_{B_\rho}|\mathbf{v}_{\epsilon_m} - \mathbf{v}_{\epsilon_\ell}|^2 dx\,dt \leq C_\eta \sum_{n=1}^{N_\eta}\int_{-\rho^2}^{0}\left[\int_{B_\rho}(\mathbf{v}_{\epsilon_m} - \mathbf{v}_{\epsilon_\ell})\cdot\mathbf{c}_n dx\right]^2 dt +$$
$$+ \eta\|(\mathbf{v}_{\epsilon_k} - \mathbf{v}_{\epsilon_\ell})\|^2_{V(Q_\rho)},$$
$$\leq C_\eta \sum_{n=1}^{N_\eta}\int_{-\rho^2}^{0}|\psi_{\epsilon_m,n}(t) - \psi_{\epsilon_\ell,n}(t)|^2\,dt + \eta\|(\mathbf{v}_{\epsilon_k} - \mathbf{v}_{\epsilon_\ell})\|^2_{V(Q_\rho)},$$

and thus Lemma 5 follows. $\square$

We remark that, unlike the situation in [20], the velocity may be unbounded, and we cannot conclude that the derivatives of u are bounded in in $\mathcal{U}^+$. It is known that, on bounded domains with smooth data there exists a solution of the Navier-Stokes equations on some interval of time $[0, T]$ that is smooth. It does not appear that we can make use of this in the case when the initial data $\mathbf{v}_0$ is smooth. Indeed, if we had smooth data we could find a solution of the approximating problem (3.2) that is smooth for some interval $[0, T_\epsilon]$, but it appears that $T_\epsilon \downarrow 0$ as $\epsilon \downarrow 0$.

References

[1] V. Alexiades and A. D. Solomon, *Mathematical Modeling of Melting and Freezing Processes*, Hemisphere Publishing, Washington, 1993.

[2] B. V. Bazalii, *A stationary two-phase Stefan problem with convective heat transport in the liquid phase*, Mat. Fiz. Nelinein Mekh. **39** (1986), 50–56.

[3] B. V. Bazalii and S. P. Degtyarev, *Classical solvability of the nonstationary Stefan problem with convection*, Soviet Math. Dokl. **33** (1986), 309–312.

[4] B. V. Bazalii and S. P. Degtyarev, *On the classical solvability of a multidimensional Stefan problem with convective motion of a viscous incompressible fluid*, Math. USSR Sb. **60** (1988), 1–17.

[5] S. Besicovitch, *On the fundamental geometric properties of linearly measurable plane set of points (I), (II), (III)*, Math. Ann. **98** (1927), 422–464; **115** (1938), 296–329; **116** (1939), 349–357.

[6] A. S. Besicovitch, *Concentrated and rarified sets of points*, Acta Math. **62** (1933), 289–300.

[7] J. Boussinesq, *Théorie analytique de la chaleur*, Gauthier Villars, Paris, 1903.

[8] H. Brézis, *On some degenerate non-linear parabolic equations*, Proceedings AMS **I, XVII** (1968).

[9] G. Caginalp, *An analysis of resolidification in welding*, Quart. Appl. Math. **44** (1987), 665–674.

[10] G. Caginalp, *A free boundary problem with moving source*, Adv. in Appl. Math. **5** (1984), 476–488.

[11] J. R. Cannon, E. DiBenedetto, and G. K. Knightly, *The steady state Stefan problem with convection*, Arch. Rat. Mech. Anal. **73** (1980), 79–97.

[12] J. R. Cannon, E. DiBenedetto, and G. K. Knightly, *The bidimensional Stefan problem with convection: the time dependent case*, Comm. Part. Diff. Equ. **8**, (1983), 1549–1604.

[13] C. J. Chang and R. A. Brown, *Natural convection in steady solidification: Finite element analysis of a two-phase Rayleigh–Bénard problem*, J. Compt. Phys. **53** (1984), 1–27.

[14] M. Chipot and J. F. Rodrigues, *On the steady-state continuous casting Stefan problem with non–linear cooling*, Quart. Appl. Math. **40** (1983), 476–491.

[15] S. P. Degtyarev, *Classical solvability of a quasistationary Stefan problem in the presence of convection*, Dokl. Akad. Nauk Ukrain. SSR Ser. A **86** (1985), 15–19.

[16] S. P. Degtyarev, *Classical solvability of a quasistationary Stefan problem in the presence of convection*, Mat. Fiz. Nelinein Mekh. **91** (1986), 65–70.

[17] S. P. Degtyarev, *Classical solvability of the multidimensional steady–state Stefan problem with convection*, Dokl. Akad. Nauk Ukrain. SSR Ser. A **90** (1986), 10–13.

[18] E. DiBenedetto, *Continuity of weak solutions to certain singular parabolic equations*, Ann. Mat. Pura Appl. (IV) **CXXX** (1982), 131–177.

[19] E. DiBenedetto and A. Friedman, *Conduction-convection problems with a change of phase*, J. Diff. Equ. **62**, 129–185.

[20] E. Dibenedetto and M. O'Leary, *Three Dimensional Conduction-Convection Problems with change of Phase*, Arch. Rat. Mech. Anal. **123**, 99–116.

[21] E. Dibenedetto and R. E. Showalter, *Implicit degenerate evolution equations and applications*, SIAM J. on Math. Anal. **12** (1981).

[22] H. Federer, *Geometric measure theory*, Springer–Verlag, New York, 1969.

[23] A. Friedman, *The Stefan problem in several space variables*, Trans. Amer. Mat. Soc. **132** (1968).

[24] M. Giaquinta, *Multiple integrals in the calculus of variations and non linear elliptic systems*, Princeton Univ. Press, 1983.

[25] E. Giusti, *Precisazione delle funzioni $H^{1,p}$ e singolarità delle soluzioni deboli di sistemi ellittici non-lineari*, Boll. UMI **2** (1969), 71–76.

[26] O. A. Ladyzhenskaya, *The mathematical theory of viscous incompressible flows*, Gordon Breach, New York, London, Paris, 1969.

[27] O. A. Ladyzhenskaya, V. A. Solonnikov, and N. N. Ural'tseva, *Linear and quasilinear equations of parabolic type*, Transl. AMS, Providence RI, 1968.

[28] L. D. Landau and E. M. Lifshitz, *Fluid Mechanics*, Pergamon Press, London, Paris, Frankfurt, 1959.

[29] S. J. Linz and M. Lücke, *Non-Oberbeck-Boussinnesq effects and barodiffusion in binary mixtures with small thermodiffusion ratio*, Phys. Review A **36** (1987), 3505–3508.

[30] A. M. Meirmanov, *The Stefan Problem*, Walter de Gruyter, Berlin, 1992.

[31] J. M. Mihaljan, *A rigorous exposition of the Boussinesq approximations applicable to a thin layer of fluid*, Astrophysical Journal **136** (1962), 1126–1135.

[32] F. Morgan, *Geometric Measure Theory*, Academic Press, New York, 1988.

[33] M. O'Leary, *Conduction-Convection Problems with change of phase*, Ph.D. Thesis, Northwestern University (1995).

[34] J. F. Rodrigues, *A steady-state Boussinesq-Stefan problem with continuous extraction*, Ann. Mat. Pura Appl. (IV) **144** (1986), 203–218.

[35] E. A. Spigel and G. Veronis, *On the Boussinesq approximation for a compressible fluid*, Astrophysical Journal **113** (1960), 442–447.

E. DiBenedetto
Department of Mathematics
Northwestern University

M. O'Leary
Department of Matematics
University of Oklahoma

WAVE-LIKE SOLUTIONS FOR GAS-SOLID REACTION IN A SEMI-INFINITE MEDIUM.

Andrea Di Liddo *

Abstract
We shall consider the pseudo-steady-state approximation of a mathematical model for the combustion of a porous solid as it reacts with a gas diffusing through its pores. We give information about the penetration and the conversion front and we show that the model exhibits waves with time-dependent velocity (constant velocity in the Neumann case).

1   Introduction.

We shall study a model for the combustion of a porous solid occupying a semi-infinite medium as it reacts irreversibly and isothermally with a gas diffusing through its pores. We shall assume that the reaction is distributed throughout the solid with power-law reaction rates.

The nondimensional equations for the gas concentration C and the solid concentration S are given by

$$
(1) \qquad
\begin{cases}
\epsilon C_t(x,t) - C_{xx}(x,t) & = \quad -C_+^p(x,t) S_+^m(x,t) \\
\\
S_t(x,t) & = \quad -C_+^p(x,t) S_+^m(x,t)
\end{cases}
\qquad x > 0,\ t > 0,
$$

where ϵ is the porosity, p and m are nonnegative constants. The notation w_+ stands for the greater of 0 and w. We shall confine ourselves to the case $\epsilon = 0$ which is a good approximation of the case $\epsilon > 0$ when ϵ is small, as it often happens in applications (the so-called *pseudo-steady-state* approximation).

Setting $X \doteq 1 - S$ (the local *conversion* of the solid) (1) becomes

$$
(2) \qquad
\begin{cases}
-C_{xx}(x,t) & = \quad -C_+^p(x,t)(1-X)_+^m(x,t) \\
\\
X_t(x,t) & = \quad C_+^p(x,t)(1-X)_+^m(x,t)
\end{cases}
\qquad x > 0,\ t > 0,
$$

We associate the initial condition

$$
(3) \qquad\qquad X(x,0) = 0,\ x > 0,
$$

and the boundary conditions

$$
(4) \qquad\qquad \alpha C(0,t) - \beta C_x(0,t) = 1, \quad C(\infty,t) = C_x(\infty,t) = 0,\ t > 0.
$$

Mathematical models for gas-solid reactions are widely used in chemical engineering. See, for instance, [1], [6], [15].

---

*Istituto per Ricerche di Matematica Applicata del CNR, Via Amendola 122 I, 70125, Bari, Italy

Problem (1) and (2) have been extensively studied in a bounded domain by Diaz and Stakgold in [2]. See also [10], [11], [12], [13], [14]. A study of traveling waves for both the parabolic and the p.s.s. approximation has been performed in [3]. Traveling waves for gas-solid reactions including a convection term have been investigated in [5]. Problem (1) has been considered in [4] in the special case of Dirichlet boundary condition for the gas at $x = 0$. The main topic of that paper is the investigation of two special phenomena which occur respectively in the case $0 \leq m < 1$ and in the case $0 \leq p < 1$. For $0 \leq m < 1$ there is a *conversion front* $x = \lambda(t)$ behind which the solid is fully converted, but no such front exist for $m \geq 1$. For $0 \leq p < 1$ there is a *penetration front* $x = \rho(t)$ ahead which the gas concentration vanishes. No such front exist for $p \geq 1$. In this paper we extend those results to more general boundary conditions (4) and we show that in some cases the solutions of (2) behave like traveling waves with time-dependent velocity. In the case of Neumann boundary condition the wave speed is constant, as already observed in [14].

Section 2 is devoted to some preliminary results about the asymptotic behaviour of the solution of (2-4), the existence of fronts and some bounds about them.

In Section 3 we show that, in the Neumann case, the solution of (2-4) is a travelling wave after a time T depending only on m and p.

In Section 4 we study the case of linear absorption of the gas, assuming general boundary conditions. In this case the solution of (2-4) has a wave-like structure with time-dependent velocity. An explicit formula for the conversion front is given.

Section 5 is concerned with the discontinuous case $p = 0$, $m = 1$. We are able to give explicit formulas for the penetration front.

2 Preliminary results.

System (2) is a quasi-monotone nondecreasing system, therefore upper and lower solutions can be defined in the usual way. See, for example, [8] and [9]. In the case of third type boundary conditions, that is for $\alpha > 0$, $(0,0)$ and $(\frac{1}{\alpha}, 1)$ are, respectively, lower and upper solution for (2-4). More delicate is the Neumann case since upper solutions with constant gas concentration do not exist. In this case upper solutions are given by the traveling waves studied in [3]. To be more precise we summarize some results given therein.

THEOREM 2.1. *For any $v > 0$ a unique solution (modulo translations) exists to system*

$$\begin{cases} u' + v(1 - w) = 0, \\ vw' = u_+^p w_+^m, \\ w(-\infty) = 0, \quad w(+\infty) = 1, \quad u(+\infty) = u'(+\infty) = 0. \end{cases}$$

Since $w(-\infty) = 0$, $u'(z) \sim -v$ as z tends to $-\infty$. Then we can choose a value of v and a suitable translation of the wave front so that $-\beta u'(0) = -\beta \frac{v}{2} \geq 1$. Since u' is nondecreasing, $-\beta u'(z) \geq 1$ for every $z < 0$. It follows that

$$(\overline{C}(x,t), \overline{X}(x,t)) \doteq (u(x - vt), 1 - w(x - vt)), \quad x \geq 0, \ t \geq 0,$$

is un upper solution for (2-4) in the case $\alpha = 0$.

Starting from a pair of lower and upper solution, existence and uniqueness of the solution of (2-4) can be proved by the monotone iterative scheme proposed in [2]. The solution (C, X) satisfies the following inequalities

$$0 \leq C(x,t) \leq \frac{1}{\alpha}, \quad 0 \leq X(x,t) \leq 1, \quad \alpha > 0,$$

$$0 \leq C(x,t) \leq u(x - vt), \quad 0 \leq X(x,t) \leq 1 - w(x - vt), \quad \alpha = 0.$$

The equation for C in (2) is elliptic, therefore the initial gas-concentration cannot be assigned since it is the solution of the equation

$$C_{xx}(x,0) = C_+^p(x,0),$$

subject to the boundary conditions (4). It is given by

$$C_0(x) = \begin{cases} \dfrac{1}{\alpha + \beta} \exp(-x) & \text{for } p = 1, \\ b^*[1 - \dfrac{x}{x_p}]_+^{\frac{2}{1-p}} & \text{for } p \neq 1, \end{cases}$$

where $x_p(1 - p)$ is the unique positive solution of the equation

$$(\alpha z + 2\beta)^{1-p} z^{1+p} - 2(1 + p) = 0.$$

Note that $x_p > 0$ if $p < 1$. In this case C_0 has a compact support $[0, x_p]$ and x_p is decreasing with respect to α and with respect to β. The value of b^* is given by

$$b^* = \frac{x_p(1 - p)}{\alpha x_p(1 - p) + 2\beta}.$$

A straightforward extension to the more general boundary condition (4) of the arguments used for Dirichlet and Neumann case in [4] and [14] gives the following results:

THEOREM 2.2. *The solution* $(C(x,t), X(x,t))$ *of (2-4) decreases in* x *for fixed* t, *and increases in* t *for fixed* x.

$$\lim_{t \to \infty} X(x,t) = 1,$$

$$\lim_{t \to \infty} C(x,t) = 1/\alpha, \quad \text{for } \alpha > 0,$$

$$\lim_{t \to \infty} C(x,t) = \infty, \quad \text{for } \alpha = 0.$$

These limits holds uniformly over compact x *sets.*

THEOREM 2.3. *Define the conversion front as*

$$\lambda(t) \doteq \sup\{x \geq 0 \mid X(x,t) = 1\},$$

and the penetration front as

$$\rho(t) \doteq \inf\{x \geq 0 \mid C(x,t) = 0\}.$$

Then, if $0 \leq m < 1$ *and* $t > \frac{1}{b^*{}^p(1-m)}$, *we have* $0 < \lambda(t) \leq \sqrt{\frac{2t}{\alpha}}$ *for* $\alpha > 0$ *and* $0 < \lambda(t) \leq \frac{t}{\beta}$ *for* $\alpha = 0$. *Moreover* $\lambda(t)$ *is increasing and* $\lim_{t \to \infty} \lambda(t) = \infty$. *If* $m \geq 1$, $X(x,t) < 1$ *for all* $x \geq 0$ *and* $t \geq 0$, *and there is no conversion front.*

If $0 \leq p < 1$, *then* $x_p < \rho(t) < \infty$, $\rho(t)$ *is increasing, and* $\rho(t) \geq \sqrt{2b^* t}$. *If* $p \geq 1$, $C(x,t) > 0$ *for all* $x \geq 0$ *and* $t \geq 0$, *and there is no penetration front.*

Proof. We prove here only the inequalities about the fronts. The other results can be proved as in [4], [14]. Note that

$$X(x,t) = A_m \left(\int_0^t C^p(x,\tau)d\tau \right),$$
(5)

where

$$A_m(t) = 1 - [1 - (1-m)t]_+^{\frac{1}{1-m}} \quad \text{for } m \neq 1,$$

$$A_1(t) = 1 - \exp(-t).$$

Then

$$X(0,t) = A_m \left(\int_0^t C^p(0,\tau)d\tau \right) \geq A_m \left(\int_0^t C^p(0,0)d\tau \right) = A_m(b^{*p}t).$$

Therefore for $0 \leq m < 1$, $X(0,t) \equiv 1$ for all $t \geq \frac{1}{b^{*p}(1-m)}$.

Integrating by parts and using the equations (2-4) we have

$$\int_0^t C(0,s)ds = \int_0^\infty xX(x,t)dx \geq \int_0^{\lambda(t)} xdx = \frac{\lambda^2(t)}{2},$$

so that

$$\lambda^2(t) \leq 2 \int_0^t C(0,s)ds \leq \frac{2t}{\alpha}, \quad \text{for } \alpha > 0.$$

For $\alpha = 0$ we have

$$\int_0^\infty X_t(x,t)dx = \int_0^\infty C_{xx}(x,t)dx = -C_x(0,t) = \frac{1}{\beta},$$
(6)

and then

$$\lambda(t) \leq \int_0^\infty X(x,t)dx \leq \frac{t}{\beta}.$$

For the penetration front we have

$$b^*t = \int_0^t C(0,0)ds \leq \int_0^\infty xX(x,t)dx = \int_0^{\rho(t)} xX(x,t)dx \leq \frac{\rho^2(t)}{2}.$$

This completes the proof of the theorem. $\square$

3 The Neumann problem.

In this section we analyze the case $\alpha = 0$, that is the Neumann case. Since no special role is played by β, we shall assume $\beta = 1$ for simplicity. We confine ourselves to the more interesting case $0 \leq m < 1$, $0 \leq p < 1$, when both fronts exist. We look for solutions of (2-4) having a wave structure for $\lambda(t) < x < \rho(t)$ and $t \geq T = T_{m,p}$, where T is the time of complete conversion of the solid at $x = 0$. Setting $C(x,t) = u(x - \lambda(t))$ and $X(x,t) = 1 - w(x - \lambda(t))$, equations (2) give

$$\begin{cases} u'' = u_+^p w_+^m, \\ w' \dfrac{d\lambda}{dt} = u_+^p w_+^m, \end{cases}$$
(7)

where we denote by $'$ the derivative with respect to $z \doteq x - \lambda(t)$. From the second equation in (7) it follows that $\frac{d\lambda}{dt}$ must be a constant k so that $\lambda(t) = k(t-T)$. For $t \geq T$ and $x \leq \lambda(t)$, C_{xx} vanishes, therefore $C(x,t) = -x + q(t)$. At $x = \lambda(t)$ we have $q(t) = u(0) + \lambda(t)$.

Equations (7) imply $u'' = kw'$ and then $u'(z) = kw(z) + u'(0) = kw(z) - 1$. At $z = \rho(t) - \lambda(t)$ we have $0 = u'(\rho(t) - \lambda(t)) = kw(\rho(t) - \lambda(t)) - 1 = k - 1$ so that $k = 1$.

System (7) is then equivalent to the first order system

(8)
$$\begin{cases} u' = w - 1, \\ w' = u_+^p w_+^m, \end{cases}$$

with the appropriate conditions at 0 and at $\rho(t) - \lambda(t)$. Since the nonlinear term $u_+^p w_+^m$ in (8) is nonLipschitzian, phase-plane analysis requires some care. Arguments similar to the ones used in [3] and [7] show that a unique orbit connecting the points $(u(0), 0)$, $(0, 1)$ exists for every $u(0) > 0$.

Dividing the two equations in (8) and integrating with respect to the independent variable between 0 and $\rho(t)$ we obtain

$$u(0) = [\frac{p+1}{(1-m)(2-m)}]^{\frac{1}{p+1}}.$$

Now we prove that $T = u(0)$. Let

$$\gamma(t) \doteq \int_0^{+\infty} X(x,t)dx,$$

(total amount of solid converted at time t). Since $\gamma(0) > 0$, from (6) we have $\gamma(t) = t$. Note that from $w(\rho(t) - \lambda(t)) \equiv 1$ it follows that $\rho(t) - \lambda(t) \equiv c = $ constant. Therefore

$$t = \gamma(t) = \lambda(t) + \int_{\lambda(t)}^{\rho(t)} X(x,t)dx$$

$$= t - T + c - \int_0^c w(z)dz,$$

and then

$$T = c - \int_0^c w(z)dz = c - \int_0^c 1 + u'(z)dz = u(0).$$

Finally note that

$$t = \int_0^{\rho(t)} X(x,t)dx \leq \rho(t).$$

We summarize the previous results in the following

THEOREM 3.1. Let $0 \leq m < 1$, $0 \leq p < 1$. The solution $(C(x,t), X(x,t))$ of the problem (2-4) with Neumann boundary conditions ($\alpha = 0$, $\beta = 1$) has the following properties:

$$\begin{cases} C(x,t) = u(x - \lambda(t)) \\ X(x,t) = 1 - w(x - \lambda(t)) \end{cases} \quad \lambda(t) \leq x \leq \rho(t), \ t \geq T,$$

where (u, w) is the solution of (8),

$$T = [\frac{p+1}{(1-m)(2-m)}]^{\frac{1}{p+1}},$$

$$\lambda(t) = t - T, \ t \geq T,$$

$$C(x,t) = -x + t, \quad for \ t \geq T, \ 0 \leq x \leq \lambda(t).$$

Moreover a positive constant c exists such that

$$\lambda(t) + c = \rho(t) \leq t, \quad for \ t \geq T.$$

4 The case $p = 1, \ 0 \leq m < 1$.

In this section we give some results in the case of linear absorption of the gas, that is for $p = 1$. Assume $0 \leq m < 1$, so that a conversion front $\lambda(t)$ exists.

We look for solutions of (2-4) having the following structure:

$$\begin{cases} C(x,t) &= g(\lambda(t))u(x - \lambda(t)) \\ X(x,t) &= 1 - w(x - \lambda(t)) \end{cases} \quad \lambda(t) \leq x, \ t \geq T,$$

where g, u, w are functions to be found and T is the time of full conversion at $x = 0$.

First of all we compute $C(x,t)$ for $t \geq T$ and $0 \leq x \leq \lambda(t)$. As in the previous section it is easy to show that $C(x,t) = a(t)x + b(t)$, where $a(t)$ and $b(t)$ can be calculated using the boundary condition (4) at $x = 0$ and the smoothness of C at the conversion front.

It results

$$a(t) = \frac{u'(0)}{\alpha u(0) - u'(0)(\beta + \alpha\lambda(t))},$$

(9)
$$b(t) = \frac{u(0) - \lambda(t)u'(0)}{\alpha u(0) - u'(0)(\beta + \alpha\lambda(t))},$$

(10)
$$g(\lambda(t)) = \frac{1}{\alpha u(0) - u'(0)(\beta + \alpha\lambda(t))}.$$

Setting $z \doteq x - \lambda(t)$, equations (2) imply

(11)
$$\begin{cases} u'' = uw_+^m, \\ w'\dfrac{d\lambda}{dt} = g(\lambda(t))uw_+^m. \end{cases}$$

The second equation in (11) implies

$$\frac{d\lambda}{dt} = kg(\lambda(t)),$$

where k is a constant.

Since $u''(z) = kw'(z)$, an integration from z to $+\infty$ gives

(12)
$$u'(z) = k(w(z) - 1),$$

so that $k = -u'(0)$ and

(13)
$$\frac{d\lambda}{dt} = -u'(0)g(\lambda(t)).$$

(11) is equivalent to the following system

(14)
$$\begin{cases} u'(z) = -u'(0)(w(z) - 1), \\[2mm] w'(z) = \dfrac{u(z)w_+^m(z)}{-u'(0)}, \end{cases}$$

with $u(+\infty) = u'(+\infty) = 0$, $w(+\infty) = 1$.

Again a phase plane analysis show that (14) has a unique solution (modulo translation).

An integration of the equation

(15)
$$\frac{dw}{du}(z) = \frac{1}{u'^2(0)} \frac{u(z)w_+^m(z)}{w(z) - 1},$$

between 0 and $+\infty$ gives

(16)
$$\frac{-u'(0)}{u(0)} = \sqrt{\frac{(1-m)(2-m)}{2}} \doteq c_m.$$

Let $\beta > 0$. The total conversion $\gamma(t)$ satisfies the equation

(17)
$$\frac{d\gamma}{dt}(t) = \frac{1 - \alpha C(0, t)}{\beta}.$$

For $t \geq T$, $C(0, t) = b(t)$ and then, from (9), (10) and (13),

$$\begin{aligned}
\frac{d\gamma}{dt}(t) &= \frac{1 - \alpha b(t)}{\beta} \\[3mm]
&= \frac{-u'(0)}{\alpha u(0) - u'(0)(\beta + \alpha\lambda(t))} = \frac{d\lambda}{dt}(t).
\end{aligned}$$

Integrating from T and $t \geq T$, it follows $\lambda(t) = \gamma(t) - \gamma(T)$.

From (12) and (16) we have

(18)
$$\begin{aligned}
\gamma(T) &= \int_0^{+\infty} X(x, T)dx = \int_0^{+\infty} 1 - w(x)dx \\[3mm]
&= \int_0^{+\infty} \frac{u'(z)}{u'(0)}dz = \frac{-u(0)}{u'(0)} = \frac{1}{c_m},
\end{aligned}$$

independent on α and β.

Now we compute T. From (5) we have

$$\int_0^T C(0, s)ds = \frac{1}{1 - m},$$

and then, integrating (17) from 0 to T, we obtain

(19)
$$\gamma(T) = \frac{T - \alpha\frac{1}{1-m}}{\beta}.$$

(18) and (19) give $T = \dfrac{\beta}{c_m} + \dfrac{\alpha}{1 - m}$, that is

(20)
$$T = \beta\sqrt{\frac{2}{(1-m)(2-m)} + \frac{\alpha}{1-m}}.$$

(10) and (16) imply

(21)
$$g(\lambda(t)) = \frac{1}{u(0)}\frac{1}{\alpha + c_m(\beta + \alpha\lambda(t))},$$

so that from (13) we have

$$\begin{cases} \dfrac{d\lambda}{dt}(t) &= \dfrac{c_m}{\alpha + c_m(\beta + \alpha\lambda(t))} \quad t \geq T, \\[2mm] \lambda(T) &= 0. \end{cases}$$

An integration between T and $t > T$ gives

$$c_m\alpha\lambda^2(t) + 2(\alpha + c_m\beta)\lambda(t) - 2c_m(t - T) = 0.$$

Solving this equation we have

(22)
$$\lambda(t) = \begin{cases} \dfrac{t - T}{\beta} & \text{for } \alpha = 0, \\[4mm] \dfrac{\sqrt{2c_m^2\alpha(t - T) + (\alpha + c_m\beta)^2} - (\alpha + c_m\beta)}{\alpha c_m} & \text{for } \alpha > 0. \end{cases}$$

An integration of (15) from 0 to x allows us to write u as a function of w. Precisely

(23)
$$u(x) = u(0)G(w(x)) \quad x \geq 0,$$

where

(24)
$$G(w) \doteq \sqrt{2c_m^2[\frac{w^{2-m}}{2-m} - \frac{w^{1-m}}{1-m}] + 1}.$$

The second equation in (14) gives

$$\begin{cases} \dfrac{dw}{dx} &= \dfrac{1}{c_m}G(w)w^m \\[2mm] & \qquad\qquad\qquad x > 0, \\[2mm] w(0) &= 0 \end{cases}$$

which can be integrated to obtain w. (23) and (21) give

(25)
$$\begin{cases} C(x,t) &= \dfrac{1}{\alpha + c_m(\beta + \alpha\lambda(t))}G(w(x - \lambda(t))) \\[2mm] & \qquad\qquad\qquad\qquad\qquad \lambda(t) \leq x, \ t \geq T. \\[2mm] X(x,t) &= 1 - w(x - \lambda(t)) \end{cases}$$

We collect the previous results in the following

THEOREM 4.1. *Let $0 \leq m < 1$, $p = 1$. The solution $(C(x,t), X(x,t))$ of the problem (2-4) is given by (25), where (u, w) is the solution of (14). G is defined in (24), T is given by (20), $\lambda(t)$ is given by (22) and*

$$C(x,t) = \frac{-c_m x + 1 + c_m\lambda(t)}{\alpha + c_m(\beta + \alpha\lambda(t))}, \quad \text{for } t \geq T, \ 0 \leq x \leq \lambda(t).$$

5   The case $p = 0$, $m = 1$.

In this section we give some explicit results about the penetration front $\rho(t)$ in the case of linear solid consumption ($m = 1$) and discontinuous absorption term ($p = 0$). Problem (2) becomes

(26)
$$\begin{cases} -C_{xx}(x,t) & = & -H(C(x,t))(1-X)(x,t) \\ \\ X_t(x,t) & = & H(C(x,t))(1-X)(x,t) \end{cases} \quad x > 0,\ t > 0,$$

where H is the Heaviside function. From (26) and the initial condition (3), it is easy to show that

$$X(x,t) = \begin{cases} 1 - \exp(-t) & \text{for } t > 0,\ x < \rho(0), \\ 1 - \exp(\theta(x) - t) & \text{for } \theta(x) \geq t,\ x > \rho(0), \end{cases}$$

where θ is the inverse of ρ. Integrating the first equation in (26) twice with respect to x and using the boundary conditions (4), it can be proved that the penetration front $\rho(t)$ satisfies the differential equation

(27)
$$\rho'(t) = \frac{1}{\alpha \rho(t) + \beta},$$

with

$$\rho(0) = \begin{cases} \dfrac{1}{\beta}, & \text{for } \alpha = 0, \\ \\ \dfrac{\sqrt{\beta^2 + 2\alpha} - \beta}{\alpha}, & \text{for } \alpha > 0. \end{cases}$$

Solving (27), we obtain

$$\rho(t) = \begin{cases} \dfrac{t+1}{\beta}, & \text{for } \alpha = 0, \\ \\ \dfrac{\sqrt{\beta^2 + 2\alpha(t+1)} - \beta}{\alpha}, & \text{for } \alpha > 0. \end{cases}$$

Moreover, from system (26), we have formulas for C. As an example, in the case $\alpha = 0$, we have $\theta(x) = \beta x - 1$, for $\frac{1}{\beta} \leq x$ and

$$C(x,t) = \frac{1}{\beta^2}[\exp(\theta(x) - t) - 1 - (\theta(x) - t)] \quad \text{for } \frac{1}{\beta} \leq x \leq \rho(t),$$

$$C(x,t) = \frac{1}{2\beta^2}[\exp(-t)\theta^2(x) + 2\exp(-\theta(x))(t - 1 - \theta(x)) + 2\exp(-t - \theta(x))(1 + \theta(x))],$$

for $0 \leq x \leq \frac{1}{\beta}$.

Concluding remarks.

In this paper we have seen that, in the cases $p = 1$ and $p = 0$, $m = 1$, the penetration and the conversion fronts behaves like t or like $\sqrt{t}$ as t tends to infinity, for Neumann or third type boundary conditions, respectively. An extension of this result to all values of m

and p, together with the study of the model with positive porosity ϵ is left to a forthcoming paper.

References

[1] R. Aris, *The Mathematical Theory of Diffusion and Reaction in Permeable Catalysts. Vol. I and II*, Clarendon Press, Oxford (1975)

[2] J. I. Diaz and I. Stakgold, "Mathematical aspects of the combustion of a solid by a distributed isothermal gas reaction" *SIAM J. Math. Anal. Appl.*, **26**, (1995) 305-328

[3] A. Di Liddo, L. Maddalena and I. Stakgold, "Traveling waves for distributed gas-solid reactions" *J. Diff. Eq.*, **113**, (1994) 452-472

[4] A. Di Liddo and I. Stakgold, "Isothermal combustion with two moving fronts" *J. Math. Anal. Appl.*, **152**, (1990) 584-599

[5] C. J. van Duijn and A. Straathof, "Traveling waves for gas-solid reactions" *Rev. Mat. Univ. Complut. Madrid*, **7**, (1994) 147-178

[6] G. F. Froment and K. B. Bischoff, *Chemical Reactor Analysis and Design*, Wiley, New York, (1979)

[7] M. A. Herrero and J. L. Vazquez, "Thermal waves in absorbing media" *J. Diff. Eq.*, **74**, (1988) 218-233

[8] C. V. Pao, *Nonlinear Parabolic and Elliptic Equations*, Plenum Press, New York and London, (1992)

[9] J. Smoller, *Shock Waves and Reaction-Diffusion Equations*, Springer-Verlag, Berlin and Heidelberg, (1983)

[10] I. Stakgold, "Conversion estimates for gas-solid reactions" *Math. Modelling*, **5**, (1984) 325-330

[11] I. Stakgold, "Partial extinction in reaction-diffusion" *Conf. Sem. Mat. Univ. Bari*, **224**, (1987)

[12] I. Stakgold, "Localization and extinction in reaction-diffusion" in K. H. Hoffmann et al. (eds.) *Free Boundary Problems: Theory and Applications*, Longman, New York, (1988) 208-221

[13] I. Stakgold, "Diffusion with Strong Absorption" in M. C. Delfour and G. Sabidussi (eds.), *Shape Optimization and Free Boudaries*, Kluwer Academic Publ., (1992) 321-346

[14] I. Stakgold, "Conversion and penetration fronts in combustion" *Proc. I World Congress of Nonlinear Analysis* , (1994)

[15] J. Szekely, J. W. Evans and H. Y. Sohn, *Gas-solid reaction*, Academic Press, New York, (1976)

Penetration of a Wetting Front in a Porous Medium with Flux Dependent Hydraulic Parameters[*]

A. Fasano[†]   P. Tani[‡]

Abstract

We consider an incompressible flow through a porous medium undergoing deformations such that: (i) the porosity and the hydraulic conductivity can be considered to be functions of the flux intensity, (ii) the macroscopic deformation can be neglected. The medium is initially dry, so that a sharp wetting front proceeds into the medium. Both gravity and capillarity are neglected. The geometry is one-dimensional.

1 Introduction

A well-known one-dimensional problem in the framework of the theory of incompressible flows through porous media is the Green-Ampt model for the penetration of a wetting front in a dry medium [3,1]. Neglecting capillarity and the action of gravity (as we shall do for the sake of brevity in the rest of the paper), the model combines Darcy's law

$$(1.1) \qquad q = -k\frac{\partial p}{\partial x}, \qquad 0 < x < s(t),\ t > 0$$

(q = volumetric velocity, k = hydraulic conductivity, p = pressure, s = thickness of the wet region) with mass conservation ($\frac{\partial q}{\partial x} = 0$) and with the boundary conditions

$$(1.2) \qquad p(0,t) = p_0(t) > 0, \qquad t > 0,$$
$$(1.3) \qquad p(s(t),t) = 0, \qquad t > 0,$$
$$(1.4) \qquad \epsilon \dot{s}(t) = q(t), \qquad t > 0.$$

(the latter expressing the fact that the wetting front moves according to Darcy's law, ϵ being the porosity). The o.d.e. for $s(t)$

$$(1.5) \qquad \epsilon s \dot{s} = k p_0(t), \qquad s(0) = 0$$

is easily derived, leading to the complete solution of the problem.

A number of generalization of this problem (still in the 1-D formulation) are known, particularly introducing some deformability of the medium (see e.g. [4], [5]). In this paper we take the rather extreme point of view of considering the physical parameters of the system (namely k and ϵ) depending on q in a prescribed way. In this fashion we want to describe media undergoing flow-induced deformations on the microscopic scale (however

---

[*]Work partially supported by Res. Contract Illycaffé S.p.a., ECMI and by the Italian MURST National Project "Problemi non lineari ...".

[†]Dipartimento di Matematica "U. Dini", Viale Morgagni 67/A, 50134 Firenze (Italy)

[‡]DEMA, Via Tevere 80/3, Sesto Fiorentino (Firenze, Italy)

negligible on the macroscopic scale) instantly affecting ϵ, k (i.e. with no relaxation). As we shall see this approach leads to a nontrivial mathematical problem.

The origin of the main difficulties lies in the fact that, being $\epsilon'(q) < 0$, the characteristics of equation (1.7) have a negative slope, so that the information travels from the free boundary to the outer surface. In this way the influence of the data is felt in a very implicit way, since p and not q is known on the boundary.

Also from the physical point of view the problem may exhibit some peculiar feature, corresponding to the possible development of singularities of the solution and which can be interpreted as a local collapse of the medium. However here we will not investigate the continuation of the solution beyond such singularities. The problem we are going to study has been considered in [6] in the case in which $p_0(t)$ is non- convex or non-concave. In this paper we utilize the same method in order to obtain a more general local existence theorem. The continuability of the solution up to the possible onset of the above mentioned singularity will be shown by means of a different argument.

For more references and for the illustration of a more general context for wetting front problems see [2].

In the sequel we take $\epsilon = \epsilon(q)$, $k = k(q)$ and we study the following free boundary problem

$$(1.6) \qquad q(x,t) = -k(q)\frac{\partial p}{\partial x}, \qquad 0 < x < s(t), \ t > 0$$

$$(1.7) \qquad \epsilon'(q)\frac{\partial q}{\partial t} + \frac{\partial q}{\partial x} = 0, \qquad 0 < x < s(t), \ t > 0,$$

$$(1.8) \qquad p(0,t) = p_0(t) > 0, \qquad t > 0,$$

$$(1.9) \qquad p(s(t),t) = 0, \qquad t > 0,$$

$$(1.10) \qquad \dot{s}(t) = \frac{q(s(t),t))}{\epsilon(q(s(t),t)}, \qquad t > 0, \ s(0) = 0.$$

On the functions $\epsilon(q)$, $k(q)$ we make the following assumptions (in agreement with physical intuition)

$$(1.11) \qquad \epsilon \in C^3, \quad \epsilon' < 0, \quad \epsilon'' \geq 0, \ \epsilon \geq \epsilon_0 > 0 \ \forall \ q > 0,$$

$$(1.12) \qquad k \in C^2, \quad k' \leq 0, \quad k'' \geq 0, \ k \geq k_0 > 0 \ \forall \ q > 0.$$

The boundary pressure $p_0(t)$ will be assumed to be such that

$$(1.13) \qquad p_0 \in C^3, \quad p_0(0) = 0, \quad \dot{p}_0(0) > 0.$$

2 Some A-Priori Results

First of all we remark that we can calculate the initial value of the function $q_0(t) = q(0,t)$. Indeed this corresponds to the initial flux intensity of the Green-Ampt model in which ϵ and k are set equal to the values $\epsilon(q_0(0))$, $k(q_0(0))$. Thus we find

$$(2.1) \qquad q_0(0) = [k(q_0(0))\epsilon(q_0(0))\dot{p}_0(0)]^{1/2},$$

from which $q_0(0) = q_0^* > 0$ can be obtained uniquely thanks to the assumptions (1.11)-(1.13).

Even though the boundary data for equation (1.7) are not given explicitly, it is important to analyze the behaviour of the solutions of the boundary value problem

$$(2.2) \qquad \epsilon'(q)\frac{\partial q}{\partial t} + \frac{\partial q}{\partial x} = 0, \qquad x > 0,\ t > 0,$$

$$(2.3) \qquad q(0,t) = q_0(t), \qquad t > 0,$$

q_0 being a positive C^1 function.

Since q is constant along the characteristics, the latter are the straight lines

$$(2.4) \qquad t = \epsilon'(q_0(\theta))x + \theta$$

parametrized by θ. The equations for the envelope are (2.4) and

$$(2.5) \qquad \epsilon''(q_0(\theta))q_0'(\theta)x + 1 = 0.$$

In other words, the solution of (2.2), (2.3) develops a singularity along the line

$$(2.6) \qquad x(\theta) = -\frac{1}{\epsilon''(q_0(\theta))q_0'(\theta)},$$

$$(2.7) \qquad t(\theta) = \frac{-\epsilon'(q_0(\theta))}{\epsilon''(q_0(\theta))q_0'(\theta)} + \theta,$$

which is defined provided that $\epsilon''(q_0(\theta))q_0'(\theta) \neq 0$.

REMARK 2.1. The curve (2.6), (2.7) lies in the half plane $x < 0$ when $q_0' > 0$. Thus in such a case it never comes into play. If however

$$(2.8) \qquad \inf_{0 < \theta < T} \epsilon''(q_0(\theta)q_0'(\theta) \equiv \lambda(T) < 0,$$

then we confine our attention to a time interval $[0, T]$ such that

$$(2.9) \qquad T\frac{\|q\|_T}{\epsilon(\|q\|_T)} < \lambda(T),$$

with $\|q\|_T = \sup_{\substack{0 < x < s(t) \\ 0 < t < T}} q(x,t)$, in which it is guaranteed that $s(t) < \lambda(T)$. □

Thus, solving (2.2), (2.3) provides a scalar field $q(x,t)$ in a region which in any case contains a neighbourhood of the origin in $\mathbb{R}_+^2$. Our first aim is to prove local existence and uniqueness in such a neighbourhood for (1.6)-(1.10).

We conclude this section showing how to exploit the information provided by the pressure boundary data. From (1.6), (1.8), (1.9) we obtain by integration

$$(2.10) \qquad p_0(t) = \int_0^{s(t)} \frac{q(x,t)}{k(q(x,t))}\,dx$$

and differentiating we get

$$(2.11) \qquad \dot{p}_0(t) = \frac{q(s(t),t)}{k(q(s(t),t))}\dot{s}(t) + \int_0^{s(t)} \frac{d}{dq}\left(\frac{q(x,t)}{k(q(x,t))}\right)\frac{\partial q}{\partial t}\,dx.$$

Now we replace $\dfrac{\partial q}{\partial t}$ by $\dfrac{-1}{\epsilon'(q)}\dfrac{\partial q}{\partial x}$ and we define the function $G(q)$ such that

(2.12) $$G'(q) = -\frac{1}{\epsilon'(q)}\frac{d}{dq}\left(\frac{q}{k(q)}\right) = -\frac{k(q) - qk'(q)}{\epsilon'(q)k^2(q)}$$

(2.13) $$G(0) = 0,$$

thus obtaining

(2.14) $$\dot{p}_0(t) = \frac{q(s(t),t)}{k(q(s(t),t))}\dot{s}(t) + G(q(s(t),t)) - G(q(0,t)).$$

Finally we use (1.10) and we introduce the functions

(2.15) $$F(q) = \frac{q^2}{k(q)\epsilon(q)},$$

(2.16) $$H(q) = F(q) + G(q),$$

arriving at the following relationship between $q(s(t),t)$ and $q(0,t) = q_0(t)$

(2.17) $$H(q(s(t),t)) = \dot{p}_0(t) + G(q_0(t)).$$

Another result which will be useful in the sequel is the relationship between the time θ at which a characteristic line of (2.2) leaves the axis $x = 0$ and the time τ at which it hits the free boundary $x = s(t)$, namely

(2.18) $$\epsilon'(q_0(\theta))s(\tau) + \theta - \tau = 0.$$

From (2.18) we obtain

(2.19) $$\frac{d\tau}{d\theta} = \frac{\epsilon''(q_0(\theta))q_0'(\theta)s(\tau) + 1}{1 - \epsilon'(q_0(\theta))\dot{s}(\tau)}.$$

3   Local Existence and Uniqueness

THEOREM 3.1. *Under assumptions* (1.11)-(1.13) *problem* (1.6)-(1.10) *has a classical solution* (q,s) *for t ranging in a suitably small time interval* $[0,T]$. *Moreover q has Lipschitz continuous first derivatives.*

Proof. We use a fixed point argument which leads to the determination of the boundary value $q_0(t)$ and hence to the existence of a solution.

For a fixed T not exceeding the time interval in Remark 2.1 we introduce the set

$$X(T,L) = \left\{ q_0 \in C([0,T]) \ \mid \ q_0(0) = q_0^*, \ 0 < q_{min} \le q_0(t) \le q_{max}, \right.$$

$$\left. \left|\frac{q_0(t') - q_0(t'')}{t' - t''}\right| \le L, \ 0 \le t' < t'' \le T \right\},$$

L being a positive constant to be determined (as well as q_{min} and q_{max}), and q_0^* is the initial value of q as in (2.1).

For a q_0 selected in $X(T,L)$ we solve (2.2), (2.3), obtaining a Lipschitz continuous solution $q(x,t)$. Of course the function $q_0'(\theta)$ is now in L_∞ and in (2.8) inf has to be replaced by ess inf.

In the region where $q(x,t)$ is defined a C^1 curve $x = s(t)$ can be determined uniquely solving (1.10). At this point we define the function $\tilde{q}(x,t)$ through the following steps.

(i) We solve
$$(3.1) \qquad H(\chi(t)) = \dot{p}_0(t) + G(q_0(t))$$

(recall (2.12)-(2.17)), noting that $F' > 0$, $G' > 0$ and consequently $H' > 0$, so that $\chi(t)$ is determined uniquely.

(ii) Imposing the condition
$$(3.2) \qquad \tilde{q}(s(t), t) = \chi(t)$$

we integrate the equation
$$(3.3) \qquad \epsilon'(q)\frac{\partial \tilde{q}}{\partial t} + \frac{\partial \tilde{q}}{\partial x} = 0, \qquad 0 < x < s(t),\ 0 < t < T,$$

which has been linearized by choosing the coefficient $\epsilon'(q)$ instead of $\epsilon'(\tilde{q})$. It is not difficult to realize the not only $\tilde{q}$ but also q is constant along the characteristics of (3.3) with the obvious consequence that such characteristics coincide with the ones of equation (2.2).

At this point we set
$$(3.4) \qquad \tilde{q}_0(t) = \tilde{q}(0, t),$$

thus defining the mapping $T : X \to C([0, T])$

$$(3.5) \qquad T q_0 = \tilde{q}_0.$$

We want to prove that for a suitable choice of T, L we have $T(X) \subset X$. The only nontrivial thing to be proved is that $\tilde{q}_0$ has a Lipschitz coefficient not greater than L.

Formally we differentiate (3.1) obtaining

$$(3.6) \qquad \frac{d\tilde{q}(s(\tau), \tau)}{d\tau} = \frac{\ddot{p}_0(\tau) + G'(q_0(\tau))\dfrac{dq_0(\tau)}{d\tau}}{H'(\tilde{q}(s(\tau), \tau))}.$$

Since (again formally)
$$(3.7) \qquad \frac{d\tilde{q}_0}{d\theta} = \frac{d\tilde{q}(s(\tau), \tau))}{d\tau}\frac{d\tau}{d\theta},$$

with θ and τ related by (2.18), and observing that we can use (2.19), combining (3.6), (3.7) we can write

$$(3.8) \qquad \frac{d\tilde{q}_0(\theta)}{d\theta} = \frac{\ddot{p}_0(\tau) + G'(q_0(\tau))\dfrac{dq_0(\tau}{d\tau}}{H'(\tilde{q}(s(\tau), \tau))}\frac{\epsilon''(q_0(\theta))\dfrac{dq_0(\theta)}{d\theta}s(\tau) + 1}{1 - \epsilon'(q(\theta))\dot{s}(\tau)}.$$

We proceed as if the functions $q_0, \tilde{q}_0$ were differentiable, transferring the bounds we will find on their derivatives to the respective Lipschitz coefficients.

Let us define the positive function $Q(t)$ via
$$(3.9) \qquad F(Q(t)) = \dot{p}_0(t),$$

remarking that, owing to (3.1), (3.2),

$$(3.10) \qquad F(\tilde{q}(s(t), t)) + G(\tilde{q}(s(t), t)) = F(Q(t)) + G(q_0(t)),$$

thus implying that $\tilde{q}(s(t), t)$ is always between $q_0(t)$ and $Q(t)$. Therefore we are lead to choose
$$(3.11) \qquad q_{\min}(T_0) = \inf_{0 < t < T_0} Q(t), \qquad q_{\max}(T_0) = \sup_{0 < t < T_0} Q(t),$$

T_0 being fixed arbitrarily (note that if $Q =$ const., i.e. $\dot{p}_0 =$ const., the problem has the trivial solution $q \equiv Q$). In this way $\tilde{q} \in [q_{min}, q_{max}]$.

Now we define

$$(3.12) \qquad L_0 = \sup_{\substack{q \in (q_{min}, q_{max}) \\ t \in (0, T_0)}} \frac{|\ddot{p}_0(t)|}{F'(q)}$$

and we take q_0 in the set $X(T, L_0)$, with $T \leq T_0$ still to be determined. Now we have

$$(3.13) \qquad \left| \ddot{p}_0(\tau) + G'(q_0(\tau)) \frac{dq_0(\tau)}{d\tau} \right| \leq L_0 H'(q_0(\tau))$$

and we can say that

$$(3.14) \qquad 0 < \frac{H'(q_0(\tau))}{H'(\tilde{q}(s(\tau), \tau))} \leq H_0(T)$$

for some constant H_0, uniformly w.r.t. the choice of q_0 in $X(T, L_0)$.

Next we note that

$$(3.15) \qquad 0 < \sigma = \frac{q_{min}}{\epsilon(q_{min})} \leq \dot{s}(\tau) \leq \frac{q_{max}}{\epsilon(q_{max})} = \Sigma$$

and, setting $E_2 = \sup_{q_{min} < q < q_{max}} \epsilon''(q)$, $e_1 = |\epsilon'(q_{max})|$,

$$0 < \frac{\epsilon''(q_0(\tau)) \frac{dq_0(\tau)}{d\tau} s(\tau) + 1}{1 - \epsilon'(q_0(\tau)) \dot{s}(\tau)} \leq \frac{E_2 L_0 T \Sigma + 1}{1 + e_1 \sigma}$$

(for T sufficiently small), so that $\hat{T}$ can be selected in such a way that

$$(3.16) \qquad \left| \frac{d\tilde{q}_0(\theta)}{d\theta} \right| \leq L_0 H_0(\hat{T}) \frac{E_2 L_0 \hat{T} \Sigma + 1}{1 + e_1 \sigma} \leq L_0,$$

since it is obvious that $\lim_{T \to 0} H_0(T) = 1$.

Thus the operator $\mathcal{T}$ maps the closed, convex and compact set $X(\hat{T}, L_0)$ into itself. In order to prove the continuity of $\mathcal{T}$ in the chosen topology, we consider two elements $q_0^{(1)}$, $q_0^{(2)}$ in $X(\hat{T}, L_0)$ and we note that, for any given θ,

$$(3.17) \qquad \tilde{q}_0^{(1)}(\theta) - \tilde{q}_0^{(2)}(\theta) = \tilde{q}^{(1)}(s^{(1)}(\tau^{(1)}), \tau) - \tilde{q}^{(2)}(s^{(2)}(\tau^{(2)}), \tau^{(2)}),$$

(the meaning of the symbols is clear). The difference on the r.h.s. coincides with

$$H^{-1}[\dot{p}_0(\tau^{(1)}) + G(q_0^{(1)}(\tau^{(1)}))] - H^{-1}[\dot{p}_0(\tau^{(2)}) + G(q_0^{(2)}(\tau^{(2)}))],$$

which is in turn estimated in terms of the norm $\|q_0^{(1)} - q_0^{(2)}\|$ and of the difference $|\tau^{(1)} - \tau^{(2)}|$. For the latter we have

$$(3.18) \qquad |\tau^{(1)}(\theta) - \tau^{(2)}(\theta)| \leq \epsilon'(q_{min}) |s^{(1)}(\tau^{(1)}) - s^{(2)}(\tau^{(2)})|.$$

Owing to the fact that both $s^{(1)}$ and $s^{(2)}$ are increasing, from (3.18) we conclude that

$$(3.19) \qquad \|\tau^{(1)} - \tau^{(2)}\| \leq C \|s^{(1)} - s^{(2)}\|$$

132

for some $C > 0$ and eventually

(3.20)
$$\|\tilde{q}_0^{(1)} - \tilde{q}_0^{(2)}\| \le C_1 T \|q_0^{(1)} - q_0^{(2)}\|,$$

which is the desived conclusion. We can also note that the operator T is contractive for small T.

So far we have proved existence in some weak sense, i.e. with q Lipschitz continuous. In order to prove that q is continuously differentiable we can slightly modify the above argument restricting the set X to the functions q_0 having Lipschitz continuous second derivative with a Lipschitz coefficient less than some M. Then, differentiating formally (3.9) w.r.t. θ, it can be shown that if T is sufficiently small and M is taken sufficiently large, then the operator T still maps the reduced set X into itself. Thus the (unique) fixed point we have found in a small neighbourhood of the origin has the required regularity. Such a regularity is automatically extended to the whole existence interval: indeed an inspection of (3.8), written for the fixed point, in connection with (2.17) reveals that any discontinuity of $\frac{dq_0}{d\theta}$ produces a sequence of discontinuity points having the origin as an accumulation point, thus contradicting the above local regularity result. □

REMARK 3.1. The preceding regularity argument can be repeated for higher order derivatives assuming more regularity on $p_0(t)$. □

From [6] we quote the continuous dependence theorem

THEOREM 3.2. *For a pair of solutions* $\{(q^{(1)}, s^{(1)}), (q^{(2)}, s^{(2)})\}$, *corresponding to the respective data* $p_0^{(1)}(t), p_0^{(2)}(t)$ *we have*

(3.21)
$$\|q_0^{(1)} - q_0^{(2)}\| \le C \|\dot{p}_0^{(1)} - \dot{p}_0^{(2)}\|$$

for some positive constant C *and in a suitabletime interval.*

4 Continuation of the Solution and some Qualitative Properties

We want to show that there are two ways in which the solution can terminate in a finite time: either $\frac{dq_0}{d\theta} \to +\infty$ (case A), or the boundary reaches the curve (2.6), (2.7) (case B).

THEOREM 4.1. *Continuation of a solution is possible to infinity, unless at some finite time* T^* *case* A *or case* B *occur.*

Proof. Suppose we have found a solution (e.g. with the method of Sect. 3) up to some time T at which the free boundary is still separated from the curve (2.6), (2.7). We know that the following equation must be valid

(4.1)
$$\frac{dq_0(\theta)}{d\theta} = \frac{\omega(\tau)\mu(\theta, \tau)}{1 - \omega\mu\epsilon''(q_0(\theta))s(\tau)},$$

where

(4.2)
$$\omega(\tau) = \frac{\ddot{p}_0(\tau) + G'(q_0(\tau))\dfrac{dq_0(\tau)}{d\tau}}{H'(q(s(\tau), \tau))},$$

(4.3)
$$\mu(\theta, \tau) = \frac{1}{1 - \epsilon'(q_0(\theta))\dot{s}(\tau)} \in (0, 1).$$

Equation (4.1) is nothing but (3.8) with $\tilde{q} = q$.

If the solution is known up to $\tau = T$, then θ reaches a value $\bar{\theta} > T$. For $\tau \in (T, \bar{\theta})$ we can write

$$(4.4) \qquad q(s(\tau), \tau) = H^{-1}(\dot{p}_0(\tau) + G(q_0(\tau))),$$

the r.h.s. being a known quantity, and with the help of (4.4) we can find the continuation of the free boundary

$$(4.5) \qquad s(\tau) = s(T) + \int_T^\tau \frac{q(s(\eta), \eta)}{\epsilon(q(s(\eta), \eta))}, d\eta.$$

Now we can construct the continuation of $q(x, t)$ by integrating (2.2) with the data (4.4), (4.5). This is equivalent to solving the first order o.d.e. (4.1) for $\theta > \bar{\theta}$. Such a procedure can be iterated to infinity, provided that we do not encounter case B, nor it happens that for some time the denominator in (4.1) tends to zero (case A). In the latter case, since for $\theta = \tau = 0$ the denominator is one, the limit of $\dfrac{dq_0}{d\theta}$ must be $+\infty$. $\qquad\square$

REMARK 4.1. Case A occurs when a branch of curve (2.6), (2.7) intersects the axis $x = 0$. Therefore both kinds of singularities have the same nature and a further continuation of the solution must include the appearance of a new boundary carrying a discontinuity of the porosity. $\qquad\square$

In [6] the cases $\ddot{p}_0 \geq 0$, $\ddot{p}_0 \leq 0$ are considered in some detail, showing in particular that $q_0(t)$ is, respectively, non-decreasing and non-increasing. Here we note that when $\ddot{p}_0 \leq 0$ we have $\omega(\tau) \leq 0$, thus excluding case A. In the opposite case ($\ddot{p}_0 \geq 0$) from (2.6) we see that case B is excluded.

A trivial example of a solution existing for all times is the steady state solution $q = q_0^*$, corresponding to $\ddot{p}_0 = 0$. It is not difficult to realize that both case A and case B can actually occur.

References

[1] J. BEAR, *Dynamics of Fluids in Porous Media*, America Elsevier, New York, 1972.

[2] A. FASANO, *Some non-standard one-dimensional filtration problems*. To appear.

[3] W. GREEN AND G. AMPT, *Studies on soil physics. The flow of air and water through soils*, J. Agric. Sci. **4**, (1911), pp. 1–24.

[4] L. PREZIOSI, D. JOSEPH, AND G. BEAVERS, *Infiltration of initially dry, deformable porous media*. To appear.

[5] K. RAJAGOPAL AND A. WINEMAN, *Developments in the mechanics of interactions between a fluid and a higly elastic solid*. To appear.

[6] P. TANI, *Fronti di saturazione in mezzi porosi con caratteristiche idrauliche variabili*. Tesi, Univ. Firenze (1994).

Clines and Material Interfaces with Nonlocal Interaction

P. C. Fife

Department of Mathematics

University of Utah

Salt Lake City, Utah 84112

To Ivar Stakgold on the occasion of his 70th birthday.

Abstract

A natural nonlocal analog of the inhomogeneous nonlinear diffusion equation $u_t = \nabla \cdot A(x)\nabla u - f(u, x)$ is the equation $u_t = \int_\Omega J(x, y - x)[u(y) - u(x)]dy - f(u, x)$, where $J > 0$. Under natural conditions on J and f, we establish here the existence of a spatially nonuniform stationary solution on the real line, satisfying given boundary conditions at $\pm\infty$, and representing a transition between two uniform states. Under additional conditions, it is unique and stable to a large range of perturbations. Because of the inhomogeneity of J and f, the solution may not be translated. Applications arise in the theories of grain boundaries and population clines.

1 Introduction

The results of this paper have bearing on continuum theories of clines in an inhomogeneous biological environment and to Allen-Cahn type theories of interfaces in inhomogeneous materials. In the first case, nonlocal migrational behavior is allowed, and in the latter, nonlocal interaction among states at different spatial locations. Our setting is one-dimensional, but when extended to higher dimensions, the results will also provide an integrodifferential analog of boundary value problems for semilinear elliptic equations (work in progress).

We begin with a description of the Allen-Cahn theory. Their original theory [3, 4, 5] was based on an order parameter $u(x, t)$ describing the relevant properties of a solid material evolving according to gradient dynamics for a free energy functional

$$(1) \qquad E[u] = \int_\Omega \left[\frac{1}{2}|\nabla u|^2 + W(u) \right] dx.$$

The second term represents the bulk free energy of the system (in their paper, it was a double well function, but in the present case it will typically have a single well for each fixed x). The first term is a penalty exacted on any spatial nonuniformity, and of course represents a local interaction in this continuum model.

The L_2-gradient dynamics for this functional yields the following evolution equation for u:

$$(2) \qquad \tau u_t = \Delta u - f(u)$$

for a positive relaxation constant τ, where $f(u) = W'(u)$.

We now make the interaction (first) part of E (1) nonlocal, replacing it by $\frac{1}{4}\int\int J(x - y)(u(x) - u(y))^2 dxdy$, where J is an even nonnegative L_1 function. Then the gradient evolution (2) is replaced by

$$(3) \qquad \tau u_t = J * u - Ku - f(u),$$

where $K = \int_\Omega J(x)dx$ and $*$ denotes convolution.

Nonlocal evolutions akin to this one have been proposed and studied before; see [14, 15, 16, 17, 18, 2, 11, 12, 19]. This specific equation was studied in [2, 9, 6].

If a basic inhomogeneity is imposed on the medium, so that its properties are described not merely by the values of u but also by the spatial position x [6], then in the above, the kernel J is a function $J(x, y - x)$, and $f = f(u, x)$. The evolution problem then is

$$(4) \qquad \tau u_t = J_x * u - K(x)u - f(u, x),$$

where we define $K(x) = \int_\Omega J(x, y - x)dy$ and $J_x * u = \int_\Omega J(x, y - x)u(y)dy$.

In all of the following, we assume

A1.

$J(x, s)$ is smooth and decreases rapidly enough as $s \to \infty$,
$J(x, s) \geq 0$,
$J(x, s) = J(x, -s)$,
$K(x) > 0$.

We specialize to the case when Ω is the real line and for some $a > 0$ the functions $J(x, s)$ and $f(u, x)$ are independent of x for $x < -a$ as well as for $x > a$. Thus the medium has a different character on the two sides of the x-interval $|x| < a$. The case $a = 0$, wherein the medium is discontinuous, is allowed, as indicated in the last section. We suppose there is a stable equilibrium of the kinetic equation $\tau u_t = -f(u, x)$ for each fixed x in these two regions, denoted for definiteness by $u = 1$ $(x > a)$ and $u = 0$ $(x < -a)$. Our purpose is to investigate the possible stationary states $u(x)$ representing a spatial transition from the state $u = 0$ to $u = 1$.

In the context of a simple model of crystalline solids, such a transition may represent an order parameter profile at the boundary between two adjoining materials with different properties but which share an order parameter field. Our prototypical example has $f(u, x) = W_u(u, x)$, where

$$(5) \qquad W(u, x) = \frac{1}{2}u^2 - \theta(x)u, \quad \theta(x) = 0 \text{ for } x < -a, \quad \theta(x) = 1 \text{ for } x > a.$$

Thus in the material on the left, W has a unique minimum at $u = 0$, and on the right it is at $u = 1$. However, our main theorem is stated in more general terms.

Inhomogeneities can also be imposed on the original Allen-Cahn model, and stationary profiles sought. This has been done, but in a population dynamical framework [7, 8, 10]. Then the transition solution may represent a cline or a spatial gradient in the population distribution of some species near the junction of two distinct habitats.

The spatial interaction described by the first two terms on the right of (4) would, in the population dynamical context, represent the tendency of the population to uniformize simply by random migration of its individuals. For example, at a local maximum of the population density, the sum of these terms would be negative, tending to drive the density downward at that point. As distinct from the model (2), the migration will generally be nonlocal in the sense that an individual can make "quick trips" over finite distances.

In this note, we establish the existence of spatially nonuniform stationary solutions of (4), typically of the form of monotone solutions which approach 0 as $x \to -\infty$ and 1 as $x \to \infty$. Their uniqueness and asymptotic stability relative to a wide class of perturbations will also be proved.

2   Comparison principles for the evolution problem

The equation (4) enjoys comparison principles very much like those in operation for (2). Some of them have been used in the context of similar equations (see most of the papers cited above following (3)). Their derivations follow along standard lines, but we review them here for reference and completeness.

We denote the right side of (4) by the (nonlinear) operator Mu. For simplicity, we set $\tau = 1$. We take Ω to be the real line R.

PROPOSITION 2.1. *Let f be continuous in x and continuously differentiable in u, uniformly for $x \in R$, and let $f(0, x) \le 0$. Let $u(x, t)$, for $0 \le t \le T$, be a bounded continuous solution of*

(6)
$$u_t \ge Mu,$$

$$u(x, 0) \ge 0,$$

continuously differentiable in t. Then $u(x, t) \ge 0$ for all $t \in [0, T]$.

Proof. For some α, β, $\epsilon > 0$ let $v(x, t) = u(x, t)e^{-\alpha t} + \epsilon(\beta t + x^2)$. Assume that v attains a negative value somewhere in the strip $\{(x, t) : t \in (0, T]\}$. Then it must attain a minimum at some finite point (x_0, t_0) in the strip, because of the term in x^2. At that point, $J_x * v - Kv > 0$ by A1. We note that

$$J_x * x^2 - Kx^2 = \int_{-\infty}^{\infty} J(x, s)s^2 ds,$$

and we choose β to be larger than this number. The inequality (6) gives a corresponding inequality for v_t. For α large enough, this last inequality implies that $v_t(x_0, t_0) > 0$, which contradicts our assumption that v has a minimum there. Hence $v(x, t) \ge 0$, and since this is true for all positive ϵ, the conclusion follows.

PROPOSITION 2.2. *Let f and u satisfy the regularity conditions of Prop. 2.1. Let $u(x, t)$ be a bounded continuous solution of (4) in the above strip, with $Mu(x, 0) \ge 0$. Then $u_t \ge 0$ in the strip. The same result holds with the inequalities reversed.*

Proof. As for example in [1], one lets $v(x, t) = u(x, t) - u(x, 0)$ and shows by Prop. 2.1 that it is nonnegative, so $u(x, t) \ge u(x, 0)$. Then for any positive δ, one sets $w(x, t) = u(x, t + \delta) - u(x, t)$ and shows it to be nonnegative, since it is so initially. Thus $u(x, t + \delta) - u(x, t) \ge 0$. Dividing by δ and passing to the limit, one obtains the conclusion.

PROPOSITION 2.3. *Let $U_0(x)$, $U_1(x)$ be bounded continuous functions satisfying $MU_0 \ge 0 \ge MU_1$, $U_0 \le U_1$. Then there exists a stationary solution $U(x)$ of (4) satisfying $U_0(x) \le U(x) \le U_1(x)$ for all x.*

Proof. Let $u_0(x, t)$ be the evolving solution of (4) with initial data U_0. By Prop. 2.2, it is monotone in t, and it can be shown with Prop. 2.1 that $u_0(x, t) \le U_1(x)$ always. Therefore $u_0(x, t)$ approaches a limit $U(x)$ pointwise. By the monotone convergence, $J_x * u_0(x, t) \to J_x * U(x)$. On a subsequence t_k depending on x, $u_t(x, t_k) \to 0$. Thus U satisfies (4).

In the following, we assume that f satisfies the hypothesis (for an example, see (5))

A2.   The function $f(u, x)$ is piecewise differentiable in x, differentiable in u uniformly for $x \in R$, and for some $a > 0$ satisfies

$$
\begin{aligned}
f(u, x) &< 0 \quad \text{for } 0 < u < 1, \ x > a; \\
f(u, x) &> 0 \quad \text{for } 0 < u < 1, \ x < -a; \\
f(0, x) \le 0 &\le f(1, x) \quad \text{for all } x; \\
f(0, x) &= 0 \quad \text{for } x < -a; \\
f(1, x) &= 0 \quad \text{for } x > a; \\
f(0, x) &< 0 \quad \text{for some } x; \quad f(1, x) > 0 \text{ for some } x.
\end{aligned}
$$

For some results, we also require

A3. The function $f(u, x)$ is differentiable in x as well as u. It and the function J satisfy

$$
\begin{aligned}
\partial f(u, x)/\partial x &\le 0, \\
s \partial J(x, s)/\partial x &\ge 0, \\
\partial f(u, x)/\partial x &< 0 \text{ for } |x| < a, \ u \in (0, 1).
\end{aligned}
$$

THEOREM 2.1. *Under assumptions A1 and A2, there exists a nonconstant solution* $u(x)$ *of*

(7) $$J_x * u - Ku - f(u, x) = 0$$

with values between 0 and 1. If in addition A3 holds, then the solution is monotone in x *and satisfies*

(8) $$u(\infty) = 1, \quad u(-\infty) = 0.$$

If $J(x, s)$ *is independent of the first variable* x, *then the solution is unique and asymptotically stable within the class of functions lying between any two of its translates.*

Proof. According to A2, the functions $U_0 = 0$, $U_1 = 1$ satisfy the hypotheses of Prop. 2.3, so we have the existence of a stationary solution $u(x)$. Moreover, U_0 and U_1 are not exact solutions (by A2), so u lies strictly between them, for at least some values of x. (It can be shown that this is true for all x.) The assumption A2 ensures that there is no constant solution between those two values, so the solutions constructed are not constant. This completes the proof of the existence statement.

Now assume A3. Let $u^*(x, t)$ satisfy (4) with $u^*(x, 0) = 0$, and let $u(x) = \lim_{t\to\infty} u^*(x, t)$. First, we shall show that $u^*(x, t)$ is monotone increasing in x for all t. Take $\tau = 1$ for simplicity. Differentiating (4), we find that the derivative $p(x, t) = u^*_x(x, t)$ satisfies

(9) $$p_t = J_x * p - Kp - f_u(u, x)p + h(x, t),$$

where

$$h(x, t) = \int J_1(x, y - x)[u(y, t) - u(x, t)]dy - f_x(x, u)$$

and $J_1(x, s) = \partial J(x, s)/\partial x$.

We have $p(x, 0) = 0$ and wish to show that p remains nonnegative. We adapt the proof of Prop. 2.1, setting $v(x, t) = p(x, t)e^{-\alpha t} + \epsilon(\beta t + x^2)$. Assume that v does not remain nonnegative. Then there will be a greatest value $t_0 \geq 0$ of t such that $v(x, t) \geq 0$ for $0 \leq t \leq t_0$, and a sequence of values $t_k \downarrow t_0$ such that $v(x, t_k) < 0$ for each k and some x depending on k. For simplicity, we take $t_0 = 0$, as the proof is similar if it is positive.

Because p is a bounded function of x for each t, we know that $v(x, t_k) > 0$ for large $|x|$, so there is a value x_k at which a negative minimum is achieved, for fixed $t = t_k$. By possibly decreasing some or all of the values t_k, we may in fact ensure that

(10) $$v_t(x_k, t_k) \leq 0.$$

We calculate

$$
\begin{aligned}
v_t &= (J_x * p - Kp - f_u(u, x)p + h(x, t))e^{\alpha t} - \alpha p e^{-\alpha t} + \epsilon\beta \\
&= J_x * v - Kv - (f_u(u, x) + \alpha)v - \\
&\quad -\epsilon\{J_x * (\beta t + x^2) - K(\beta t + x^2) - (\alpha + f_u)(\beta t + x^2)\} + he^{-\alpha t} + \epsilon\beta.
\end{aligned}
$$

At any of the points (x_k, t_k), we have $v < 0$, $J_x * v - Kv > 0$. Choose α and β so large that $\alpha + f_u(...) > 0$ and $\beta > J_x * x^2 - Kx^2 \equiv \beta_0$, as in Prop. 2.1. We then have

(11) $$v_t(x_k, t_k) > \epsilon(\beta - \beta_0) + h(x_k, t_k)e^{-\alpha t}.$$

Because of our assumptions A3, we know that $h(x, 0) \geq 0$ (in case $t_0 > 0$, the second inequality in A3 is needed at this point), so that by continuity we can ensure that this function is larger than an arbitrarily small negative number by choosing $t = t_k$ sufficiently small. Therefore from (11), $v_t(x_k, t_k) > 0$, which contradicts (10).

Therefore v remains positive, and since ϵ is arbitrary, p does as well, so that $u^*(x, t)$ is monotone increasing in x (as well as in t). Hence the limit function $u(x)$ is monotone. The limits $u(\pm\infty)$ therefore exist. Letting $|x| \to \infty$ in (7), we see that the limits must be zeros of f, which does not depend on x for large $|x|$. By A2, the only possibilities are (8).

For any $\delta > 0$, let $w(x, t)$ be the solution of (4) with initial data $w(x, 0) = u(x - \delta)$. The assumption that $f(u, x)$ is nonincreasing in x implies that $Mw(x, 0) \geq 0$, so that by Prop. 2.1, $w(x, t)$ lies, for all t, between $u(x - \delta)$ and $u(x)$. Moreover, it is monotone increasing in t so approaches a monotone stationary solution $u_1(x)$ as $t \to \infty$.

Now assume that $J = J(x - y)$ is independent of the first argument. Then $K(x)$ is a constant, which we take to be 1. Since $\int_{-\infty}^{\infty} u'(x)[J * u - u]dx = 0$ (direct verification), we know that

$$(12) \qquad \int_{-\infty}^{\infty} f(u(x), x)u'(x)dx = \int_0^1 f(u, x(u))du = 0,$$

where $x(u)$ is the inverse function to $u = u(x)$. Similarly, the same equation holds with the symbol u in the first integral replaced by u_1 and the symbol $x(u)$ in the second replaced by $x_1(u)$, where x_1 is the inverse function generated by $u_1(x)$. However, by its construction $x_1(u) \geq x(u)$. This fact, together with the third part of A3, ensure that

$$\int_0^1 f(u, x(u))du > \int_0^1 f(u, x_1(u))du$$

unless $x(u) = x_1(u)$. Therefore this last equation holds. This in turn implies that the solutions u and u_1 are the same. A similar argument was used in [7].

By a sweeping argument, this establishes uniqueness of monotone solutions $u(x)$ in the class of solutions that lie between $u(x)$ and $u(x - \delta)$. A simple extension yields the same result in the class of functions between $u(x + \delta)$ and $u(x - \delta)$. It also establishes the asymptotic stability of u relative to the class of perturbations bounded in this way. This completes the proof.

3 Clines and discontinuities

In applications to clines, it is important to include functions f typified by

$$f(u, x) = (1 - 2\theta(x))u(1 - u),$$

where the function θ is as in (5), and we require $\theta'(x) > 0$ for $|x| < a$. This function f satisfies all the hypotheses in A2 and A3 except the last one in A2. However, our results still hold; in the proof of the theorem, we simply construct the function $U_0 = U_0(x)$ so that it has nontrivial compact support in $\{x > a\}$, is nonnegative, and is sufficiently flat. The function U_1 is changed in a similar manner.

Throughout the paper, we have assumed that $a > 0$, but in fact the results also hold for $a = 0$; this case represents a discontinuity in the underlying medium, such as happens when two homogeneous materials adjoin at the point $x = 0$. To obtain these stationary solutions, one lets $u_a(x)$ be those constructed above, and takes the limit as $a \to 0$. Since all the u_a are monotone, a limit along a subsequence exists everywhere and produces a solution of (7) with $a = 0$.

We have not mentioned the regularity of the solution $u(x)$ of (7), but it may or may not be continuous, even for $a > 0$. Such phenomena are studied in [2].

Acknowledgment

I am grateful to the I. Newton Institute, University of Cambridge, where this research was performed.

References

[1] D. G. Aronson and H. F. Weinberger, Multidimensional nonlinear diffusion arising in population genetics, *Advances in Math. 30*, 33-76 (1978).

[2] Peter W. Bates, Paul C. Fife, Xiaofeng Ren, and Xuefeng Wang, Traveling waves in a convolution model for phase transitions, *Arch. Rat. Mech. Anal.*, to appear.

[3] J. W. Cahn, Theory of crystal growth and interface motion in crystalline materials, *Acta Metallurgica 8*, 554-562 (1960).

[4] J. W. Cahn and S. M. Allen, A microscopic theory for domain wall motion and its experimental verification in Fe-Al alloy domain growth kinetics, *J. de Physique 38 Colloque C7*, 51-54 (1977).

[5] J. W. Cahn and S. M. Allen, A microscopic theory for antiphase boundary motion and its application to antiphase domain coarsening, *Acta. Met. 27*, 1085-1095 (1979).

[6] P. C. Fife, An integrodifferential analog of semilinear parabolic PDE's, to appear in *Topics in Partial Differential Equations and Applications*, P. Marcellini, G. Talenti, and E. Vesentini, eds., Marcel Dekker.

[7] P. C. Fife and L. A. Peletier, Nonlinear diffusion in population genetics, *Arch. Rat. Mech. Anal. 64*, 93-109 (1977).

[8] P. C. Fife and L. A. Peletier, Clines with variable selection and migration, *Proc. Roy. Soc. Lon. B214*, 99-123 (1981).

[9] Paul C. Fife and Xuefeng Wang, A convolution model for phase transitions: the generation and propagation of internal layers in higher space dimensions, preprint.

[10] W. H. Fleming, A selection-migration model in population genetics, *Jour. Math. Biol. 2*, 219-233 (1975).

[11] M. A. Katsoulakis and P. E. Souganidis, Generalized motion by mean curvature as a macroscopic limit of stochasic Ising models with long range interactions and Glauber dynamics, *Commun. Math. Phys. 169*, 61-97 (1995).

[12] M. A. Katsoulakis and P. E. Souganidis, Interacting particle systems and generalized front propagation, *Arch. Rat. Mech.*, in press.

[13] A. de Masi, E. Orlandi, E. Presutti, and L. Triolo, Glauber evolution with Kac potentials, I. Mesoscopic and macroscopic limits, interface dynamics. Preprint.

[14] A. de Masi, T. Gobron, and E. Presutti, Travelling fronts in non local evolution equations, *Arch. Rat. Mech. Anal.*, to appear.

[15] A. de Masi, E. Orlandi, E. Presutti, and L. Triolo, Motion by curvature by scaling nonlocal evolution equations, *J. Stat. Physics 73*, 543-570 (1993).

[16] A. de Masi, E. Orlandi, E. Presutti, and L. Triolo, Stability of the interface in a model of phase separation, *Proc. Roy. Soc. Edinburgh 124A* , 1013-1022 (1994).

[17] A. de Masi, E. Orlandi, E. Presutti, and L. Triolo, Uniqueness of the instanton profile and global stability in non local evolution equations, *Rend. Math., 14*, 693-723 (1994).

[18] O. Penrose, A mean field equation of motion for the dynamic Ising model, *J. Stat. Phys. 63*, 975-986 (1991).

[19] P. E Souganidis, Interface dynamics in phase transitions, in Proceedings of the International Congress of Mathematicians 1994.

The Seamount Problem *

R. P. Gilbert[†]   Yongzhi S. Xu[‡]

Abstract

This paper deals with inverse acoustics problems in a shallow ocean. The problem we investigate is the non-homogeneities in the seabed caused by a sea-mount or some object lying on the sea-floor. This problem resembles our earlier work on the unidentified submersible problem.

1 Formulation of the problem

One of the pressing problems in underwater acoustics today is formulating the direct and inverse problems for acoustic waves in a shallow ocean with an interactive seabed. A primary reason for this interest is the desire to investigate inhomogeneities in the seabed. These inhomogeneities might be caused by objects buried in the seabed, such as submerged wreckage, or mineral deposits, etc.. Such investigations give rise to a class of problems known as inverse problems. Gilbert and Xu have investigated inverse problems in a finite depth ocean with a reflecting seabed and reported the results in a sequence of papers [5, 6, 7, 8, 9, 10, 11]. The methodology used in these papers was first to obtain an operator which produced the far field from an incident ray scattered off the target. Then the inverse problem was formulated as an extremal problem. As a first step in this direction we use the far-field representation to investigate an irregularity in the sea-bottom. Such a methodology should be applicable to an ocean with a Biot type seabed once a suitable fundamental singular solution is available. These extremal methods depend on a quick forward solver such as produced by an integral representation. We consider an ocean with a reflecting sea-bottom, where a sea-mount in a constant depth ocean as pictured in the Figure ??. The acoustic pressure, generated by a point source at the given location $\vec{x} := (x_0^1, z_0) = (x_0, y_0, z_0)$, satisfies

$$
\begin{align}
(1) \qquad & \Delta p + k^2 p = -\delta(\vec{x} - \vec{x_0}), \ \vec{x} \in R_h^3 \setminus \bar{\Omega}, \\
(2) \qquad & p = 0 \qquad \text{at } z = h \\
(3) \qquad & \frac{\partial p}{\partial z} = 0 \qquad \text{at } z = 0, \ a \leq r \leq \infty \\
(4) \qquad & \frac{\partial p}{\partial \nu} = 0 \ \text{on } \mathcal{M},
\end{align}
$$

and the out-going radiation condition. Here we assume that $k \neq \frac{(2n+1)\pi}{2h}$, $n = 0, 1, \ldots,$,

$$ R_h^3 = \{(r, \theta, z) : 0 \leq r < \infty, \ 0 \leq \theta \leq 2\pi, \ 0 \leq z \leq h\}, $$

---

*This research was supported in part by the National Science Foundation through grant BES-9402539.

[†]Department of Mathematical Sciences, University of Delaware, Newark, DE 19716.

[‡]Department of Mathematics, University of Tennessee, Chattanooga, TN 37403.

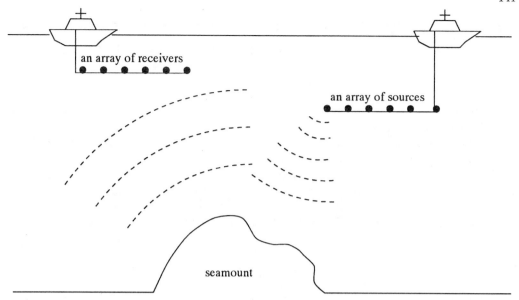

an array of receivers

an array of sources

seamount

FIG. 1. *One ship has an array of transmitters, and the other an array of receivers.*

Ω is the sea-mount, and $\mathcal{M}$ is the surface of the sea-mount,

$$\mathcal{M} := \{(r, \theta, z) : z := f(r, \theta), 0 \le \theta \le 2\pi, 0 \le r \le a\};$$

a is a positive constant that the sea-bottom is flat for $r > a$. The sea-mount is denoted by $D := \{(r, \theta, z) : 0 < z < f(r, \theta), 0 \le \theta < 2\pi, \ 0 \le r \le a\}$.

For the constant depth ocean without a sea-mount, the solution to (1), (2), (3), (4) is the Green's function for the Helmholtz equation in R_h^3, which has the form

$$(5) \qquad G(\vec{x}, \vec{x_0}) = \sum_{n=-\infty}^{\infty} (-1)^n [g(z - z_0 + 2nh) - g(z + z_0 + 2nh)],$$

where

$$(6) \qquad g(z) := \frac{\exp\{ik(\rho^2 + z^2)^{1/2}\}}{4\pi(\rho^2 + z^2)^{1/2}}, \ \rho^2 = |x' - x_0^1|^2.$$

The solution of problem (1), (2), (3), (4) can be represented as

$$(7) \qquad p(\vec{x}, \vec{x_0}) = G(\vec{x}, \vec{x_0}) + \int_{\mathcal{M}} \left\{ G(\vec{x}, \vec{y}) \frac{\partial p_{sc}(\vec{y})}{\partial \nu_y} - p_{sc}(\vec{y}) \frac{\partial G(\vec{x}, \vec{y})}{\partial \nu_y} \right\} ds_y,$$

for $\vec{x} \in R_h^3 \setminus \bar{\Omega}$; where $p_{sc}(\vec{y})$ is the unique solution of the integral equation

$$(8) \qquad p_{sc}(\vec{y}) + 2 \int_{\mathcal{M}} p_{sc}(\vec{x}) \frac{\partial G(\vec{y}, \vec{x})}{\partial \nu_y} ds_x = -2 \int_{\mathcal{M}} G(\vec{x}, \vec{y}) \frac{\partial}{\partial \nu_y} G(\vec{y}, \vec{x_0}) ds, \quad \vec{y} \in \mathcal{M}$$

and

$$(9) \qquad \frac{\partial}{\partial \nu} p_{sc}(\vec{y}) = -\frac{\partial}{\partial \nu_y} G(\vec{y}, \vec{x_0}), \ \vec{y} \in \mathcal{M}.$$

The inverse problem is: *Given $p(\vec{x}, \vec{x_0})$ for all $\vec{x} \in \Gamma$, $\Gamma := \{(r, \theta, z) : z = d = \text{constant}\}$,* and $\vec{x}_0 \in \Gamma_2$, $\Gamma_2 := \{(r, \theta, z) : z = d_2 = \text{constant}\}$, *determine the sea-mount* $\mathcal{M}$. Here we assume that Γ_1 and Γ_2 are strictly above the sea-mount, i.e., $\max_{r,\theta}\{z | z = f(r, \theta)\} < \min\{d_1, d_2\}$.

2 Uniqueness of the sea-mount problem

We assume that both Γ_1 (the receiving plane) and Γ_2 (the source location plane) are above the sea-mount. That is, $\mathcal{M}$ is disjoint with Γ_j, $j = 1, 2$. Our proof is inspired by that for uniqueness in homogeneous media, in particular by [19].

THEOREM 2.1. *Assume that D_1 and D_2 are two sea-mounts with rigid boundary $\mathcal{M}_1$ and $\mathcal{M}_2$ such that the corresponding solutions of problem (1), (2), (3), (4) coincide on Γ_1 for all $\vec{x}_0 \in \Omega$, where Ω is the unbounded component of $R_h^3 (\bar{D}_1 \cup \bar{D}_2)$, then $D_1 = D_2$.*

Proof. Suppose that $D_1 \neq D_2$. Then without loss of generality we can assume there exists $\vec{x}^* \in \partial\Omega$ such that $\vec{x}^* \in \mathcal{M}_1$ and $\vec{x}^* \notin \mathcal{M}_2$. We choose $\epsilon > 0$ such that

$$
(10) \qquad \vec{x}_n := \vec{x}^* + \frac{\epsilon}{n}\nu(\vec{x}^*), \quad (n = 1, 2, \ldots)
$$

is contained in Ω, where ν is the unit normal vector. Consider the solution $u_{n,j}$ to the problem (1), (2), (3), (4) with $\vec{x}_0$ replaced by $\vec{x}_n$ corresponding to the sea-mount D_j, $(j = 1, 2)$. By assumption,

$$
(11) \qquad u_{n,1} = u_{n,2} \quad \text{on} \quad \Gamma_1.
$$

In view of $u_{n,1} = u_{n,2} = 0$ at $z = h$, and $u_{n,j}(j = 1, 2)$ are out-going, we know that $u_n := u_{n,1} - u_{n,2} \equiv 0$ in the region between Γ_1 and the surface $z = h$. Owing to the real analyticity of solutions of the Helmholtz equation, it follows that $u_n \equiv 0$ in Ω. Hence, $u_{n,1} = u_{n,2}$ in Ω. Consider $u_n = u_{n,2}$ as the *wave* corresponding to sea-mount D_2. We know that the

$$
(12) \qquad \frac{\partial u_n}{\partial \nu} = -\frac{\partial G(\cdot, \vec{x}_n)}{\partial \nu} \quad \text{on} \ \mathcal{M}_2
$$

are uniformly bounded with respect to the maximum norm on $\mathcal{M}_2$. It follows from the continuous dependence of boundary value of exterior Neumann problem that the $\frac{\partial u_n}{\partial \nu}$ are uniformly bounded with respect to the maximum norm on closed subsets of $R_h^3 \backslash \bar{D}_2$. In particular,

$$
(13) \qquad \left| \frac{\partial u_n}{\partial \nu}(\vec{x}^*) \right| \leq C
$$

for all n and some positive constant C.

On the other hand, consider $u_n = u_{n,1}$ as the *wave* corresponding to the sea-mount D_1. From the boundary condition on $\mathcal{M}_1$,

$$
\begin{aligned}
\left| \frac{\partial u_n}{\partial \nu}(\vec{x}^*) \right| &= \left| \frac{\partial G}{\partial \nu}(\vec{x}^*, \vec{x}_n) \right| \\
&= \left| \frac{\partial}{\partial \nu} \frac{e^{ik|\vec{x}^* - \vec{x}_n|}}{4\pi|\vec{x}^* - \vec{x}_n|} + \frac{\partial}{\partial \nu}\Phi_1(\vec{x}^*, \vec{x}_n) \right| \\
&= \left| \frac{1 - ik|\vec{x}^* - \vec{x}_n|}{4\pi|\vec{x}^* - \vec{x}_n|} + \frac{\partial}{\partial \nu}\Phi_1(\vec{x}^*, \vec{x}_n) \right| \to \infty, \quad \text{as} \quad n \to \infty,
\end{aligned}
$$

where $\Phi_1(\vec{x}^*, \vec{x}_n)$ has continuous derivatives at $\vec{x}^* = \vec{x}_n$ [9]. This is a contradiction to (13). Therefore, $D_1 = D_2$. $\square$

In order to eliminate the requirement that incident waves must arrive from all directions, we need the following lemmas.

LEMMA 2.1. *Let D be a bounded domain with C^2 boundary and $R_h^3 \backslash D$ is connected. D is located strictly below Γ_1, i.e., $\max\{x|(r,\theta,z) \in \bar{D}\} < d_1$. Let $G(\cdot, \vec{x}_0)$ be the Green's function with source at $\vec{x}_0$,*

$$(14) \qquad \mathcal{H} := \Big\{ \frac{\partial G}{\partial \nu}(\cdot, \vec{x}_0) - iG(\cdot, \vec{x}_0) : x_0 \in \Gamma_1 \Big\}.$$

Then $\mathcal{H}$ is complete in $L^2(\partial D)$.

Proof. Assume $\varphi \in L^2(\partial D)$ satisfies

$$(15) \qquad \int_{\partial D} \overline{\varphi(\vec{y})} \Big\{ \frac{\partial}{\partial \nu_y} G(\vec{y}, \vec{x}_0) - iG(\vec{y}, \vec{x}_0) \Big\} ds(\vec{y}) = 0,$$

for all $\vec{x}_0 \in \Gamma_1$. Then the combined single- and double-layer potential

$$(16) \qquad u(\vec{x}) := \int_{\partial D} \overline{\varphi(\vec{y})} \Big\{ \frac{\partial}{\partial \nu_y} G(\vec{y}, \vec{x}) - iG(\vec{y}, \vec{x}) \Big\} ds(\vec{y}), \quad x \in R_h^3 \backslash \partial D,$$

satisfies the Helmholtz equation in $R_h^3 \backslash \partial D$, the out-going radiation condition as $r \to \infty$, and

$$(17) \qquad u(\vec{x})\Big|_{\Gamma_1} = u(\vec{x})\Big|_{z=h} = 0.$$

This implies that $u \equiv 0$ in $R_h^3 \backslash \bar{D}$, as discussed in the proof of Theorem 3.1.

Because of the polar singularity of

$$(18) \qquad G(\vec{x}, \vec{y}) = \frac{e^{ik|\vec{x}-\vec{y}|}}{4\pi|\vec{x}-\vec{y}|} + \bar{\Phi}_1(\vec{x}, \vec{y}),$$

where $\Phi_1(\vec{x}, \vec{y})$ is continuous at $\vec{x} = \vec{y}$, by letting $\vec{x} \to \vec{y}$, we obtain the boundary integral equation

$$(19) \qquad \varphi + \mathbf{K}\varphi - i\mathbf{S}\varphi = 0 \quad \text{on} \quad \partial D.$$

Here

$$(20) \qquad \mathbf{K}\varphi(\vec{x}) : = 2 \int_{\partial D} \frac{\partial G}{\partial \nu_y}(\vec{y}, \vec{x})\varphi(\vec{y}) ds(\vec{y}),$$

$$(21) \qquad \mathbf{S}\varphi(\vec{x}) : = 2 \int_{\partial D} G(\vec{y}, \vec{x})\varphi(\vec{y}) ds(\vec{y}).$$

The operator $\mathbf{I} + \mathbf{K} - i\mathbf{S}$ is invertible for $k \neq \frac{(2n+1)\pi}{2h}, (n = 0, 1, 2, \ldots)$ (See [9]!) and its inverse is a bounded linear operator in $L^2(\partial D)$. Hence, we have from (19) $\varphi = 0$ on ∂D and the completeness of $\mathcal{H}$ is proved.

LEMMA 2.2. *Let D be a bounded domain with C^2 boundary ∂D such that $R_h^3 \backslash \bar{D}$ is connected. D is located strictly below Γ_1. Let $u \in C^2(D) \cap C^1(\bar{D})$ be a solution of the Helmholtz equation. Then there exists a sequence v_n in the span of*

$$V := \quad \text{span} \quad \{G(\cdot, \vec{x}_0) : \vec{x}_0 \in \Gamma_1\}$$

such that

$$(22) \qquad v_n \to u, \quad \nabla v_n \to \nabla u, \quad \text{as} \quad n \to \infty,$$

uniformly on compact subsets of D.

Proof. By Lemma 2.1, there exists a sequence $\{v_n\}$ in V, such that

$$(23) \qquad \left\|\frac{\partial v_n}{\partial \nu} - iv_n - \left(\frac{\partial u}{\partial \nu} - iu\right)\right\|_{L^2(\partial D)} \to 0, \quad \text{as} \quad h \to \infty.$$

Notice that $w_n := u_n - u$ satisfies the Helmholtz equation in D and the impedance boundary condition

$$(24) \qquad \frac{\partial w_n}{\partial \nu} - iw_n = \frac{\partial v_n}{\partial \nu} - iv_n - \left(\frac{\partial u}{\partial \nu} - iu\right) \quad \text{on} \quad \partial D.$$

We represent w_n as a sequence of single-layer potentials

$$(25) \qquad w_n(\vec{x}) := \int_{\partial D} G(\vec{y}, \vec{x})\psi_n(\vec{y})ds(\vec{y}), \ \vec{x} \in D, \ n = 1, 2, \ldots$$

where the $\psi_n \in C(\partial D)$ are density functions. Letting $\vec{x}$ approach a point on ∂D, we obtain the boundary integral equations

$$(26) \qquad \psi_n + \mathbf{K}'\psi_n - i\mathbf{S}'\psi_n = 2f_n, \quad \text{on} \quad \partial D,$$

where

$$(27) \qquad \mathbf{K}'\psi_n := 2 \int_{\partial D} \frac{\partial G}{\partial \nu_x}(\vec{y}, \vec{x})\psi_n(\vec{y})ds(\vec{y})$$

and

$$(28) \qquad f_n = \frac{\partial v_n}{\partial \nu} - iv_n - \left(\frac{\partial u}{\partial \nu} - iu\right).$$

For $k \neq (2n+1)\pi/2h, n = 0, 1, 2, \ldots$, the integral operator $(\mathbf{I} + \mathbf{K}' - i\mathbf{S})$ is invertible and has a bounded inverse $(\mathbf{I} + \mathbf{K}' - i\mathbf{S})^{-1}$,

$$(29) \qquad \psi_n = 2(\mathbf{I} + \mathbf{K}' - i\mathbf{S})^{-1}f_n.$$

Hence, for $\vec{x} \in D$, we have, using the Schwarz inequality on (25),

$$(30) \quad |v_1(\vec{x}) - u(\vec{x})| = |w_n(\vec{x})| \leq 2|\partial D|^{\frac{1}{2}} \sup_{\vec{y} \in \partial D} |G(\vec{y}\vec{x})| \cdot \|(\mathbf{I} + \mathbf{K}' - i\mathbf{S})^{-1}f_n\|_{L^2(\partial D)},$$

$$|\nabla v_n(\vec{x}) - \nabla u(\vec{x})| = |\nabla w_n(\vec{x})|$$
$$(31) \qquad \leq 2|\partial D|^{\frac{1}{2}} \sup_{\vec{y} \in \partial D} \|\nabla_{\vec{x}} G(\vec{y}, \vec{x})\| \|(\mathbf{I} + \mathbf{K}' - i\mathbf{S})^{-1}f_n\|_{L^2(\partial D)}.$$

Now the L^2 convergence of (31) follows from the uniform convergence of v_n and ∇v_n to u and ∇u.

THEOREM 2.2. *Assume that D_1 and D_2 are two sea-mounts with rigid boundaries $\mathcal{M}_1$ and $\mathcal{M}_2$, such that the corresponding solutions of (1), (2), (3), (4) coincide on Γ_1 for all $x_0 \in \Gamma_2$, then $D_1 = D_2$.*

Proof. We need to prove that under the assumption of the Theorem, the solutions corresponding to D_1 and D_2 coincide on Γ_1 for all $\vec{x}_0 \in \Omega$, the unbounded component of $R_h^3\backslash(\bar{D}_1 \cup \bar{D}_2)$. First, for any $\vec{x}_0 \in \Omega$, we consider two exterior Neumann problems for out-going solutions of the Helmholtz equation

$$(32) \qquad \Delta p_j^s + k^2 p_j^s = 0, \quad \text{in} \quad R_h^3\backslash\bar{D}_j, \ (j = 1, 2),$$

with boundary conditions

(33)
$$\frac{\partial p_j^s}{\partial \nu} + \frac{\partial G}{\partial \nu}(\cdot, \vec{x}_0) = 0, \quad \text{on} \quad \partial D_j, \ (j = 1, 2).$$

We want to show that $p_1^s = p_2^s$ in Ω. We choose a bounded C^2 domain D such that $R_h^3 \backslash \bar{D}$ is connected, $\bar{D}_1 \cup \bar{D}_2 \subset D$ and $\vec{x}_0 \notin \bar{D}$. Then by Lemma 2.2, there exists a sequence $\{v_n\}$ in $V = \text{span} \{G(\cdot, \vec{x}) : \vec{x} \in \Gamma_1\}$; such that

(34)
$$\nabla v_n \to \nabla G(\cdot, \vec{x}_0), \ n \to \infty$$

uniformly on $\bar{D}_1 \cup \bar{D}_2$.

In view of that v_n are linear combinations of point-source waves from sources on Γ_1, the assumption of the Theorem follows that the corresponding solutions $v_{n,1}^s$ and $v_{n,2}^s$ for the sea-mounts D_1 and D_2 coincide in Γ_1. Using the same argument as that in the proof of Theorem (2.1) it follows that

(35)
$$v_n^s := v_{n,1}^s = v_{n,2}^s \quad \text{in} \quad \Omega.$$

Moreover,

(36)
$$\frac{\partial v_n^s}{\partial \nu} + \frac{\partial v_n}{\partial \nu} = 0 \quad \text{on} \quad \partial D_j \cap \partial \Omega, \ (j = 1, 2).$$

As a consequence of the continuous dependence of the solution to the exterior Neumann problem on the boundary condition, along with the boundary condition (32) and the convergence (34), it follows that

(37)
$$v_n^s \to p_j^s, \quad n \to \infty,$$

uniformly on compact subsets of Ω for $(j = 1, 2)$. Therefore, it must hold that $p_1^s = p_2^s$ in Ω. By Theorem 2.1, we conclude that $D_1 = D_2$. $\square$

3 A linearized algorithm for construction of sea-mount

Let us consider the following linearized algorithm to find the shape of the sea-mount. Let

(38)
$$\delta p_n = p - p_n, \text{ and } \delta f_n = f - f_n, \ (n = 0, 1, 2, \ldots),$$

where $f_0(r, \theta)$ is the initial guess for $f(r, \theta)$,

$$D_n : = \{(r, \theta, z) := 0 < z < f_n(r, \theta), 0 \le \theta \le 2\pi, 0 < r < a\}$$
$$M_n : = \{(r, \theta, z) : z := f_n(r, \theta), 0 \le \theta \le 2\pi, 0 \le r \le a\},$$

Substituting (38) into (1), (2), (3), (4) and neglecting terms of $O(\delta^2)$ and higher we have

(39) $\qquad \Delta p_n + k^2 p_n = -\delta(\vec{x} - \vec{x}_0), \quad \text{where} \vec{x} \in R_h^3 \backslash \bar{D}_n$

(40) $\qquad\qquad p_n = 0 \qquad\qquad \text{at } z = h$

(41) $\qquad\qquad \frac{\partial p_n}{\partial z} = 0 \qquad\qquad \text{at} z = 0, a \le r \le \infty$

(42) $\qquad\qquad \frac{\partial p_n}{\partial \nu} = 0 \qquad\qquad \text{on } M_n, \ (n = 0, 1, \ldots),$

146

and

(43) $$\Delta \delta p_n + k^2 \delta p_n = 0 \qquad \text{where } \vec{x} \in R_h^3 \setminus \bar{D}_n$$

(44) $$\delta p_n = 0 \qquad \text{at } z = h$$

(45) $$\frac{\partial \delta p_n}{\partial z} = 0 \qquad \text{at } z = 0, 0 \le r \le a$$

(46) $$\frac{\partial \delta p_n}{\partial \nu} = -\left(\frac{\partial^2}{\partial \nu^2} p_n\right) \delta f_n, \quad \text{on } \mathcal{M}_n.$$

We can now use single-layer potentials to obtain a relation between δp_n and δf_n in (46). Let us represent δp_n as

(47) $$\delta p_n(\vec{x}) := \int_{\mathcal{M}_n} G(\vec{x}, \vec{y}) \phi(\vec{y}) ds_y, \ \vec{x} \in R_h^3 \setminus \bar{D}_n.$$

Then $\phi(\vec{y})$ satisfies

(48) $$\phi(\vec{x}) - 2 \int_{\mathcal{M}_n} \frac{\partial G(\vec{x}, \vec{y})}{\partial \nu_x} \phi(\vec{y}) ds_y = -2 \frac{\partial^2}{\partial \nu^2} p_n \delta f_n \text{ on } \mathcal{M}_n,$$

and

(49) $$\int_{\mathcal{M}_n} G(\vec{x}, \vec{y}) \phi(\vec{y}) ds_y = \delta p_n(\vec{x}) := p(\vec{x}) - p_n(\vec{x}), \text{ for } \vec{x} \in \Gamma_1.$$

Now we can conclude an iterative algorithm for solving the inverse problem as follows:
1. Initial guess $f_0(r, \theta)$; for $n = 0$,
2. Solve for $p_n(\vec{x})$ from (43,44 45,46).
3. Let $\delta p_n(\vec{x}) = p(\vec{x}) - p_n(\vec{x})$, Solve $\phi(\vec{y}) = \phi_n(\vec{y})$ for $\vec{y} \in \mathcal{M}_n$ from (49).
4. For chosen $\epsilon_n > 0$, set

$$\delta f_n = \min \left\{ \epsilon_n, \left[\phi(\vec{x}) - 2 \int_{\mathcal{M}_n} \frac{\partial G(\vec{x}, \vec{y})}{\partial \nu_x} \phi(\vec{y}) ds_y \right] \left[-2 \frac{\partial^2}{\partial \nu^2} p_n \right]^{-1} \right\}.$$

5. Let $f_{n+1} = f_n + \delta f_n$.
Repeat Steps 2 through 5 for $(n = 1, 2, \ldots)$ solving for $p_n, \delta p_n, \phi_n, \delta f_n$ respectively until $|\delta f_n| < \epsilon$ for some chosen ϵ.

Step 3 in the above algorithm solves an ill-posed integral equation, inherited from the original ill-posedness of the inverse problem. A proper regularization method must be adapted in order to solve (49). With this in mind we first discuss some properties of the integral operators $\mathbf{T}$ and $\mathbf{T}_N$ defined by

(50) $$\mathbf{T}\phi(\vec{x}) = \int_{\mathcal{M}} G(\vec{x}, \vec{y}) \phi(\vec{y}) ds_y, \ \vec{x} \in \Gamma_1,$$

(51) $$\mathbf{T}_n \phi(\vec{x}) = \int_{\mathcal{M}_n} G(\vec{x}, \vec{y}) \phi(\vec{y}) ds_y, \ \vec{x} \in \Gamma, \ (n = 0, 1, 2, \ldots).$$

We will need the following spaces that are weighted in $x' = (x_1, x_2) \in R^2$.

$$L^{2,s}(R^2) := \left\{ u : (1 + |x'|^2)^{s/2} u \in L^2(R^2) \right\},$$

$$H^{1,s}(R^2) := \left\{ u : D^\alpha u \in L^{2,s}(R^2), |\alpha| < 1 \right\},$$

where we use the multi-index notation $\alpha = (\alpha_1, \alpha_2)$, $|\alpha| = |\alpha_1| + |\alpha_2|$ and $D^\alpha = \frac{\partial^\alpha}{\partial x_1^{\alpha_1} \partial x_2^{\alpha_2}}$; L^2 denote the space of square-integrable functions. We use $L^2(\mathcal{M})$, $H^1(\mathcal{M})$, $L^2(\mathcal{M}_n)$ and $H^1(\mathcal{M}_n)$ to denote the usual Hilbert spaces and Sobolev spaces.

In view of the normal mode expansion of $G(\vec{x}, \vec{y})$

$$(52) \quad G(\vec{x}, \vec{y}) = \frac{i}{4} \sum_{n=0}^{\infty} \sum_{m=-\infty}^{\infty} \frac{\epsilon_m \phi_n(z) \phi_n(z')}{\|\phi_n\|^2} H_m^{(1)}(k a_n r) J_m(k a_n r') \cos[m(\theta - \theta')], \quad r' < r,$$

where $\vec{x} = (r, \theta, z)$, $\vec{y} = (r', \theta', z')$ and a_n is pure imaginary except for a finite number of n's, $\epsilon_0 = 1$, and $\epsilon_m = 2$ for $m \neq 0$; we know that $G(\vec{x}, \vec{y})$ is real analytic in $\vec{x}$ for any $\vec{y} \in \mathcal{M}$; and for some constant C,

$$|G(\vec{x}, \vec{y})| < C|x'|^{-1/2},$$

$$|D^\alpha G(\vec{x}, \vec{y})| < C|x'|^{-1/2}, \quad |\alpha| \leq 2,$$

uniformly for $\vec{y} \in \mathcal{M}$ as $|x'| \to \infty$. This follows

THEOREM 3.1. *(1) The operator $\mathbf{T}$ is compact from $L^2(\mathcal{M})$ into $H^{1,-s}(\Gamma_1)$ for $s > 1/2$.*
(2) The operator $\mathbf{T}_n$ is compact from $L^2(\mathcal{M}_n)$ into $H^{1,-s}(\Gamma_1)$ for $s > 1/2$.

THEOREM 3.2. *The operator $\mathbf{T}$ is injective and has dense range provided that the mixed boundary valued problem*

$$(53) \qquad\qquad \Delta u + k^2 u = 0, \ \vec{x} \in \bar{D},$$

$$(54) \qquad\qquad u = 0 \qquad \text{on } \mathcal{M},$$

$$(55) \qquad\qquad \tfrac{\partial u}{\partial z} = 0 \qquad \text{at } z = 0, 0 \leq r \leq a,$$

$$(56)$$

has no nontrivial solution.

Proof. We first prove that $\mathbf{T}\phi = 0$ follows $\phi = 0$. Consider

$$(57) \qquad\qquad u(\vec{x}) = \int_{\mathcal{M}} G(\vec{x}, \vec{y}) \phi(\vec{y}) ds_y, \ \vec{x} \in R_h^3.$$

$u(\vec{x})$ satisfies (43) in $R_h^3 \backslash \bar{D}$, $u(\vec{x}) = 0$ for $\vec{x} \in \Gamma_1 \cup \Gamma_h$ where $\Gamma_h := \{(r, \theta, z) : z = h\}$; and $u(\vec{x})$ satisfies the out-going radiation condition. Then $u(\vec{x}) = 0$ for $\vec{x} \in \{(r, \theta, z) : d \leq z \leq h\}$; hence $u(\vec{x}) = 0$ for $\vec{x} \in R_h^3 \backslash D$.

Define

$$(58) \qquad\qquad \mathbf{K}\phi(\vec{x}) := 2 \int_{\mathcal{M}} \frac{\partial}{\partial \nu_x} G(\vec{x}, \vec{y}) \phi(\vec{y}) ds_y, \ \vec{x} \in \mathcal{M}.$$

The jump relation of the normal derivative of $u(\vec{x})$ on $\mathcal{M}$ implies

$$(59) \qquad\qquad \phi - \mathbf{K}\phi = 0, \ \text{on } \mathcal{M}.$$

A typical way similar to that in ([3], p128) follows $\phi = 0$ on $\mathcal{M}$, provided the problem (56) has no nontrivial solution. This concludes that $\mathbf{T}$ is injective.

Now we show that if $(\psi, \mathbf{T}\phi)_{L^{2,-s}(\Gamma_1)} = 0$ for all $\phi \in L^2(\mathcal{M})$, then $\psi = 0$. That is, we need to show that $\mathbf{T}^*\psi = 0$ on $\mathcal{M}$ follows $\psi = 0$ on Γ; where $\mathbf{T}^* : L^{2,-s}(\Gamma_1) \to L^2(\mathcal{M})$ is the adjoint operator of $\mathbf{T}$:

$$(60) \qquad\qquad \mathbf{T}^*\psi(\vec{y}) = \int_\Gamma G(\vec{x}, \vec{y}) \psi(\vec{x})(|x'|^2 + 1)^{-s/2} ds_x, \ \vec{y} \in \mathcal{M}.$$

148

Now consider

(61)
$$v(\vec{y}) := \int_\Gamma G(\vec{x}, \vec{y})\psi(\vec{x})(|x'|^2 + 1)^{-s/2} ds_x, \quad \vec{y} \in R_h^3.$$

$v(\vec{y})$ is a solution of the problem (56); hence $v(\vec{y}) = 0$ in Ω. But $v(\vec{y})$ satisfies the Helmholtz equation in $R_h^3 \setminus \Gamma_1$. So $v(\vec{y}) = 0$ on $\{(r, \theta, z) : 0 \leq z \leq d\}$. Define

(62)
$$\mathbf{S}\psi(\vec{y}) := 2 \int_\Gamma \frac{\partial}{\partial \nu_y} G(\vec{x}, \vec{y})\psi(\vec{x})(|x'|^2 + 1)^{-s/2} ds_x, \quad \vec{y} \in \Gamma_1.$$

The jump relation of the normal derivative of $v(\vec{y})$ on Γ_1 implies

(63)
$$\psi + \mathbf{S}\psi = 0, \quad \text{on } \Gamma_1.$$

Now we can conclude similar to the discussion for $u(\vec{x})$ that $\psi = 0$ on Γ_1. $\square$

Based on Theorem 3.3 and 3.4, we may apply the Tikhonov regularization to step 3., that is, we solve

(64)
$$\alpha\phi_\alpha + \mathbf{T}_n^*\mathbf{T}_n\phi_\alpha = \mathbf{T}_n^*(p - p_n)$$

with some regularization parameter $\alpha > 0$ instead of (49). The regularity of discrepancy principle for the Tikhonov regularization (see, for example, [3], Th. 4.16, p99) follows

THEOREM 3.3. *If $\delta p_n \in \mathbf{T}(L^2(\mathcal{M}))$, then*

(65)
$$\phi_\alpha = (\alpha\mathbf{I} + \mathbf{T}_n^*\mathbf{T}_n)^{-1}\mathbf{T}_n^*(p - p_n)$$

approaches $\mathbf{T}_n^{-1}(\delta p_n)$ as $\alpha \to 0$. $\square$

References

[1] Ahluwalia,D. and Keller, J.: *Exact and asymptotic representations of the sound field in a stratified ocean* in *Wave Propagation and Underwater Acoustics*, Lecture Notes in Physics 70, Springer, Berlin, 1977.

[2] Burton, A.J.,Miller, G.F. *The application of integral equation methods to the numerical solutions of some exterior boundary-value problems*, Proc. Royal Soc. London Ser.A, 323(1971), 201-210.

[3] Colton, David and Kress, Rainer: *Inverse Acoustic and Electromagnetic Scattering Theory*, Springer-Verlag, (1993).

[4] James Corones, Zhiming Sun:*Transient Reflection and transmission problem for fluid-saturated porous Media*, Invariant Imbedding and Inverse Problems,SIAM, Philadelphia.

[5] Gilbert, R.P. and Xu, Y.: *Starting fields and far fields in ocean acoustics*, Wave Motion 11(1989)507-524.

[6] Gilbert, R.P. and Xu, Y.: *Dense sets and the projection theorem for acoustic harmonic waves in homogeneous finite depth oceans*, Math. Methods Appl. Scis. 12(1989)69-76.

[7] Gilbert, R. P.; Xu, Y.: *The propagation problem and far-field patterns in a stratified finite-depth ocean*, Math. Methods in the Appl. Sciences, 12,(1990), 199-208.

[8] Gilbert, R.P. and Xu, Y.: *Acoustic waves and far-field patterns in two dimensional oceans with poro-elastic seabed*, Proceeding of Their IMACS Symposium on Computational Acoustics, Harvard University, Massachusetts, (1991).

[9] Gilbert, R.P., Xu, Y., and Thejll, P. *An approximation scheme for the three-dimensional scattered wave and its propagating far-field pattern in a finite depth ocean*, ZAMM, 72 (1992)10, 459-480.

[10] Gilbert, R.P. and Xu, Y.: *Generalized Herglotz functions and inverse scattering problems in finite depth oceans* in *Inverse Problems*, SIAM, (1992).

[11] Gilbert, R.P., Xu, Y.: *An inverse problem for harmonic acoustics in stratified oceans*, J. Math. Anal. Appl. 17(1) (1993), 121-137.

[12] Gilbert, R.P., Khashanah, K., and Swart, B.: *Computation of the Acoustic Modal Solutions in a Shallow Ocean with an Elastic Seabed* (to appear) Istanbul Jour. Math. (1995).

[13] Gilbert, R.P., Khashanah, K., and Xu, Y.: *An unidentified inclusion problem for a shallow ocean I: construction of the* GREEN*'s function*, in Computational Acoustics **2**(1993), 1-19.

[14] Gilbert, R.P. and Wood, D. H.:*A transmutation approach to underwater sound propagation*, Wave Motion, 8, 5, 383- 397(1986)

[15] Gilbert, R.P. and Zhongyan Lin: *On the conditions for uniqueness and existence of the solution to an acoustic inverse problem: I Theory*, J. computational Acoustics, Vol.1, No.2(1993) 229-247

[16] Gilbert, R.P. and Zhongyan Lin: *An acoustic inverse problem: Numerical experiments*, J. Computational Acoustics,(to appear 1995).

[17] Gilbert, R.P. and Zhongyan Lin: *Acoustic waves in shallow inhomogeneous ocean with a layer of sediment*, Acustica

[18] Gilbert, R.P.,Zhongyan Lin, and Buchanan, J.L.:*Acoustic waves in shallow inhomogeneous ocean with an interactive seabed*, Proceedings of the Conference on Porous Media, St-Etienne, France.

[19] Kirsch, A. and Kress, R.: *Uniqueness in inverse obstacle scattering*, Inverse Problems, 9(1993), 285-299.

[20] Schenck, H.A.: *Improved integral formulation for acoustic radiation problems*, J. Acoust. Soc. Amer., 44(1968), 41-58.

[21] Xu, Y.: *Direct and Inverse Scattering in Shallow Oceans*, Ph.D. Thesis, University of Delaware, 1990.

On the Existence of Asymmetric Minimizers for the One-Dimensional Ginzburg-Landau Model of Superconductivity

by

Stuart P. Hastings

And

William C. Troy

This paper is dedicated to Ivar Stakgold on the occasion of his seventieth birthday.

1   Introduction

A simple application of the Ginzburg-Landau theory [6] is to the one dimensional problem describing a film of superconductor material subjected to a tangential external magnetic field. In an early paper Marcus [9] assumes that all quantities are functions only of the transverse coordinate, and he shows that, in reduced variables, the energy difference between normal and superconductor states is given by

$$\mathcal{G} = \frac{1}{d} \int_{-\frac{1}{2}d}^{\frac{1}{2}d} \left[\tilde{\phi}^2(\tilde{\phi}^2 - 2) + \frac{2(\tilde{\phi}')^2}{k^2} + 2\tilde{\phi}^2\tilde{a}^2 + 2(\tilde{a}' - h_e)^2 \right] dy. \tag{1}$$

Here $\tilde{\phi}$ is the order parameter, which measures the density of superconducting electrons, and $\tilde{a}$ is the magnetic field potential. Also h_e is the external magnetic field, d is the thickness of the film, and k is the dimensionless material constant distinguishing different superconductors, i.e. $k > \frac{1}{\sqrt{2}}$ for type II super-conductors and $k < \frac{1}{\sqrt{2}}$ for type I superconductors (see also [5], [10]). The minimizer requirement, which implies that $\mathcal{G}$ be stationary with respect to general first order variations of the functions ϕ and a, leads to the boundary value problem

$$\tilde{\phi}'' = k^2 \tilde{\phi}(\tilde{\phi}^2 + \tilde{a}^2 - 1) \tag{2}$$

$$\tilde{a}'' = \tilde{\phi}^2 \tilde{a} \tag{3}$$

$$\tilde{\phi}'(\pm\frac{d}{2}) = 0 \quad , \quad \tilde{a}'(\pm\frac{d}{2}) = h_e. \tag{4}$$

It is routine to prove that $\mathcal{G}$ has a smooth minimizer satisfying (2)-(4), for any positive h_e. Further, Marcus gives arguments to show that a non-trivial minimizer of $\mathcal{G}$ also satisfies $\tilde{\phi} > 0$ in $[-d, d]$, and this is the only kind of solution we consider. A solution of (2)-(3)-(4) is called symmetric if

$$\tilde{\phi}'(0) = 0 \quad \text{and} \quad \tilde{a}(0) = 0. \tag{5}$$

It follows from (2) and (3) that, if $\tilde{\phi}'(0) = \tilde{a}(0) = 0$, then $\tilde{\phi}$ is an even function and $\tilde{a}$ is odd. If (5) is not satisfied, then the solution is called asymmetric. The existence of at least one symmetric solution has been investigated by Odeh [10], Wang and Yang [13] and also Yang [14]. See also Bolley and Helffer [1]-[4]. The possibility of asymmetric solutions was recognized by Marcus [9], who, however, conjectured that they probably did not exist. However, Odeh gave criteria for asymmetric solutions to appear by bifurcation as h_e increases, and in [4], Bolley and Helffer show that these criteria are satisfied. None of these authors, however, address the question of whether the solutions to the differential equations are global minimizers, which is crucial to the physical relevance of the solution. Numerical studies, such as the bifurcation studies

of Seydel ([11],[12]), and also more recent numerical and theoretical work of Kwong [8], predict that a range of the parameters exists for which multiple symmetric solutions, and also asymmetric solutions, are possible. In a previous paper [7] we have proved the existence of multiple symmetric solutions. Stated in terms of (2)-(4), our main results were the following:

Theorem 1 *(See [7]).*
  (i) *Let $k > \frac{1}{\sqrt{2}}$. If d is sufficiently large, then there is a range of h_e for which at least three symmetric solutions exist.*
  (ii) *Let $k \in (0, \frac{1}{\sqrt{2}})$. If d is sufficiently large, then there is a range of h_e for which at least two symmetric solutions exist.*

Our interest in this paper is in a global minimizer, not just solutions of (2)-(3)-(4). It can be shown that if the product $h_e d$ is sufficiently large, then the global minimizer of $\mathcal{G}$ is the trivial solution $\tilde{\phi} = 0, \tilde{a}(x) = h_e x$, where $\mathcal{G} = 0$. (See [8] , where the details are given for the symmetric case.) For our purposes, the following simple observation is essential.

Theorem 2 *Suppose that $h_e d < 2\sqrt{3}$. Then the global minimizer of (1) is a non-trivial solution of (2)-(3)-(4).*

Proof. In (1) we substitute the test functions $\tilde{\phi}(x) = c, a(x) = h_e x$. We find that

$$\mathcal{G} = c^2(c^2 - 2 + \frac{h_e^2 d^2}{6}).$$

The result follows by taking c sufficiently small (for a fixed $h_e d < \sqrt{12}$), giving a negative value of $\mathcal{G}$.

Our main result is the following:

Theorem 3 *There is a $\bar{k} \in (0, \frac{1}{\sqrt{2}})$ such that if $k > \bar{k}$ is fixed, then for sufficienty large d, the global minimizer of (1) is asymmetric for some interval of values of h_e.*

2 Proof of Theorem 3

At this point it is convenient to make a change of variables, to those used, for example, by Kwong in [8]. We set

$$t = \frac{2x}{d}, \ K = \frac{k^2 d^2}{4}, \ r = \frac{1}{k^2}, \ h = \frac{h_e d}{2}. \tag{6}$$

Then the problem (2)-(3)-(4) becomes

$$\phi'' = K\phi(\phi^2 + a^2 - 1), \tag{7}$$

$$a'' = rK\phi^2 a, \tag{8}$$

$$\frac{d\phi}{dt}(\pm 1) = 0, \ \frac{da}{dt}(\pm 1) = h. \tag{9}$$

(The results in [7] were stated in terms of these variables.) We assume that $\phi'(0) = a(0) = 0$, so that the solution is symmetric about $t = 0$. Thus it suffices only to consider the interval $0 \leq t \leq 1$. We take an initial condition

$$\phi(0) = \beta, \ \phi'(0) = 0, \ a(0) = 0, \ a'(0) = \alpha. \tag{10}$$

Kwong [8] proved that for each $\beta \in (0, 1)$ there is a unique $\alpha = \alpha(\beta) > 0$ such that $\phi'(1) = 0$. Also, $\alpha(\beta)$ is continuous in β. This determines a unique value of $h = a'(1)$, which we denote by $h(\beta)$. Further, $h_0 = \lim_{\beta \to 0+} h(\beta)$ exists, so h extends continuously to $[0, 1]$, with $h(1) = 0$. Our goal is to estimate the maximum of the function h in $[0, 1]$, showing that this maximum is less than $\sqrt{3}$ for sufficiently large K. By Theorem 2, this will imply the existence of asymmetric minimizers.

We will see that this is not difficult for small r. However, it is interesting to determine if the phenomenon of asymmetric minimizers extends into the range to type I superconductors, which is to say the region $r > 2$. This will require more delicate estimates.

2.1 Initial steps in estimating region of symmetric solutions

We make repeated use of the well known fact that the "energy" function

$$\frac{\phi'^2}{K} + \frac{a'^2}{rK} - \frac{\phi^4}{2} + \phi^2 - a^2\phi^2$$

is constant. From (10),

$$\frac{\phi'^2}{K} + \frac{a'^2}{rK} \equiv \frac{\alpha^2}{rK} + \beta^2 - \frac{\beta^4}{2} + \frac{\phi^4}{2} - \phi^2 + a^2\phi^2 \tag{11}$$

on $[0, 1]$. We set

$$Q(x) = \beta^2 - \frac{\beta^4}{2} + \frac{\phi(x)^4}{2} - \phi(x)^2 + a^2\phi(x)^2.$$

We recall that the function $\alpha(\beta)$ is decreasing. In [7] it was further shown that

$$1 \leq \frac{\alpha(0)}{\sqrt{K}} \leq 1 + \frac{3}{\sqrt{K}}.$$

Thus, it follows from (11) that

$$\frac{a'(1)^2}{K} \leq 1 + O(1/K) + rQ(1) \tag{12}$$

as $K \to \infty$, uniformly in β. We distinguish two cases, according to whether $Q(1) \leq 1$ or (ii) $Q(1) > 1$.

Let

$$h_{sym} = max_{0 \leq \beta \leq 1} h(\beta).$$

If $Q(1) \leq 1$, we get from (12) that $\frac{a'(1)^2}{K} \leq 1 + r + O(1/K)$ for large K. This gives the desired inequality $h_{sym} < \sqrt{3}\sqrt{K}$ for large K if $r < 2$.

If $Q(1) > 1$, then there is a largest $x_0 \in (0,1)$ such that $Q(x_0) = 1$, and $Q > 1$ on $(x_0, 1]$. From (11),

$$a'(x_0) < \sqrt{rK} + o(\sqrt{K})$$

as $K \to \infty$. Also, $Q \geq 1$ implies that $a > 1$, so $\phi'' > 0$ on $[x_0, 1]$, and we can consider $a'(x)$ as a function of $\phi'(x)$. From (7)-(8),

$$\frac{da'(x)}{d\phi'(x)} = \frac{ra\phi}{\phi^2 + a^2 - 1}. \tag{13}$$

Further, on $[x_0, 1]$, because $Q \geq 1$, we find that

$$\phi^2 + a^2 - 1 \geq \sqrt{(1 - a^2)^2 + 1} \equiv D(a). \tag{14}$$

Next, we multiply (7) by ϕ', integrate, and obtain $\phi'(x) \geq -\sqrt{K}$ on $[0, 1]$. Using this at x_0, with (13) and (14), gives $a'(1) \leq \sqrt{rK} + \sqrt{K}\Gamma r$, where

$$\Gamma = \max_{a \geq 0} \frac{a}{\sqrt{(1 - a^2)^2 + 1}}. \tag{15}$$

This shows that $h_{sym} < \sqrt{3}\sqrt{K}$ for small r.

2.2 Extending the range of r.

We extend the proof in the previous section. We will make use of (11), (13), and (14), and we find it necessary to do some detailed numerical computations. Our goal is to prove that

$$\limsup_{K \to \infty} \{ \sup_{0 \leq \beta < 1} \frac{a'(1, \alpha(\beta), \beta)}{\sqrt{K}} \} < \sqrt{3}. \tag{16}$$

We recall some additional facts from [7] which hold for any solution of (7)-(8) such that $\phi \in (0, 1]$, $\phi' < 0$, $a' > 0$, $a'' > 0$ on $(0, 1)$, and $\phi'(1) = 0$. For such a solution, let $H(x) = \phi^2 + a^2 - 1$. There is a unique $\hat{x} \in (0, 1)$ such that $H(\hat{x}) = 0$, and $H' > 0$, $H'' > 0$ on $(\hat{x}, 1]$. Also, if $s = \frac{a'}{a}$, then $s' < 0$ and $s \geq \sqrt{rK}\phi$ on $(0,1]$. Finally, $-\beta\sqrt{K} \leq \phi' < 0$ on $(0,1)$.

Lemma 1 *Let $\{K_N\}$ and $\{\beta_N\} \subseteq (0, 1)$ be positive sequences, with $\lim_{N \to \infty} K_N = \infty$.*
  (i) If $\lim_{N \to \infty} \sqrt{K_N}\phi(1, \alpha(\beta_N), \beta_N) = 0$ then $\limsup_{N \to \infty} \frac{a'(1)}{\sqrt{K_N}} \leq \sqrt{1 + \frac{r}{2}}$.
  (ii) If there is a $\delta > 0$ such that

$$\sqrt{K_N}\phi(1, \alpha(\beta_N), \beta_N) \geq \delta$$

for sufficiently large N, then

$$\lim_{N \to \infty} \frac{\alpha(\beta_N)}{\sqrt{K_N}} = 0. \tag{17}$$

Proof: In [7] we saw that $\frac{\alpha(\beta_N)^2}{K_N} \leq 1 + \frac{3}{K_N}$ for all N, and using (11) we quickly obtain (i). For (ii), if we have $\alpha(\beta_N) \geq \mu\sqrt{K_N} > 0$, where μ is independent of N, for infinitely many N's, then $a' > \mu\sqrt{K_N}$, $a \geq \mu\sqrt{K_N}x$ on $(0,1]$. Hence $a(x) \geq 1$ for $x \geq \frac{1}{\mu\sqrt{K_N}}$. Integrating gives $a \geq 1 + \mu\sqrt{K_N}(x - \frac{1}{\mu\sqrt{K_N}})$. Substituting this into (7), using the hypothesis of (ii), and integrating, leads to the conclusion that $\phi'(1) > 0$, which is impossible. This proves Lemma 1.

We wish to prove (16) for an interval $(0, \bar{r}]$ of r's with $\bar{r} > 2$. For definiteness, we take $\bar{r} = 2.01$. (It is not hard to show that (16) does not hold for large r.) Assume now that for each K, β is chosen so that $a'(1)$ is as large as possible.

If (16) is false, then there must be a sequence $\{K_N\}$ tending to infinity such that $\lim_{N \to \infty} \frac{a'(1)}{\sqrt{K_N}} \geq \sqrt{3}$. We will assume from now on that all our solutions come from the corresponding solutions of (7)-(9). We see that in case (i) of Lemma 1, (16) holds if $r < 4$. Now suppose that case (ii) holds for some small $\delta > 0$. We therefore have (17). If we have a subsequence of $\{K_N\}$ for which $Q(1) \leq 1$, then along this subsequence it follows from (11) and (17) that $\frac{a'(1)^2}{K} \leq r + O(\frac{1}{K_N})$, and we have (16) if $r < 3$. We therefore will assume that $Q(1) > 1$, and define x_0 as above, so that $Q(x_0) = 1$ and $Q > 1$ on $(x_0, 1]$. Also, as above, $\hat{x}$ will denote the unique zero of ϕ''.

In the next key result we again assume that a' is a function of ϕ'. As noted above, this is valid in $[\hat{x}, 1]$:

Lemma 2 *Suppose that for some $\lambda > 0$, $\frac{da'}{d\phi'} \leq \lambda$ on $[x_0, 1]$.*
  (i) *If, for some $P > 0$, $(\lambda^2 + r)(r + \frac{\alpha^2}{K}) < P^2 r$, then*

$$\frac{a'(1)}{\sqrt{K}} \leq P. \tag{18}$$

  (ii) *If $P > 0$, $(\lambda^2 + r)(r + \frac{\alpha^2}{K}) \geq P^2 r$, and*

$$G(r, \lambda, \alpha) \equiv \frac{-P\lambda + \sqrt{(r + \frac{\alpha^2}{K})(r + \lambda^2) - P^2 r}}{\lambda^2 + r} \leq \frac{\phi'(x_0)}{\sqrt{K}} < 0, \tag{19}$$

then once again, (18) holds.

Proof: Recall that $Q(x_0) = 1$. Integrating $\frac{da'}{d\phi'} \leq \lambda$ on $[x_0, 1]$ and using (11) gives

$$\frac{a'(1)}{\sqrt{K}} \leq \sqrt{r + \frac{\alpha^2}{K} - r\frac{\phi'(x_0)^2}{K}} - \lambda\frac{\phi'(x_0)}{\sqrt{K}}.$$

We then find that (18) holds if

$$(\lambda^2 + r)\frac{\phi'(x_0)^2}{K} + 2P\lambda\frac{\phi'(x_0)}{\sqrt{K}} + P^2 - r - \frac{\alpha^2}{K} \geq 0.$$

Both (i) and (ii) in our Lemma follow from the last inequality.

We now consider three different ranges of β, which are chosen somewhat arbitrarily:

$$0 < \beta \le .1, \quad .1 < \beta \le 1 - \frac{1}{\sqrt{K}}, \quad \text{and} \quad 1 - \frac{1}{\sqrt{K}} < \beta \le 1.$$

Of these, the most difficult is the last, and so we now turn our attention to that situation. We apply Lemma 2 with P chosen slightly less than $\sqrt{3}$. We choose $P = 1.7$ for convenience. From (13) and (14) we see that $\frac{da'}{d\phi'} \le r$. Suppose that case (ii) of Lemma 2 holds, as it will if $r = \bar{r}$. Then (17) and Lemma 2 (ii) imply (18) for sufficiently large $K = K_N$ if $\frac{\phi'(x_0)}{\sqrt{K}} \ge -.14$. Hence we assume that for large N, $\frac{\phi'(x_0)}{\sqrt{K}} < -.14$. As in [7], we use the inequality $\phi'' \le K\phi(\phi^2 - 1)$, multiplying this by ϕ', which is negative, and integrating over $[0, x_0]$. From this and our assumption that $\beta > 1 - \frac{1}{\sqrt{K}}$, we find that $\phi(x_0) \le .9$ for large K.

Now use (11) and (12) again to conclude that $\frac{da'}{d\phi'} \le \lambda = 1.1\, r\phi(x_0)$. This allows us to obtain (16) from (17) and part (i) of Lemma 2 if $\phi(x_0) \le 0.407$. There remains to consider the interval $.407 < \phi(x_0) \le .9$. We now suppose that $\sigma \in (0, \sqrt{2})$ is chosen independent of K or β.

Lemma 3 *For sufficiently large K, and any $\beta \in [1 - \frac{1}{\sqrt{K}}, 1]$,*

$$\frac{a'(\hat{x})}{a(\hat{x})} \le \sqrt{rK(1 + \frac{\sigma}{\sqrt{K}}\phi'(\hat{x}))}, \quad \text{and} \quad \phi(\hat{x}) \le \sqrt{1 + \frac{\sigma}{\sqrt{K}}\phi'(\hat{x})}.$$

Proof: At $\hat{x}$, $a^2 = 1 - \phi^2$. Also, $\sup_{1 - \frac{1}{\sqrt{K}} \le \beta \le 1} \alpha(\beta) \to 0$ as $K \to \infty$. Next, solve (11) for $a'(\hat{x})^2$, divide both sides of the resulting expression by $a(\hat{x})^2 = 1 - \phi^2(\hat{x})$, and find the maximum value of the resulting right hand side for given $\phi'(\hat{x}) < 0$ over $0 < \phi(\hat{x}) < 1$. This leads to the first inequality. Finally, we multiply (7) by ϕ', integrate, use $\phi' < 0$ and $\beta \to 1$ as $K \to \infty$ to arrive at the second inequality.

Lemma 4 *Suppose that for some $\eta > 0$ and $\mu > 0$, $\frac{\phi'(x_0)}{\sqrt{K}} \le -\eta$ and $\frac{a'(\hat{x})}{a(\hat{x})} \le \mu\sqrt{K}$. Then*

$$T(\phi(\hat{x})) \ge S(\phi(x_0)), \tag{20}$$

where $T(\phi) = \phi + \frac{\eta}{2\mu}\ln(1 - \phi^2)$ and $S(\phi) = \phi + \frac{\eta}{2\mu}\ln(\frac{1}{2\phi^2} + 1 - \frac{\phi^2}{2})$.

Proof: Recall that $\frac{a'}{a}$ is decreasing, so that $\frac{a'}{a} \le \mu\sqrt{K}$ in $[\hat{x}, x_0]$. From this it follows that $x_0 - \hat{x} \ge \frac{1}{\mu\sqrt{K}}\ln(\frac{a(x_0)}{a(\hat{x})})$. Since $\phi'' > 0$ on $[\hat{x}, x_0]$, we have $\phi' < -\eta\sqrt{K}$ on this interval, and an integration leads to

$$\phi(x_0) \le \phi(\hat{x}) - \eta\sqrt{K}(x_0 - \hat{x}) \le \phi(\hat{x}) - \frac{\eta}{\mu}\ln(\frac{a(x_0)}{a(\hat{x})}). \tag{21}$$

Further, from $Q(x_0) = 1$ we obtain that

$$a(x_0)^2 \ge F(\phi(x_0)) \equiv \frac{1}{2\phi(x_0)^2} + 1 - \frac{\phi(x_0)^2}{2}. \tag{22}$$

Then (20) follows from (21) and the equality $a(\hat{x}) = \sqrt{1 - \phi(\hat{x})^2}$. This proves Lemma 4.

We now outline the rest of the proof of Theorem 3. We choose a specific bound ϵ for $\frac{\alpha}{\sqrt{K}}$, assume that (19) does not hold, and get a contradiction to (20) for K so large that $\frac{\alpha^2}{K} < \epsilon^2$. For definiteness, take $\epsilon = .01$. The η and μ in Lemma 4 are constant in each of a finite set of subintervals $[\phi_a, \phi_b]$ of the range $.407 \leq \phi(x_0) \leq .9$. For each of these subintervals we will see that S is increasing in $[\phi_a, \phi_b]$ for the given choice of η and μ, giving the bound $S(\phi(x_0)) \geq S(\phi_a)$. Our contradiction will come by using Lemma 3 to show that if $\phi(x_0) \in [\phi_a, \phi_b]$, then $T(\phi(\hat{x})) < S(\phi_a)$.

To carry out this plan we will use the functions D, F, and G defined in (14), (22), and (19), respectively. From (22) we obtain that $a(x_0)^2 \geq F(\phi_b)$. Note that the function $\frac{a}{D(a)}$ has its maximum, Γ in (15) , at $a = 2^{1/4}$ and decreases for $a \geq 2^{1/4}$. Now let

$$Z(r, \phi) = \begin{cases} \frac{r\phi\sqrt{F(\phi)}}{D(\sqrt{F(\phi)})} & \text{if } F(\phi) \geq \sqrt{2}, \\ \Gamma r\phi & \text{otherwise .} \end{cases}$$

Also, let

$$\nu(r, \phi, \alpha) = G(r, Z(r, \phi), \alpha) - \frac{.05}{\sqrt{rF(.400)}} .$$

Lemma 5 *If* $\frac{\phi'(x_0)}{\sqrt{K}} \leq G(r, Z(r, \phi_b), \alpha)$, *then* $\frac{\phi'(\hat{x})}{\sqrt{K}} \leq \nu(r, \phi_b, \alpha)$.

Proof: We need a lower bound for $x_0 - \hat{x}$. Recall that $H(\hat{x}) = 0, H(x_0) \geq 1, H'' > 0$, and $H' > 0$ on $(\hat{x}, x_0]$. Thus there is a unique $\tilde{x} \in (\hat{x}, x_0)$ such that $H(\tilde{x}) = \frac{1}{2}$. On $[\tilde{x}, x_0]$, $F(\phi) < F(.4)$ and $\frac{a'}{a} \leq \sqrt{rK}$, using Lemma 3. Further, for sufficiently large K,

$$H' = 2aa' + 2\phi\phi' \leq H'(x_0) \leq 2a(x_0)a'(x_0) \leq 2\sqrt{rK}a(x_0)^2 \leq 2F(.4)\sqrt{rK},$$

because $\phi(x_0) \geq .407, \beta \geq 1 - \frac{1}{\sqrt{K}}$, and $Q(x_0) = 1$. An integration gives $x_0 - \tilde{x} \geq (4F(.4)\sqrt{rK})^{-1}$. Also, on this interval, $\phi'' = K\phi H \geq K\phi/2 \geq .2K$. The conclusion of Lemma 5 follows.

We now let $\eta = -G(r, Z(r, \phi_b), \alpha)$ and $\mu = \sqrt{r}\sqrt{1 - \sigma\eta}$, so that from Lemma 3, the hypotheses of Lemma 4 hold. The function T in Lemma 4 is increasing in the interval $I = [0, -\frac{\eta}{2\mu} + \sqrt{1 + (\frac{\eta}{2\mu})^2}]$. In each of the subintervals $\phi_a \leq \phi(x_0) \leq \phi_b$, we must check that $\sqrt{1 + \sigma\nu(r, \phi_b, \alpha)}$ lies in I. This implies, from Lemmas 3 and 5, that $T(\phi(\hat{x})) \leq T(\sqrt{1 + \sigma\nu(r, \phi_b, \alpha)})$.

Hence, we have a contradiction of Lemma 4 if $T(\sqrt{1 + \sigma\nu(r, \phi_b, \alpha)}) < S(\phi_a)$. The proof is completed by choosing a specific σ, say $\sigma = 1.4141$, a subdivision of $[.407, .9]$ into subintervals and checking this inequality in each subinterval. It turns out that seven subintervals, $[.407, .59]$, $[.59, .64]$, $[.64, .67]$, $[.67, .70]$, $[.70, .74]$, $[.74, .8]$, and $[.80, .9]$, are sufficient. We omit these details for the sake of brevity.

There remain to discuss the other two intervals of β described above, that is, $0 < \beta \leq .1$ and $.1 < \beta \leq 1 - \frac{1}{\sqrt{K}}$. For the first, we use (13)-(15) to estimate λ. Assuming that $Q(1) > 1$, and defining x_0 as before, we have $\phi(x_0) \leq .1$ so that $\lambda < .17$. Then the conditions of Lemma 2, part (i) hold for large N with $P = 1.7$, giving (16). The case $.1 < \beta \leq 1 - \frac{1}{\sqrt{K}}$ hinges on

Lemma 6

$$\lim_{K \to \infty} \left(\sup_{.1 \le \beta \le 1 - \frac{1}{\sqrt{K}}} \{\sqrt{K}\phi(1, \alpha(\beta), \beta)\} \right) = 0.$$

Proof:. Suppose that for some $\nu > 0$, $\sqrt{K}\phi(1, \alpha(\beta), \beta) \ge \nu$ for large K (or for a sequence of $K's$ tending to infinity). We claim that for small $\delta > 0$ and large K, $\phi(\delta) \ge \delta$. If not, then there is a unique $x_\delta \in (0, \delta)$ such that $\phi(x_\delta) = \delta$. By the mean value theorem there is a $\gamma_\delta \in (0, x_\delta)$ such tht $\phi'(\gamma_\delta) = \frac{\phi(x_\delta) - \phi(0)}{x_\delta} \le -\frac{1}{20\delta}$, for small $\delta > 0$. If $\hat{x} \ge 2\delta$ then $\phi' \le -\frac{1}{20\delta}$ over $[\delta, 2\delta]$ and an integration gives $\phi(2\delta) < 0$ for small δ, a contradiction. If $\hat{x} \le 2\delta$, then $\phi''(2\delta) \ge 0$ so that $a(2\delta) \ge \sqrt{1 - 4\delta^2}$ for small $\delta > 0$. This and the mean value theorem imply that $a'(\zeta) \ge \frac{1}{3\delta}$ for some $\zeta \in (0, 2\delta)$. Thus $a' \ge \frac{1}{3\delta}$ for $x \ge 2\delta$, since $a'' > 0$, and an integration shows that $a(3\delta) > 1$. For $x \ge 3\delta$ a further integration leads to $\phi'' \ge \frac{\sqrt{K}\nu}{9\delta^2}(x - 3\delta)^2$, since $\sqrt{K}\phi(1) \ge \nu$. A final integration shows that $\phi'(1) > 0$, a contradiction which proves our contention that $\phi(\delta) \ge \delta$. Next, we obtain a lower bound for a at a fixed point in $(0, \delta)$. We choose the point $\frac{\delta}{10}$, and suppose that $a(\frac{\delta}{10})^2 < \frac{1}{2\sqrt{K}}$. Then integrate (7), using $\phi \ge \delta$ in $[0, \delta]$, to obtain $\phi(\frac{\delta}{10}) < 0$ for large K, a contradiction. Hence we obtain $a(\frac{\delta}{10}) \ge \frac{1}{\sqrt{2}} K^{-\frac{1}{4}}$. Further integrations of (7) again lead to $\phi'(\delta) > 0$; hence $\phi'(1) > 0$, a contradiction. This completes the proof.

References

[1] Bolley, C. and Helffer, B., *Rigorous results for the G.L. equations associated to a superconducting film in the weak κ limit*, Preprint Ecole Centrale de Nantes (May 1384).

[2] Bolley, C., *Modélisation du champ de retard à la condensation d'un supraconducteur par un probléme de bifurcation M^2 AN*, Vol. 26, n°2, p. 175-287, 1392.

[3] Bolley, C. and Helffer, B., *An application of semi-classical analysis to the asymptotic study of the supercooling field of a superconducting material*, Ann. Inst. Henri Poincaré, Phys. Theéorique **58**, n°2 (1393), 189-233.

[4] Bolley, C. and Helffer, B., *Rigorous results on G.L. models in a film submitted to an exterior parallel magnetic field*, Preprint, Ecole Centrale de Nantes (April 1393).

[5] Chapman, S.J., Howison, S.D., and Ockendon, J.R., *Macroscopic models for superconductivity*, SIAM Review **34** (1392), 529-560.

[6] Ginzburg, V. L. and Landau, L. D., *On the theory of superconductors*, Soviet Phys., JETP **20** (1950), 1064.

[7] Hastings, S.P., Kwong, M.K., and Troy, W.C., *The existence of multiple solutions for a Ginzburg-Landau type model of superconductivity*, European J. Appl. Math., to appear.

[8] Kwong, M.K., *On the one-dimensional Ginzburg-Landau BVPs*, Journal of Differential and Integral Equations **8** (1995), 1395-1405.

158

[9] Marcus, P., *Exact solutions of the Ginzburg-Landau equations for slabs in tangential magnetic fields*, Reviews of Modern Physics (1364), 294-299.

[10] Odeh, F., *Existence and bifurcation theorems for the Ginzburg-Landau equations*, J. Math. Phys. **8** (1367), 4351.

[11] Seydel, R., *From equilibrium to chaos; Practical bifurcation and stability analysis*, Elsevier Publ. Co., 1388.

[12] Seydel, R., *Branch switching in bifurcation problems in ordinary differential equations*, Numer. Math. **41** (1383), 93-116.

[13] Wang, S., and Yang, Y., *Symmetric superconducting states in thin films*, SIAM J. Appl. Math. **52** (1392), 614-629.

[14] Yang, Y., *Boundary value problems of the Ginzburg-Landau equations*, Proc. Roy. Soc. of Edin. **114A** (1990), 355-365.

Applications of Boundary Element Methods to a Class of Nonlinear Boundary Value Problems in Mechanics

George C. Hsiao[*]

Abstract

This paper is concerned with a class of nonlinear problems in mechanics which can be treated by boundary element methods. This class includes both the contact and the elasto-plastic interface problems in elasticity. Emphasis will be placed upon the variational formulations, and mathematical foundations of the solution procedures. An iterative scheme for strongly monotone operator equations is introduced, and some numerical results are presented for a nonlinear problem arising from irrotational compressible flow past obstacles.

1 Introduction

Boundary element methods (BEMs) are generally methods for treating problems with linear differential equations. The essence of the method hinges on the existence of fundamental solutions (or more generally parametrices) of the associated linear differential operators. Nevertheless, there are various nonlinear problems to which BEMs are applicable from both theoretical and computational points of view. The purpose of this paper is to explore the applicability of the BEM to nonlinear boundary-value problems in mechanics. Through a collection of simple model problems, we discuss the basic common features of the BEM approaches. Hopefully, these discussions may shed some light and provide systematic guidelines for newcomers to the field.

In general, to treat nonlinear problems by the BEM depends upon the nature of the nonlinearity involved as well as where the nonlinearity appears. Examining the existing literature in the field, we can classify them into three groups of problems: (a) semilinear equation problems, (b) nonlinear boundary condition problems, and (c) linear-nonlinear interface problems. Problems in group (a) are most commonly handled by putting the nonlinear term on the right-hand side of the equation as a given forcing term. Then based on some kind of iterative method (e.g., the monotone iteration scheme [32]), the corresponding problem may be treated in the same way as a linear problem by BEMs with a Newtonian potential over the domain. In this paper, we shall not discuss this group of problems; for these problems we do not believe that BEMs are the most effective methods, except when the domain is unbounded. (For those readers with a genuine interest, we refer them to [14], [26], [9], [6] and the recent paper [25].) In the next two sections, we shall concentrate only on the problems in groups (b) and (c).

As will be seen, in some cases, the nonlinear problems can be reduced to operator equations of the strongly monotone type. Hence in Section 4, we will present an iterative scheme. This scheme is particularly designed for this kind of nonlinear problem. To

---

[*]Department of Mathematical Sciences, University of Delaware, Newark, DE. 19716-2553, USA.

160

conclude the paper, in Section 4, we present some numerical results for the nonlinear problem arising from irrotational compressible flow past obstacles. These computations are based on the iterative scheme introduced in the previous section, and may serve as the purpose for demonstrating the effectiveness of the scheme.

2 Nonlinear Boundary Condition Problems

This class of problems consists of linear differential equations but with nonlinear boundary conditions. To simplify the presentation, we further characterize the problems according to the nonlinearity: (i) nonlinear Robin conditions—this leads to nonlinear boundary integral equations of the Hammerstein type with compact perturbation and strongly monotone operator equations, (ii) boundary conditions of the Signorini type with or without friction—this leads to boundary variational inequalities and the Yosida approximations, and (iii) boundary optimization—this leads to shape optimal control, free boundary and inverse problems. In the following, we shall confine ourselves only to (i) and (ii). For (iii), we refer to [1] for optimal design problems in ship hydrodynamics, [17] for inverse scattering of elastic waves, and the monograph by [4] for inverse acoustic problems in general.

In the following, let Ω be a bounded domain in $I\!R^3$ with smooth boundary Γ. We consider the Lamé equation for the displacement field $\mathbf{u}$ in linear elasticity

$$(1) \qquad -\Delta^*\mathbf{u} := -\mu\Delta\mathbf{u} - (\lambda+\mu)\nabla\ div\ \mathbf{u} = \mathbf{f} \quad \text{in} \quad \Omega.$$

Here μ and λ are the given Lamé constants for the isotropic material and they are subject to the restrictions, $\mu > 0$ and $3\lambda + 2\mu > 0$; $\mathbf{f}$ is a given body force. For (1), we have the fundamental tensor given explicitly by

$$(2) \qquad \mathbf{E}(x,y) = \frac{\lambda+3\mu}{8\pi\mu(\lambda+2\mu)}\left\{\frac{1}{|x-y|}\mathbf{I} + \left(\frac{\lambda+\mu}{\lambda+3\mu}\right)\frac{1}{|x-y|^3}(x-y)(x-y)^t\right\},$$

where the superscript ^{t} denotes the operation of taking the transpose of the vector or matrix. Before we consider various nonlinear boundary conditions, let us also introduce here the stress operator, stress and strain tensors, respectively

$$(3) \quad \mathbf{T}_n[\mathbf{u}]|_\Gamma := \sigma(\mathbf{u})\cdot\mathbf{n}|_\Gamma, \quad \sigma(\mathbf{u}) := \lambda\ div\ \mathbf{u}\ \mathbf{I} + 2\ \mu\ \epsilon(\mathbf{u}), \quad \epsilon(\mathbf{u}) := \frac{1}{2}(\nabla\mathbf{u}+\nabla\mathbf{u}^t).$$

Here and in the equal, $\mathbf{n}$ denotes the unit normal vector directed to the exterior of Ω.

2.1 Nonlinear Robin Conditions

We begin with the nonlinear problem consisting of (1) together with the nonlinear Robin condition

$$(4) \qquad \mathbf{T}_n[\mathbf{u}] + \phi(x,\mathbf{u}) = \mathbf{g}(x) \quad \text{on} \quad \Gamma$$

for given functions ϕ and $\mathbf{g}$ satisfying some regularity conditions to be specified. To reduce the boundary value problem to an equivalent boundary integral equation (BIE), we use the direct method based on the Betti representation formula

$$(5) \qquad \mathbf{u}(x) := \int_\Gamma \mathbf{E}(x,y)\tau(y)ds_y - \int_\Gamma (\mathbf{T}_{n_y}[\mathbf{E}(x,y)])^t v(y)ds_y + \mathbf{u}_f, \quad x\in\Omega$$

in terms of the Cauchy data $\tau = \mathbf{T}_n[\mathbf{u}]|_\Gamma$ and $v = \mathbf{u}|_\Gamma$, where $\mathbf{u}_f$ is the particular solution of (1) given by the Newtonian potential $\mathbf{u}_f = \int_\Omega \mathbf{E}(x,y)\mathbf{f}(y)dy$. Then, carrying the limit transition to the surface points for the stress operator, we arrive at the BIE of the form

$$(6) \qquad V\tau - (\tfrac{1}{2}I + K)v = -\mathbf{u}_f|_\Gamma \quad \text{on} \quad \Gamma.$$

The boundary integral operators V and K are respectively the weakly singular and singular operators given by

$$(7) \qquad V\tau(x) := \int_\Gamma \mathbf{E}(x,y)\tau(y)ds_y, \quad Kv(x) := \int_\Gamma (\mathbf{T}_{n_y}[\mathbf{E}(x,y)])^t v(y)ds_y, \quad x \in \Gamma.$$

Then from the boundary condition (4), we obtain the nonlinear BIE for v

$$(8) \qquad A(v) := Tv + \Phi(v) = \mathbf{g} + V^{-1}(\mathbf{u}_f|_\Gamma) \quad \text{on} \quad \Gamma.$$

Here $T := V^{-1}(\tfrac{1}{2}I + K)$ is the Steklov-Poincaré operator and Φ is the Nemyckiĭ operator defined by

$$\Phi(v)(x) := \phi(x, v(x)), \ x \in \Gamma.$$

In the formulation of (8), we have tacitly employed the fact that V is invertible in the underlying space under consideration. Indeed, in the present case, it can be shown that $V : H^s(\Gamma) \to H^{s-1}(\Gamma)$ is an isomorphism for any $s \in R$, and is $H^{1/2}(\Gamma) - elliptic$ ([27], [5]).

Existence, uniqueness as well as error estimates for approximate solutions of (8) may be established in the same way as for the linear equations, provided that A is Lipschitz continuous and strongly monotone on some appropriate Sobolev space. These, of course, require the nonlinear function ϕ to fulfill certain conditions. Note that the Steklov-Poincaré operator T has rigid motions as eigenvectors. If for any $\omega \in H^{1/2}(\Gamma)$, let $\mathbf{w} \in H^1(\Omega)$ satisfying $-\Delta^*\mathbf{w} = \mathbf{0}$ in Ω with $\mathbf{w}|_\Gamma = \omega$, then the generalized Betti formula of the first kind implies that

$$(9) \qquad < T\omega, \omega >_\Gamma = \int_\Omega \sigma(\mathbf{w}) : \epsilon(\mathbf{w})dx,$$

where $< \cdot, \cdot >_\Gamma$ denotes the duality pairing between the dual $H^{-1/2}(\Gamma)$ and the Sobolev space $H^{1/2}(\Gamma)$. Hence, one can show that

$$\exists \gamma > 0 : \ < T\omega, \omega >_\Gamma + ||\omega||^2_{L^2(\Gamma)} \geq \gamma ||\omega||^2_{H^{1/2}(\Gamma)} \quad \forall \omega \in H^{1/2}(\Gamma)$$

by using Korn's inequality and the trace theorem. This motivates us to see what kinds of conditions ϕ should satisfy. We now make the following assumptions.

(H1) Carathéodory conditions: The nonlinear function $\phi_i(\cdot, \mathbf{v})$ is measurable on Γ for all $\mathbf{v} \in R^3$ and $\phi_i(x, \cdot)$ is continuous in R^3 for almost all $x \in \Gamma$.

(H2) Lipschitz condition: The nonlinear functions $\phi_i(x, \cdot), i = 1, 2, \cdots$, have continuous first order partial derivatives in R^3 for almost all $x \in \Gamma$. Also, there exists a constant $C > 0$ such that $\frac{\partial}{\partial v_j}\phi_i(x, \mathbf{v})$ satisfies the Caratheódory conditions (H1) and $|\frac{\partial}{\partial v_j}\phi_i(x, \mathbf{v})| \leq C$ for all $\mathbf{v} \in R^3$ and for almost all $x \in \Gamma$.

(H3) Strong monotonicity: There is a constant $c > 0$ such that
$\sum_{i=1,j=1}^{3} \frac{\partial}{\partial v_j}\phi_i(x, \mathbf{v})\xi_i\xi_j \geq c\sum_{i=1}^{3} \xi_i^2$ for all $\mathbf{v}, \xi \in R^3$ and for almost all $x \in \Gamma$.

Our main results concerning the operator A are summarized in the following theorem.

THEOREM 2.1. *Under the assumptions* (H1)–(H3), *the operator* $A : H^{1/2}(\Gamma) \to H^{-1/2}(\Gamma)$ *is Lipschitz continuous, i.e.,*

$$\exists \, \Lambda > 0 : \quad \|A(v) - A(\nu)\|_{H^{-1/2}(\Gamma)} \leq \Lambda \|v - \nu\|_{H^{1/2}(\Gamma)} \quad \forall v, \nu \in H^{1/2}(\Gamma),$$

and is strongly monotone, i.e.,

$$\exists \, \gamma > 0 : \quad < A(v) - A(\nu), v - \nu >_\Gamma \, \geq \gamma \|v - \nu\|^2_{H^{1/2}(\Gamma)} \quad \forall v, \nu \in H^{1/2}(\Gamma).$$

As a consequence of the theorem, one may show that for given $g \in H^{-1/2}(\Gamma)$ and $f \in H^{-1}(\Omega)$, (8) has a unique solution $v \in H^{1/2}(\Gamma)$, and obtain a lemma of Céa's type for the Galerkin approximations of the solution (see, e.g., [13]). We remark that the nonlinear Robin problem considered here for the Lamé equation is a direct generalization of [29] for the Laplace equation. Other results for the Laplacian can also be found in [28], [3], and [7].

2.2 Boundary Conditions of Signorini Type

Next let us consider (1) together with boundary conditions of the Signorini type, the contact problems. The physical situation for the contact problems can be simply described as follows. An elastic body occupies a region Ω in R^3 with boundary Γ. A part of the boundary Γ_c is in contact with a rigid support and is subject to some sort of contact conditions to be specified. The complementary part $\Gamma \setminus \Gamma_c$ shall be subject to conditions of classical type. If we assume that $\Gamma \setminus \Gamma_c = \bar{\Gamma}_u \cup \bar{\Gamma}_\sigma$, then typical classical boundary conditions are

$$(10) \qquad \mathbf{u}|_{\Gamma_u} = \mathbf{0} \quad \text{on} \quad \Gamma_u, \quad \mathbf{T}_n[\mathbf{u}]|_{\Gamma_\sigma} = \mathbf{g} \quad \text{on} \quad \Gamma_\sigma.$$

Now on the contact surface Γ_c, there are two types of contact conditions, one with friction and the other one without friction. For simplicity, we consider here only the conditions without friction:

$$(11) \qquad \tau_T = \mathbf{0} \quad \text{on} \quad \Gamma_c \quad \text{and} \quad u_N \geq 0, \quad \tau_N \geq 0, \quad u_N \tau_N = 0 \quad \text{on} \quad \Gamma_c.$$

Here the stress vector $\tau|_{\Gamma_c} = \mathbf{T}_n[\mathbf{u}]|_{\Gamma_c}$ is decomposed into normal and tangential components :

$$\tau_N = \tau \cdot \mathbf{n} \quad \text{and} \quad \tau_T = \tau - \tau_N \mathbf{n}.$$

Analogously to τ_N and τ_T, u_N and $\mathbf{u}_T$ denote the normal and the tangential components of the displacement vector $\mathbf{u}|_{\Gamma_c}$.

Contact problems such as the one consisting of (1),(10),(11), are usually studied within the framework of the theory of variational inequalities over the domain. The variational formulation of the problem (1),(10),(11) reads: *Given* $\mathbf{f} \in L^2(\Omega)$ *and* $\mathbf{g} \in H^{-1/2}(\Gamma_\sigma)$, *find* $\mathbf{u} \in \mathcal{K}_\Omega := \{\mathbf{v} \in H^1(\Omega) : \mathbf{v}|_{\Gamma_u} = \mathbf{0}, \, v_N \geq 0 \quad \text{on} \quad \Gamma_c\}$ *such that*

$$(12) \qquad a_\Omega(\mathbf{u}, \mathbf{v} - \mathbf{u}) := \int_\Omega \sigma(\mathbf{u}) : \varepsilon(\mathbf{v} - \mathbf{u}) dx \geq \ell(\mathbf{v} - \mathbf{u}) \quad \forall \, \mathbf{v} \in \mathcal{K}_\Omega,$$

where $\ell(\mathbf{v} - \mathbf{u}) := \int_\Omega \mathbf{f} \cdot (\mathbf{v} - \mathbf{u}) dx + \int_{\Gamma_\sigma} \mathbf{g} \cdot (\mathbf{v} - \mathbf{u}) dx$. We note that the variational inequality (12) contains the bilinear form $a_\Omega(\mathbf{u}, \mathbf{v})$ over the domain Ω. However, from the Betti formula, we see that

$$(13) \qquad a_\Omega(\mathbf{u}, \mathbf{v}) = \int_\Gamma \mathbf{T}_n[\mathbf{u}] \cdot \mathbf{v} ds, \quad \text{if} \quad -\Delta^* \mathbf{u} = 0.$$

It is this relation that allows one to replace the variational inequality (12) over the domain by a boundary variational inequality.

To apply the BEM, the first step is to transform (1) into a homogeneous equation by introducing a particular solution $\mathbf{u}_f$ defined by

$$-\Delta^* \mathbf{u} = \mathbf{f} \quad \text{in} \quad \Omega, \quad \mathbf{u}_f|_\Gamma = 0 \quad \text{on} \quad \Gamma.$$

Then by setting $\mathbf{w} = \mathbf{u} - \mathbf{u}_f$, we may replace (12) by the variational inequality for $\mathbf{w} \in \mathcal{K}_\Omega$ such that

$$(14) \qquad a_\Omega(\mathbf{w}, \mathbf{v} - \mathbf{w}) \geq \int_{\Gamma_c} (\mathbf{g} - \mathbf{T}_n[\mathbf{u}_f]) \cdot (\mathbf{v} - \mathbf{w}) ds \quad \forall \quad \mathbf{v} \in \mathcal{K}_\Omega.$$

Since $-\Delta^* \mathbf{w} = 0$ in Ω, (13) implies that

$$(15) \qquad a_\Omega(\mathbf{w}, \mathbf{v} - \mathbf{w}) = \int_\Gamma \chi \cdot (\mathbf{v} - \mathbf{w}) \, ds$$

with $\omega := \mathbf{w}|_\Gamma, \chi := \mathbf{T}_n[\mathbf{w}]|_\Gamma$, and $\nu := \mathbf{v}|_\Gamma$.

Next, from the Betti representation formula of $\mathbf{w}$(cf. (5)), we obtain two BIEs to relate the Cauchy data ω and χ, namely,

$$(16) \qquad \omega = V\chi + (\tfrac{1}{2}I - K)\omega \quad \text{and} \quad \chi = (\tfrac{1}{2}I + K')\chi + W\omega \quad \text{on} \quad \Gamma.$$

The first equation in (16) is similar to (6) while the second one is obtained from the representation formula by applying the stress operator to $\mathbf{w}$ and taking its limit on Γ. Here the boundary integral operators K' and W for $x \in \Gamma$, are given by

$$(17) \quad K'\chi(x) := \int_\Gamma \mathbf{T}_{n_x}[\mathbf{E}(x,y)]\chi(y)ds_y, \quad W\omega(x) := -\mathbf{T}_{n_x} \int_\Gamma (\mathbf{T}_{n_y}[\mathbf{E}(x,y)])^t \omega(y)ds_y.$$

The first equation in (16) is equivalent to $\chi = V^{-1}(\tfrac{1}{2}I + K)\omega$. Substituting it into the second equation in (16), we obtain a Steklov- Poincaré operator $T : H^{1/2}(\Gamma) \to H^{-1/2}(\Gamma)$ in the form

$$(18) \qquad T\omega := (\tfrac{1}{2}I + K')V^{-1}(\tfrac{1}{2}I + K)\omega + W\omega = \chi.$$

Then from (15), we now define the boundary bilinear form

$$b_\Gamma(\omega, \nu - \omega) :=< T\omega, \nu - \omega >_\Gamma \quad \text{for} \quad \omega, \nu \in \mathcal{K}_\Gamma,$$

and (14) may be replaced by a boundary variational inequality. Here $\mathcal{K}_\Gamma$ is the corresponding convex set defined by $\mathcal{K}_\Gamma := \{\nu \in H^{1/2}(\Gamma) : \nu|_{\Gamma_u} = 0, \quad \nu_N \geq 0 \text{ on } \Gamma_c\}$. The main result can be summarized in the following.

THEOREM 2.2. *Given* $\mathbf{f} \in L^2(\Omega)$ *and* $\mathbf{g} \in H^{-1/2}(\Gamma_\sigma)$, *there exists a unique solution* $\omega \in \mathcal{K}_\Gamma$ *of the boundary variational inequality:*

$$b_\Gamma(\omega, \nu - \omega >_\Gamma \geq \int_{\Gamma_\sigma} (\mathbf{g} - \mathbf{T}_n[\mathbf{u}_f]) \cdot (\nu - \omega)ds \quad \forall \quad \nu \in \mathcal{K}_\Gamma.$$

Existence, uniqueness and approximate solutions can be established in the same manner as in [18], [24], and [22]. The latter includes also contact problems with friction. See also the works of [31], [19], [20], [15], [16], [21], and the monograph [2].

To conclude this section, we now replace the unilateral conditions in (11) by the perturbed conditions

$$u_{\epsilon N} + \epsilon\, \tau_{\epsilon N} \geq 0,\ \tau_{\epsilon N} \geq 0,\ (u_{\epsilon N} + \epsilon\, \tau_{\epsilon N})\tau_{\epsilon N} = 0 \quad \text{on} \quad \Gamma_c,$$

where $0 < \epsilon$ is a small parameter. It is not difficult to see that these perturbed unilateral constraints together with the condition $\tau_T = 0$ on Γ_c are equivalent to the nonlinear Robin condition in the form:

$$(19) \qquad \mathbf{T}_n[\mathbf{u}_\epsilon] = \left[sup\left(0, -\epsilon^{-1} u_{\epsilon N}\right)\right]\mathbf{n} =: -\epsilon^{-1}\, \Phi(\mathbf{u}_\epsilon) \quad \text{on} \quad \Gamma_c.$$

Hence, in principle, the perturbed problem defined by (1),(10) and (19) for $\mathbf{u}_\epsilon$, could be analyzed as (1) and (6) previously. The solution of the perturbed problem, if exists, is referred to as the Yosida approximation of the original unperturbed problem. Error estimates concerning the Yosida approximation are available for the Dirichlet-Signorini problem for the Laplacian in [30], and [3].

3 Linear-Nonlinear Interface Problems

In recent years, combined methods of finite and boundary elements have received increasing attention in computational mechanics (see, e.g., [23]). The methods are efficient and can be implemented on parallel multi-processor computers by taking the advantage of modern computer architectures. To illustrate the essence of the methods we consider here an elasto-plastic interface problem in elasticity. This is a typical linear-nonlinear interface problem, which can be treated by these methods.

To describe the problem, let Ω be a bounded, simply connected domain in R^3 with boundary Γ. We assume that Ω is decomposed into two subdomains, an inner domain Ω_e which is completely contained in Ω with boundary Γ_0 and an annular domain $\Omega_p := \Omega \setminus \bar{\Omega}_e$. The annular region Ω_p bounded by Γ_0 and Γ is occupied by an elasto–plastic material while the bounded region Ω_e inside Γ_0 consists of linear elastic material. For the elasto–plastic material Ω_p, we assume that the Hencky–Mises stress–strain relation holds, namely, if $\mathbf{u}$ denotes the displacement field of the material, then

$$\sigma_p(\mathbf{u})(x) = \left[\kappa(x) - \frac{2}{3}\tilde{\mu}(x, G(\mathbf{u})(x))\right] \text{div } \mathbf{u}(x)\mathbf{I} + 2\tilde{\mu}(x, G(\mathbf{u})(x))\epsilon(\mathbf{u})(x),$$

where $\kappa : \Omega_p \to R$ is the bulk modulus, $\tilde{\mu} : \Omega_p \times R^+ \to R$ the Lamé function, and $G(\mathbf{u}) := \epsilon^*(u) : \epsilon^*(u)$ with $\epsilon^*(\mathbf{u}) := \epsilon(\mathbf{u}) - \frac{1}{3}(\text{div } \mathbf{u})\mathbf{I}$ being the deviator of the small strain tensor. Here the functions κ and $\tilde{\mu}$ are supposed to be continuous, and $\tilde{\mu}(x, \cdot)$ to be continuously differentiable in R^+ such that the following estimates hold:
(H4) $0 < \kappa_0 < \kappa(x) \leq \kappa_1$ for all $x \in \bar{\Omega}_p$.

(H5) $0 < \mu_0 \leq \tilde{\mu}(x, \eta) \leq \frac{3}{2}\kappa(x)$ and $0 < \mu_1 \leq \tilde{\mu}(x, \eta) + 2\left(\frac{\partial}{\partial\eta}\tilde{\mu}(x, \eta)\right)\eta \leq \mu_2$
for all $(x, \eta) \in \bar{\Omega}_p \times R^+$.

For the linear elastic material in Ω_e, we assume the validity of the standard Hooke's law for the isotropic material as in (3).

The interface problem can be formulated as follows: *For given body force* $\mathbf{f}$ *defined on* Ω_p, *find the displacement field* $\mathbf{u}$ *satisfying*

$$-\Delta^*\mathbf{u} := -\operatorname{div}\sigma(\mathbf{u}) = 0 \quad in \quad \Omega_e; \quad -N(\mathbf{u}) := -\operatorname{div}\sigma_p(\mathbf{u}) = \mathbf{f} \quad in \quad \Omega_p,$$

$$\text{(20)} \qquad\qquad\qquad \mathbf{u}|_\Gamma = 0 \quad on \quad \Gamma,$$

$$\mathbf{u}^- = \mathbf{u}^+, \ T_n[\mathbf{u}]^- = T_n[\mathbf{u}]^+ \quad on \quad \Gamma_0.$$

In the formulation, for any function $\mathbf{v}$ defined on $\Omega_e \cup \Omega_p$, we denote by $\mathbf{v}^\mp$ its limits on Γ_0 from Ω_e and Ω_p, respectively; in particular, we have

$$T_n[\mathbf{u}]^- = \sigma(\mathbf{u}) \cdot \mathbf{n}|_{\Gamma_0} \quad and \quad T_n[\mathbf{u}]^+ = \sigma_p[\mathbf{u}] \cdot \mathbf{n}|_{\Gamma_0},$$

where $\mathbf{n}$ denotes the outward unit normal to the interface Γ_0.

The basic idea of the combined method is the reduction of the problem for $\mathbf{u}$ in the elastic domain Ω_e to a BIE on Γ_0, and the transformation of the original boundary value problem (20) to an equivalent problem, the so called *nonlocal boundary value problem*, for $\mathbf{u}$ in Ω_p together with an appropriate BIE for $\tau = T_n[\mathbf{u}]^-$ on the interface Γ_0. Specifically, if we represent the solution $\mathbf{u}$ in Ω_e by

$$\text{(21)} \qquad \mathbf{u}(x) = \int_{\Gamma_0} \mathbf{E}(x,y)\tau(y)ds_y - \int_{\Gamma_0} (T_{n_y}[\mathbf{E}(x,y)])^t\mathbf{u}^+(y)ds_y, \quad x \in \Omega_e,$$

then we arrive at a nonlocal boundary value problem in Ω_p: Find $\mathbf{u}$ and τ satisfying

$$\text{(22)} \quad -N(\mathbf{u}) = \mathbf{f} \text{ in } \Omega_p; \quad \mathbf{u}|_\Gamma = 0 \text{ on } \Gamma, \quad and \quad T_n[\mathbf{u}]^+ = (\frac{1}{2}I + K')\tau + W\mathbf{u}^+ \text{ on } \Gamma_0$$

together with the nonlocal boundary condition,

$$\text{(23)} \qquad\qquad V\tau - \left(\frac{1}{2}I + K\right)\mathbf{u}^+ = 0 \quad on \ \Gamma_0.$$

We observe that if one can express τ explicitly in terms of $\mathbf{u}^+$, then the solution $\mathbf{u}$ in Ω_p would be the only unknown, which may be determined completely by (22) alone. However, in general both $\mathbf{u}$ in Ω_p and τ on Γ_0 are the unknowns and they are related implicitly by the coupled system of equations (22) and (23).

We will now consider the variational solutions $\mathbf{u}$ and τ of the nonlocal boundary problem (22)–(23). Let us first introduce the subspace $H^1_\Gamma(\Omega_p)$ of $H^1(\Omega_p)$ defined by

$$H^1_\Gamma(\Omega_p) = \{\mathbf{v} \in H^1(\Omega_p) : \mathbf{v}|_\Gamma = 0\}.$$

The variational formulation of the nonlocal problem (22)–(23) then reads: *Given* $\mathbf{f} \in H^1_\Gamma(\Omega_p)$, *find* $(\mathbf{u}, \tau) \in H^1_\Gamma(\Omega_p) \times H^{-1/2}(\Gamma_0)$ *such that*

$$\text{(24)} \qquad\qquad A_{\Omega_p}(\mathbf{u}, \mathbf{v}) + B_{\Gamma_0}((\mathbf{u}, \tau), (\mathbf{v}, \chi)) = \int_{\Omega_p} \mathbf{f} \cdot \mathbf{v}dx$$

for all $(\mathbf{v}, \chi) \in H^1_\Gamma(\Omega_p) \times H^{-1/2}(\Gamma_0)$. *Here* A_{Ω_p} *and* B_{Γ_0} *are, respectively, the semilinear and bilinear forms defined by*

$$A_{\Omega_p}(\mathbf{u}, \mathbf{v}) := \int_{\Omega_p} \sigma_p(\mathbf{u}) : \varepsilon(\mathbf{v})dx$$

and

$$B_{\Gamma_0}((\mathbf{u},\tau),(\mathbf{v},\chi)) := \langle W\mathbf{u}^+, \mathbf{v}^+\rangle_{\Gamma_0} + \langle \left(\frac{1}{2}I + K'\right)\tau, \mathbf{v}^+\rangle_{\Gamma_0} - \langle \chi, \left(\frac{1}{2}I + K\right)\mathbf{u}^+\rangle_{\Gamma_0} + \langle \chi, V\tau\rangle_{\Gamma_0}.$$

The variational formulation (24) is a special case of the one considered in [10] and [11]. Now let $\mathcal{H}^*$ denote the dual of $\mathcal{H} := H^1_\Gamma(\Omega_p) \times H^{-1/2}(\Gamma_0)$. Then one may define a nonlinear operator $\mathcal{A} : \mathcal{H} \to \mathcal{H}^*$ on $\mathcal{H}$ by putting

(25)
$$[\mathcal{A}(\mathbf{u},\tau),(\mathbf{v},\chi)] := A_{\Omega_p}(\mathbf{u},\mathbf{v}) + B_{\Gamma_0}((\mathbf{u},\tau),(\mathbf{v},\chi))$$

and rewrite (24), equivalently, in the form of an operator equation for the unknowns $(\mathbf{u},\tau)$:

(26)
$$\mathcal{A}(\mathbf{u},\tau) = \mathcal{F},$$

where $\mathcal{F} \in \mathcal{H}^*$ is defined by $[\mathcal{F},(\mathbf{v},\tau)] := \int_{\Omega_p} \mathbf{f}\cdot\mathbf{v}dx$. From the mapping properties of the boundary integral operators, it is not difficult to see that the operator $\mathcal{A}$ is completely dominated by the semilinear form A_{Ω_p} in (24). Based on the general theory of monotone operators, the following existence and uniqueness result was established in [13].

THEOREM 3.1. *Under the assumptions* (H4)–(H5) *on the bulk modulus* κ *and on the Lamé function* $\tilde{\mu}$, *we have the following:* (a) *The operator* $\mathcal{A}$ *defined by* (25) *is both strongly monotone and Lipschitz continuous on* $\mathcal{H}$. (b) *Given* $\mathbf{f} \in L^2(\Omega_p)$, *there exists a unique solution* $(\mathbf{u},\tau) \in \mathcal{H}$ *of the operator equation* (26).

We remark that there is a great amount of literature concerning the coupling of the BEM and the FEM for nonlinear problems. Most of them are dealing with problems of strongly monotone type (cf. the recent monograph on this subject [13] and the bibliography therein). For non-monotone type problems, see [12] and [8].

4   An Iterative Scheme and Some Numerical Results

In this section we present an iterative scheme which is particularly designed for the operator equations of strongly monotone type. Let $\mathcal{H}$ be a Hilbert space with its dual $\mathcal{H}^*$. Consider the nonlinear operator equation

(27)
$$\mathbf{A}(\mathbf{u}) = \mathbf{f},$$

for given $\mathbf{f} \in \mathcal{H}^*$, where $\mathbf{A} : \mathcal{H} \longrightarrow \mathcal{H}^*$ is Lipschitz continuous and strongly monotone. Then (27) may be solved by the following interative scheme.

- Begin with an initial guess $\mathbf{u}_0 \in \mathcal{H}$.

- Define a sequence $\{\theta^{n+1/2}\}$:

$$B(\theta^{n+1/2},\mathbf{v}) = <\mathbf{A}(\mathbf{u}^n) - \mathbf{f}, \mathbf{v} >_{\mathcal{H}^*\times\mathcal{H}}, \quad n = 0,1,\cdots,$$

where B is any symmetric $\mathcal{H} - elliptic$ continuous *bilinear* form.

- Up-date the iteration:

$$\mathbf{u}^{n+1} = \mathbf{u}^n - \omega\theta^{n+1/2}, \quad n = 0,1,\cdots$$

for a suitable relaxation parameter $\omega > 0$.

In the algorithm, we may eliminate $\theta^{n+1/2}$ and express $\mathbf{u}^{n+1}$ directly in terms of $\mathbf{u}^n$ so that

$$\mathbf{u}^{n+1} = \mathbf{u}^n - \omega \mathbf{B}^{-1}(\mathbf{A}(\mathbf{u}^n) - \mathbf{f}).$$

By using standard fixed-point arguments, it can be shown that there is a $\tilde{\omega} > 0$ such that $\mathbf{u}^n \to \mathbf{u}$ in $\mathcal{H}$ for each $\omega \in (0, \tilde{\omega})$. Here $\mathbf{B}$ is the corresponding operator associated with the bilinear form $B(\cdot, \cdot)$, and hence, plays the role of the preconditioner in the iterative scheme.

We have applied the present scheme to a nonlinear problem arising from the irrotational compressible flow past an obstacle in the plane. Fig. 1 below gives the convergence history where $\epsilon := max\ |u^{n+1} - u^n|/max\ |u^n|$ is the relative error for the stream function u, and n the total number of iterations. Details and more numerical results are available. For interested readers, we refer to the recent paper [8].

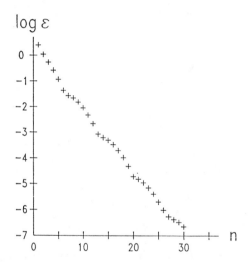

FIG. 1. *Convergence history.*

References

[1] T. S. Angell, G. C. Hsiao, and R. E. Kleinman, *An optimal design problem for submerged bodies*, Math. Meth. Appl. Sci. 8 (1986), pp. 50–76.

[2] H. Ante and P. D. Panagiotopulos, *The Boundary Integral Approach to Static and Dynamic Contact Problems*, Birkhäuser Verlag, Basel-Boston-Berlin, 1992.

[3] M. Bach and H. Schmitz, *A boundary element method for some potential problems with monotone boundary condition*, IBM Technical Report TR 75.92.02, IBM Germany, Heidelberg Scientic Center, 1992.

[4] D. Colton and R. Kress, *Inverse Acoustic and Electromagnetic Scattering Theory*, Springer-Verlag, Berlin, 1992.

[5] R. Dautray, J. L. Lions, *Mathematical Analysis and Numerical Methods for Science and Technology*, volume 4, Springer-Verlag, Berlin- Heidelberg- New York-London-Paris- Tokyo-Hong-Kong, 1990.

[6] Y. Deng, G. Chen, W. M. Ni, and J. Zhow, *Boundary element- monotone iterative scheme for semilinear elliptic partial differential equations*, preprint 1994.

[7] P. P. B. Eggermont and J. Sarannen, *L^p estimates of boundary integral equations for some nonlinear boundary value problems*, Numer. Math., 58 (1990), pp. 465–478.

[8] M. Feistauer, G. C. Hsiao, R. E. Kleinman, and R. Tezaur, *Analysis and numerical realization of coupled BEM and FEM for numerical exterior problems*, in Inverse Scattering and Potential Problems in Mathematical Physis, Verlag Peter Lang, Frankurt am Main, 1995.

[9] T. M. Fisher, G. C. Hsiao, and L. W. Wendland, *Singular perturbations for the exterior three-dimensional slow viscous flow problem*, J. Math. Anal. Appl., 110 (1985), pp. 583–603.

[10] G. N. Gatica and G. C. Hsiao, *The coupling of boundary element and finite element methods for a nonlinear exterior boundary value problem*, Zeit. Anal. Anwend. (ZAA), 8 (4) (1989), pp. 377–387.

[11] G. N. Gatica and G. C. Hsiao, *On a class of variational formulations for some nonlinear interface problems*, Rendiconti di Matematica VII, 10, (1990), pp. 681–715.

[12] G. N. Gatica and G. C. Hsiao, *The coupling of boundary integral and finite element methods for monotone nonlinear problems*, Numer. Funct. Anal. and Optimiz., 13 (5 & 6), (1992), pp. 431–447.

[13] G. N. Gatica and G. C. Hsiao, *Boundary-Field Equation Methods for a Class of Nonlinear Problems*, Pitman Research Notes in Mathematics, Pitman Publishing, London, 1995, in press.

[14] R. P. Gilbert and G. C. Hsiao, *On Dirichlet's problem for quasi-linear elliptic equations*, in Lecture Notes in Mathematics 430, Springer-Verlag, Berlin-Heidelberg-New York, 1974.

[15] J. Gwinner and E. P. Stephan, *A boundary element procedure for contact problems in plane linear elastostatics*, M^2AN, 27 (1993), pp. 457–480.

[16] J. Gwinner and E. P. Stephan, *Boundary element convergence for a variational inequality of the second kind*, in Parametric Optimization and Related Topics III, P. Lang, 1993.

[17] P. Hähner and G. C. Hsiao, *Uniqueness theorems in inverse obstacle scattering of elastic waves*, Inverse Problems, 9 (1993), pp. 525–534.

[18] H. Han, *The boundary finite element method for Signorini problems*, in Lecture Notes in Mathematics 1297, Springer-Verlag, Berlin-Heidelberg, 1987.

[19] H. Han, *A direct boundary element method for Signorini problems*, Math. Comp., 55 (1990), pp. 115-128.

[20] H. Han, *A boundary element methods for Signorini problems in three dimensions*, Numer. Math., 60, (1991), pp. 63–75.

[21] H. Han, *The boundary element method for solving variational inequalities*, in Contemporary Mathematics 163, AMS, 1994.

[22] H. Han and G. C. Hsiao, *The boundary element method for a contact problem*, in Theory and Applications of Boundary Element Methods, Tsinghua University Press, Peking, China, 1988.

[23] G. C. Hsiao, *The coupling of BEM and FEM-a brief review*, in Boundary Elements X, vol 1, Springer-Verlag, Berlin, 1988.

[24] G. C. Hsiao, *On the progeny of Fichera's boundary integral method*, in Probelmi Attuali Dell'analisi e Della Fisica Matematica, Universitá di Roma "La Sapienza", 1993.

[25] G. C. Hsiao, *Nonlinear problems via boundary element methods*, in IABEM Symposium on Boundary Integral Methods for Nonlnear Problems, Siena, Italy, 1995.

[26] G. C. Hsiao and R. C. MacCamy, *Singular perturbations for the two-dimensional viscous flow problems*, in Lecture Notes in Mathematics 942, Singer-Verlag, Berlin, 1982.

[27] G. C. Hsiao, W. L. Wendland, *On a boundary integral method for some exterior problems in elasticity*, Trudy Tbilisi Univ., Mat. Mekh. Astronom, No. 18 (1985), pp. 31–60.

[28] K. Ruotsalainen and J. Sarannen, *On the collocation method for a nonlinear boundary integral equation*, J. Comput. Appl. Math., 28 (1989), pp. 339–348.

[29] K. Ruotsalainen and W. L. Wendland, *On the boundary element method for some nonlinear boundary value problems*, Numer. Math., 53 (1988), pp. 299-314.

[30] H. Schmitz, G. Schneider, and W. L. Wendland, *Boundary element methods for problems involving unilateral boundary conditions*, in Nonlinear Computational Mechanics, Springer-Verlag, 1991.

[31] W. Spann, *Fehlerabschätzungen zur Randelementmethode beim Signorini-Problem für die Laplace-Gleichung*. PhD thesis, Universität München, München, Germany, 1989.

[32] I. Stakgold, *Green's Functions and Boundary Value Problems*, John Wiley & Sons, New York, 1979.

On the Metastable and Supersensitive Motion of Shocks for Viscous Conservation Laws in One Space Dimension

Jacques G.L. Laforgue* †   Robert E. O'Malley, Jr.‡§

1   The Supersensitive Steady-State Burgers' Equation

Consider the nonlinear two-point boundary value problem

(1)
$$\begin{cases} \epsilon u_{xx} = u u_x, & -1 \le x \le 1 \\ u(-1) = \alpha, & u(1) = -1 \end{cases}$$

for various values of α, in the limit as the viscosity parameter ϵ tends to zero through *positive* values. The boundary values actually provide upper and lower bounds, i.e., we are guaranteed the existence of a solution that remains between them (cf, e.g., Howes (1976)). That the solution for each α is unique might be determined by a phase-plane analysis or by direct integration. Since $2\epsilon u_x = u^2 - A^2$ for some real constant A^2, separation of variables provides when $A \ne 0$ a so-called "general solution"

(2)
$$u(x, \epsilon) = -A \tanh\left(A(x - \beta)/2\epsilon\right) = -A \left[\frac{1 - e^{-A(x-\beta)/\epsilon}}{1 + e^{-A(x-\beta)/\epsilon}}\right]$$

where the real or purely imaginary *amplitude* A and the *zero location* β might be determined by the boundary values, i.e., the relations

(3)
$$e^{A\beta/\epsilon} = \left(\frac{A-1}{A+1}\right) e^{A/\epsilon} = \left(\frac{A+\alpha}{A-\alpha}\right) e^{-A/\epsilon}.$$

We note that alternative solutions (2) could be expressed in terms of the tangent or hyperbolic cotangent. Further, since (2) is unchanged when A is replaced by $-A$, we shall take $\Re A \ge 0$. Presuming the last equation has a non-trivial asymptotic solution $A(\epsilon)$, $\beta(\epsilon)$ will follow immediately by taking a logarithm.

We shall clarify the situation by solving three separate cases, depending on the sign of $\alpha - 1$, taking into account asymptotically exponentially small terms. For $\alpha = 1$, we expect

---

* Research supported in part by the Consejo de Investigación de la Universidad De Oriente.

† Departamento de Matemáticas, Universidad De Oriente, Apartado 245, Cumaná 6101A, Venezuela.

‡ The research was finished while this author was visiting the Isaac Newton Institute for Mathematical Sciences at the University of Cambridge, U.K., and the Université Claude Bernard Lyon 1, France. Support from these institutions, the CNRS, and the U.S. National Science Foundation is gratefully acknowledged.

§ Department of Applied Mathematics, University of Washington, Seattle, Washington 98195, U.S.A.

a symmetric solution with $\beta(\epsilon) = 0$ and an $A(\epsilon)$ satisfying $A = 1 + (A+1)e^{-A/\epsilon}$. Successive approximations indeed provides the series

$$A(\epsilon) = 1 + 2e^{-1/\epsilon} + O(\frac{1}{\epsilon}e^{-2/\epsilon}),$$

roughly in powers of $e^{-1/\epsilon}$, and thereby the asymptotic solution

$$(4) \qquad u(x, \epsilon) = \frac{-1 + e^{-x/\epsilon}}{1 + e^{-x/\epsilon}} + O(e^{-1/\epsilon}).$$

Note the solution's discontinuous limit

$$u(x, \epsilon) \sim \begin{cases} 1, & x < 0 \\ -1, & x > 0 \end{cases}$$

as $\epsilon \to 0$ and the nonuniform convergence which occurs in an $O(\epsilon)$-thick shock or transition layer about the midpoint $x = 0$. For $\alpha > 1$, we instead use successive approximations in $A = \alpha + \frac{(A+\alpha)(A+1)}{A-1}e^{-2A/\epsilon}$ to get

$$A(\epsilon) = \alpha + \frac{2\alpha(\alpha+1)}{\alpha-1}e^{-2\alpha/\epsilon} + O(\frac{1}{\epsilon}e^{-4\alpha/\epsilon}),$$

$$\beta(\epsilon) = 1 + \frac{\epsilon}{\alpha}\log\left(\frac{\alpha-1}{\alpha+1}\right) + O(\epsilon e^{-2\alpha/\epsilon})$$

and, thereby, the solution

$$(5) \qquad u(x, \epsilon) = \alpha\left[\frac{(\alpha-1) - (\alpha+1)e^{-\alpha(1-x)/\epsilon}}{(\alpha-1) + (\alpha+1)e^{-\alpha(1-x)/\epsilon}}\right] + O(e^{-2\alpha/\epsilon})$$

which has the constant limit α for $x < 1$ and nonuniform convergence in an $O(\epsilon)$-thick terminal boundary layer. Note that higher order terms break down at $\alpha = 1$. Finally, for $\alpha < 1$, we likewise find the unique solution

$$(6) \qquad u(x, \epsilon) = \frac{-(1-\alpha) + (1+\alpha)e^{-(1+x)/\epsilon}}{(1-\alpha) + (1+\alpha)e^{-(1+x)/\epsilon}} + O(e^{-2/\epsilon}),$$

which has the limit -1 for $x > -1$ and an initial layer of nonuniform convergence near the left endpoint. In particular, for $\alpha = -1$, the constant solution is obtained. For each fixed α, then, the limiting solution away from $x = \pm 1$ is equal to an endvalue α or -1, unless $\alpha = 1$ when the solution jumps at $x = 0$ from 1 to -1. One anticipates nature not being so abrupt.

Interior shock layer solutions become less rare if we introduce

$$(7) \qquad \alpha_+(\epsilon) = 1 + e^{-a/\epsilon}$$

for some constant a with $0 < a < 1$ (i.e., for $\alpha(\epsilon)$ very near 1). We now get the amplitude

$$A_+(\epsilon) = 1 + e^{-a/\epsilon} + 4e^{-(2-a)/\epsilon} + O(\frac{1}{\epsilon}e^{-2/\epsilon} + e^{-(4-3a)/\epsilon}),$$

the zero location

$$\beta_+(\epsilon) = 1 - a - \epsilon\log 2 + O(e^{-a/\epsilon} + \epsilon e^{-(2-2a)/\epsilon}),$$

and the asymptotic solution

$$(8) \qquad u_+(x, \epsilon) \sim \frac{-2 + e^{-(x-1+a)/\epsilon}}{2 + e^{-(x-1+a)/\epsilon}} + O(e^{-a/\epsilon})$$

with the limit

$$\begin{cases} 1, & x < 1 - a \\ -1, & x > 1 - a \end{cases}$$

and a shock layer centered at $\beta_+(\epsilon)$.

Likewise, for

$$(9) \qquad \alpha_-(\epsilon) \equiv 1 - e^{-a/\epsilon},$$

we obtain

$$A_-(\epsilon) = 1 + 4e^{-(2-a)/\epsilon} + O(e^{-(4-3a)/\epsilon} + e^{-2/\epsilon}),$$

$$\beta_-(\epsilon) = -1 + a + \epsilon \log 2 + \epsilon O(e^{-a/\epsilon} + e^{-2(1-a)/\epsilon}),$$

and

$$(10) \qquad u_-(x, \epsilon) = \frac{-1 + 2e^{-(x+1-a)/\epsilon}}{1 + 2e^{-(x+1-a)/\epsilon}} + O(e^{-2(1-a)/\epsilon}).$$

By varying a from 0 to 1, the layer for $u_\pm$ thereby moves from the endpoint ± 1 to 0, i.e., an asymptotically negligible perturbation in the initial value moves the shock by an $O(1)$ amount. Laforgue and O'Malley (1993) called this significant phenomenon "super-sensitivity," while Carr et al (1995) described the related reaction-diffusion situation as a "numerical nightmare." For $a > 1$, the perturbation is too negligible to alter the symmetric limiting solution. The shock layer, in all cases, remains well-described by the shifting hyperbolic tangent profile, which is known as the Taylor shock in the fluid mechanics literature (cf. Crighton (1986)).

2 The Viscous Shock Equation

Now consider a variety of viscous conservation laws

$$(11) \qquad u_t + \big(f(u)\big)_x = \epsilon u_{xx}, \quad -1 \le x \le 1, \quad t \ge 0$$

where the function f is smooth and such that

$$(12) \qquad f(u) < f(-1) \equiv f(1) \text{ for } |u| < 1 \text{ and } -A_+ \equiv f'(-1) < 0 < f'(1) \equiv A_-.$$

To get a single interior shock, in contrast to the more common endpoint layers, we impose the special constant boundary values

$$(13) \qquad u(\pm 1, t) = \mp 1$$

(corresponding to the maxima of f) and smooth initial values $u(x, 0)$ such that $u(\pm 1, 0) = \mp 1$ and, say, $u_x(x, 0) < 0$. The nontrivial task of constructing an asymptotic solution which develops a shock in finite time is described, e.g., in Il'in (1992), based on applying the method of characteristics to the limiting first-order partial differential equation. The shock layer behavior will, naturally, be described via the stretched steady-state problem

$$(14) \qquad \big(f(\phi)\big)_\eta = \phi_{\eta\eta}, \quad \phi(\pm\infty) = \mp 1$$

obtained by using a spatial variable $\eta = \frac{x-\hat{x}}{\epsilon} = O(\frac{1}{\epsilon})$, stretched about the interior shock location $\hat{x}$. To fix the solution of this translation-invariant problem, we ask that $\phi(0) = 0$ and obtain the implicit representation

$$(15) \qquad \eta = \int_0^\phi \frac{dv}{f(v) - f(\pm 1)}.$$

The unique profile $\phi(\eta)$ could readily be computed numerically, e.g., by carefully integrating the initial value problem $\phi_\eta = f(\phi) - f(\pm 1)$, $\phi(0) = 0$ backward and forward to $\eta = \pm\infty$. Moreover, elementary estimates can be used to display the exponential tails, i.e.

$$(16) \qquad \phi(\eta) \pm 1 \sim \pm L_\pm e^{\mp A_\pm \eta} \text{ as } \eta \to \pm\infty,$$

where

$$L_\pm \equiv \exp\left[\mp \int_0^{\mp 1} \left(\frac{A_\pm}{f(\mp 1) - f(s)} - \frac{1}{1 \pm s}\right) ds\right] > 0$$

(cf. Laforgue and O'Malley (1995b)), as well as higher-order estimates. Corresponding to $x_\pm = \pm 1$ are large values $\eta_\pm = O(\frac{1}{\epsilon})$, so (16) implies that $\phi(\eta)$ fails to satisfy the endvalues for u by asymptotically exponentially small amounts. We will find that (12) is sufficient to guarantee the continued existence of a time-varying asymptotic approximation $\phi(\eta)$ with $\eta \equiv \frac{x-\hat{x}(t)}{\epsilon}$, featuring an interior shock layer at a slowly-moving location $\hat{x}(t)$ with constant outer limits elsewhere. The asymptotic profile ϕ is fundamental to finding the asymptotic solution $u(x,t)$.

In the special case of Burgers' equation, $f = \frac{1}{2}u^2$, $\phi(\eta) = -\tanh(\eta/2)$, and $A_\pm = 1$. Moreover, the solution can then be independently obtained by using the Cole-Hopf transformation to convert the given nonlinear convection equation to the linear heat equation and by solving the resulting initial-boundary value problem via a Fourier series (cf. Laforgue and O'Malley (1995a)). One finds that the limiting solution is asymptotically given by the slowly moving profile

$$(17) \qquad -\tanh\left(\frac{x - x_\epsilon(t)}{2\epsilon}\right)$$

which features an $O(\epsilon)$-thick shock layer centered at

$$(18) \qquad x_\epsilon(t) \sim \epsilon \log\left(\frac{1 + \tanh\left(\frac{x_\epsilon(t_\epsilon)}{2\epsilon}\right) e^{-2(t-t_\epsilon)/\epsilon e^{1/\epsilon}}}{1 - \tanh\left(\frac{x_\epsilon(t_\epsilon)}{2\epsilon}\right) e^{-2(t-t_\epsilon)/\epsilon e^{1/\epsilon}}}\right)$$

for $t \geq$ some finite t_ϵ, so the profile (17) has the outer limits

$$\begin{cases} 1 & \text{for } x < x_\epsilon(t) \\ -1 & \text{for } x > x_\epsilon(t). \end{cases}$$

Here, we use the initial shock location

$$x_\epsilon(t_\epsilon) \sim \frac{1}{2}\int_{-1}^1 u(s,0)ds + \frac{\epsilon^2}{2}\left(u_x(1,0) - u_x(-1,0)\right) + \cdots$$

and the time

$$t_\epsilon \sim \min_{-1 \leq x \leq 1}\left(\frac{-2}{u_x(x,0)}\right) + \cdots$$

of initial shock formation which were fully determined asymptotically by Il'in. From (17)–(18), we find that the shock location $x_\epsilon(t)$ moves monotonically, but exponentially slowly, from $x_\epsilon(t_\epsilon)$ to the midpoint $x_\epsilon(\infty) = 0$ over an asymptotically exponentially long t interval. An impatient person might erroneously let short-term computations suggest that the asymptotic solution remains stationary, missing the *metastability* encountered. The special case of (11) with $f'(u) = u^n$ and $n > 1$ is studied by Crighton (1992) and Harris (1995) as the "modified Burgers' equation."

For the general viscous shock equation (11), we shall use the Burgers' solution to suggest by analogy a limiting asymptotic solution

$$\phi(\eta) \text{ for } \eta \equiv \frac{x - x_\epsilon(t)}{\epsilon},$$

using the specific profile ϕ which is uniquely defined by (15), with the shock location $x_\epsilon(t)$ defined by

(19) $$u(x_\epsilon(t), t) = 0.$$

We shall seek a solution of the initial-boundary value problem in the form

(20) $$u(x, t, \epsilon) = \phi(\eta) + v(\eta, t, \epsilon)$$

which requires the correction term v to satisfy the nonlinear boundary value problem defined by the ordinary differential equation

(21) $$v_{\eta\eta} - f'(\phi + v)v_\eta - \big(f'(\phi + v) - f'(\phi)\big)\phi_\eta = -\frac{dx_\epsilon}{dt}(\phi_\eta + v_\eta) + \epsilon\frac{dv}{dt}$$

on $\eta_- \equiv -\frac{1}{\epsilon}(1 + x_\epsilon) \leq \eta \leq \eta_+ \equiv \frac{1}{\epsilon}(1 - x_\epsilon)$ and by the three conditions

$$v(\eta_\pm) = \mp 1 - \phi(\eta_\pm) \sim \mp L_\pm e^{\mp A_\pm \eta_\pm} \text{ and } v(0) = 0.$$

Because the forcing term $-\frac{dx_\epsilon}{dt}\phi_\eta$ and the boundary values $v(\eta_\pm)$ are all expected to be aymptotically negligible, we shall estimate the small solution v by that of the corresponding linearized problem

(22) $$\begin{cases} (w_\eta - f'(\phi)w)_\eta = -\frac{dx_\epsilon}{dt}\phi_\eta \\ w(\eta_\pm) = \mp L_\pm e^{\mp A_\pm \eta_\pm}, \ w(0) = 0 \end{cases}$$

in which t simply remains a parameter. Integrating with $w(0) = 0$ implies that w must have the form

(23) $$w(\eta) = \lambda(t)c_0(\eta) - \frac{dx_\epsilon}{dt}c_1(\eta)$$

for $c_j(\eta) \equiv \int_0^\eta \phi^j(s)e^{\int_s^\eta f'(\phi(r))dr}ds$, $j = 0$ and 1, and the constant of integration $\lambda(t)$. Since $c_1(\eta_\pm) \sim \mp c_0(\eta_\pm) \sim \frac{-1}{A_\pm}$, the limiting endpoint conditions thereby imply the two limiting linear algebraic equations

(24) $$-A_\pm L_\pm e^{\mp A_\pm \eta_\pm} \sim \lambda(t) \pm \frac{dx_\epsilon}{dt}$$

for the unknowns $\lambda(t)$ and $\frac{dx_\epsilon}{dt}$. Elimination then provides the differential equation

$$\frac{dx_\epsilon}{dt} \sim \frac{1}{2}\left(A_- L_- e^{-\frac{A_-}{\epsilon}} e^{\frac{-A_- x_\epsilon}{\epsilon}} - A_+ L_+ e^{-\frac{A_+}{\epsilon}} e^{\frac{A_+ x_\epsilon}{\epsilon}} \right)$$

for the motion of the shock. Note that this limiting equation has the unique rest point

$$\hat{x}_\epsilon(\infty) = \frac{A_+ - A_-}{A_+ + A_-} - \frac{\epsilon}{A_+ + A_-} \log\left(\frac{A_+ L_+}{A_- L_-}\right)$$

located within $(-1, 1)$ and that the differential equation can be conveniently rewritten as

$$(25) \qquad \frac{dx_\epsilon}{dt} \sim k\left[e^{-A_-(x_\epsilon - \hat{x}_\epsilon(\infty))/\epsilon} - e^{A_+(x_\epsilon - \hat{x}_\epsilon(\infty)/\epsilon)}\right]$$

for the asymptotically negligible constant $k \equiv \frac{1}{2} A_\pm L_\pm e^{-A_\pm(1 \mp \hat{x}_\epsilon(\infty))/\epsilon} > 0$.

Further, since one term on the right of (25) is asymptotically dominant, the sign of $\frac{dx_\epsilon}{dt}$ is fixed, and the shock will move monotonically from $x_\epsilon(t_\epsilon)$ to its steady state $x_\epsilon(\infty) = \hat{x}_\epsilon(\infty)$ as t increases. With the limiting solution $\phi(\eta)$, linearization of (25) further implies that

$$(26) \qquad u(x, t, \epsilon) \sim \phi\left(\frac{x - \hat{x}_\epsilon(\infty)}{\epsilon}\right) + O(e^{-\omega_\epsilon t})$$

as $t \to \infty$ for the decay constant $\omega_\epsilon \equiv \frac{1}{\epsilon}(A_+ + A_-)k$. Thus, the asymptotically negligible size of k explains the observed sluggish approach to the steady-state solution. Note that the steady state is completely independent of the initial condition. These important metastability conclusions concerning long time shock movement have been confirmed by a number of careful numerical computations (cf. Laforgue and O'Malley (1995b)). Reyna and Ward (1995) obtained analogous results using a projection method, based on a natural solvability condition.

Solving (24) for $\lambda(t)$ finally provides the asymptotically negligible estimate

$$
\begin{aligned}
(27) \qquad w(\eta) \sim &\frac{1}{2} A_+ L_+ e^{-A_+(1-x_\epsilon)/\epsilon}(c_1(\eta) - c_0(\eta)) \\
&- \frac{1}{2} A_- L_- e^{-A_-(1+x_\epsilon)/\epsilon}(c_0(\eta) + c_1(\eta)) \\
= &O(e^{-A_+(1-x_\epsilon(t_\epsilon))/\epsilon} + e^{-A_-(1+x_\epsilon(t_\epsilon))/\epsilon})
\end{aligned}
$$

for the correction term v in (20). Being more careful, let us rewrite (21) in the form

$$(28) \qquad v_{\eta\eta} - (f'(\phi)v)_\eta = -\frac{dx_\epsilon}{dt}\phi_\eta + H(v)$$

for

$$H(v) \equiv -\frac{dx_\epsilon}{dt}v_\eta + \epsilon\frac{dv}{dt} + (f'(\phi + v) - f'(\phi))v_\eta + (f'(\phi + v) - f'(\phi) - f''(\phi)v)\phi_\eta.$$

Since $v(0) = 0$, integration implies that $v(\eta)$ must satisfy the integral equation

$$(29) \qquad v(\eta) = \lambda(t)c_0(\eta) - \frac{dx_\epsilon}{dt}c_1(\eta) + \int_0^\eta e^{\int_s^\eta f'(\phi(r))dr} \int_0^s H(v(\rho))d\rho ds$$

with a constant of integration $\lambda(t)$ and functions c_0 and c_1 as in (23). Imposing the boundary conditions at $\eta_\pm$ allows us to again uniquely obtain expressions for the unknowns $\lambda(t)$ and $\frac{dx_\epsilon}{dt}$. Equation (29) can be solved iteratively for v. Because the last term is asymptotically exponentially small, this justifies our preceding formal analysis and makes the limit $w(\eta)$ the natural first iterate. A numerical method of solution can be developed by slowly-translating the asymptotic profile $\phi(\eta)$ in order to minimize the correction v (cf. Garbey and O'Malley (1995)).

To illustrate the supersensitivity of the asymptotic solution, we restrict attention to the multiply-perturbed Burgers' problem

(30)
$$\begin{cases} u_t + u u_x = \epsilon u_{xx} + b_f e^{-a/\epsilon} \\ u(\pm 1, t, \epsilon) = \mp 1 + 2 b_\pm e^{-a/\epsilon} \end{cases}$$

where a again satisfies $0 < a < 1$ and $u(x, 0, \epsilon)$ is as previously prescribed. Now it is natural to set

(31)
$$u(x, t, \epsilon) = -\tanh(\eta/2) + v(\eta, t, \epsilon),$$

again using the stretched variable $\eta = (x - x_\epsilon(t))/\epsilon$ where $u(x_\epsilon(t), t, \epsilon) = 0$. This implies the nonlinear boundary value problem

$$\begin{cases} v_{\eta\eta} + \phi_{\eta\eta} - (v + \phi)(v_\eta + \phi_\eta) + (v_\eta + \phi_\eta)\frac{dx_\epsilon}{dt} - \epsilon\frac{dv}{dt} + \epsilon b_f e^{-a/\epsilon} = 0 \\ v(\eta_\pm, t, \epsilon) = \mp 1 - \phi(\eta_\pm) + 2 b_\pm e^{-a/\epsilon}, \ v(0, t, \epsilon) = 0 \end{cases}$$

for v, with $\phi(\eta) \equiv -\tanh(\eta/2)$. We next approximate v by the solution w of the linearized problem

$$\begin{cases} w_{\eta\eta} + (w\tanh(\eta/2))_\eta = -\epsilon b_f e^{-a/\epsilon} + \frac{1}{2}\mathrm{sech}^2\frac{\eta}{2}\frac{dx_\epsilon}{dt} \\ w(\eta_\pm, t) = 2 b_\pm e^{-a/\epsilon} \mp 2 e^{\mp\eta_\pm}, \ w(0, t) = 0 \end{cases}$$

on the algebraically-long interval $\eta_- \leq \eta \leq \eta_+$. Any solution w of the differential equation with $w(0) = 0$ must have the form

$$w(\eta, \sigma) = \frac{dx_\epsilon}{dt}\tanh^2(\eta/2) + \lambda(t, \epsilon)\left(\frac{\eta}{2}\mathrm{sech}^2\frac{\eta}{2} + \tanh\frac{\eta}{2}\right)$$
$$-\epsilon b_f e^{-a/\epsilon}\left(\frac{\eta^2}{4}\mathrm{sech}^2\frac{\eta}{2} + \eta\tanh\frac{\eta}{2} - \tanh^2\frac{\eta}{2}\right).$$

Here $\frac{dx_\epsilon}{dt}$ and the constant of integration λ must be determined asymptotically by the endconditions

$$2 b_\pm e^{-a/\epsilon} \mp 2 e^{\mp\eta_\pm} \sim \frac{dx_\epsilon}{dt} \pm \lambda - \epsilon b_f e^{-a/\epsilon}(\pm\eta_\pm - 1).$$

They specify an asymptotically negligible correction term v and, more critically, require the shock layer profile $\phi\left(\frac{x - x_\epsilon(t)}{\epsilon}\right)$ to move according to the equation

(33)
$$\frac{dx_\epsilon}{dt} \sim e^{-1/\epsilon}\left(e^{-x_\epsilon/\epsilon} - e^{x_\epsilon/\epsilon}\right) + b(\epsilon)e^{-a/\epsilon}$$

where $b(\epsilon) \equiv b_+ + b_- + b_f(1 - \epsilon)$. Equation (33) can be readily analyzed to provide a solution which converges sluggishly to the steady state with a shock located at $1 - a$ if $b(\epsilon) > 0$, at $a - 1$ if $b(\epsilon) < 0$, and at 0 if $b(\epsilon) = 0$.

References

[1] J. Carr, D.B. Duncan, and C.H. Walshaw, *Numerical Approximation of a Metastable System*, IMA J. Numer. Anal., to appear.

[2] D.G. Crighton, *The Taylor Internal Structure of Weak Shock Waves*, J. Fluid Mech., 173(1986), pp. 625–642.

[3] D.G. Crighton, Nonlinear Acoustics, *Modern Methods in Analytical Acoustics*, D.G. Crighton, A.P. Dowling, J.E. Ffowcs-Williams, M. Heckl, and F.G. Leppington, editors, Springer-Verlag, Berlin, 1992, pp. 648–670.

[4] M. Garbey and R.E. O'Malley, Jr., work in progress.

[5] S.E. Harris, *Sonic Shocks Governed by the Modified Burgers' Equation*, to appear.

[6] F.A. Howes, *Singular Perturbations and Differential Inequalities*, Memoirs, Amer. Math. Soc., 168(1976).

[7] A.M. Il'in, *Matching of Asymptotic Expansions of Solutions of Boundary Value Problems*, Amer. Math. Soc., Providence, 1992.

[8] J.G.L. Laforgue and R.E. O'Malley, Jr., *Supersensitive Boundary Value Problems*, Asymptotic and Numerical Methods for Partial Differential Equations with Critical Parameters, H.G. Kaper and M. Garbey, editors, Kluwer, Dordrecht, 1993, pp. 215–223.

[9] ——, *Shock Layer Movement for Burgers' Equation*, SIAM J. Appl. Math., 55(1995), pp. 332–347.

[10] ——, *Viscous Shock Motion for Advection-Diffusion Equations*, Studies in Appl. Math. 95(1995), pp. 147–170.

[11] L.G. Reyna and M.J. Ward, *On the exponentially slow motion of a viscous shock*, Comm. Pure Appl. Math., 48(1995), pp. 79–120.

ON EXPANSION METHODS
FOR INVERSE AND RECOVERY PROBLEMS
FROM PARTIAL INFORMATION

M. Zuhair Nashed[*]

October 20, 1995

Abstract

In this paper we delineate the role of reproducing kernel Hilbert spaces in expansion methods for inverse problems associated with recovery problems from partial or indirect information. This is illustrated by applications to sampling theorems in signal analysis and by the problem of reconstructing a function from its moments with respect to a given family of functions. We also describe a new approach to numerical solution of integral equations of the first kind using sampling expansions.

1 Introduction and Motivation

Series expansions and integral representations of functions and operators play a fundamental role in the analysis of *direct problems* of applied mathematics; witness the role of Fourier series, power series, and eigenfunction expansions associated with self-adjoint linear operators, and the role of integral representations in potential theory, boundary value problems, complex analysis and other areas. We refer, but not only on this important occasion, to the classic books by Stakgold [20], [21].

Expansion methods play also a useful role in certain *inverse problems*. To motivate our presentation, we first consider two simple recovery problems for a function f belonging to a function space V of dimension n. Let $\{u_1, u_2, \cdots, u_n\}$ be a basis for V. Then f has the unique representation $f(x) = \sum_{i=1}^{n} c_i u_i(x)$. The form of the coefficients, c_i, and the ease of their computation depend on the type of *recovery problem* considered, and the choice of a basis that would be most convenient for computation.

(i) **Recovery of f from inner products (moments):** If we know $\alpha_j := \langle f, u_j \rangle$, then the c_i's are the solutions of the system

$$\sum_{i=1}^{n} c_i \langle u_i, u_j \rangle = \alpha_j, \quad j = 1, \cdots, n.$$

---
[*]Department of Mathematical Sciences, University of Delaware, Newark, DE 19716.

In particular, if $\{u_1, \cdots, u_n\}$ is an orthonormal basis, then $c_j = \langle f, u_j \rangle$ and so we have the Fourier expansion

(1)
$$f(x) = \sum_{j=1}^{n} \langle f, u_j \rangle u_j(x).$$

(ii) **Recovery of a function from its values on a subset of its domain.** If we know $f(t_j), j = 1, \cdots, n$, then the c_i's are uniquely determined from the system

$$\sum_{i=1}^{n} c_i u_i(t_j) = f(t_j), \quad j = 1, \cdots, n.$$

In particular, if we choose $u_i(t_j) = \delta_{ij}$, the Kronecker delta, then we obtain the expansion

(2)
$$f(t) = \sum_{i=1}^{n} f(t_i) u_i(t),$$

which may be viewed as a Fourier-type expansion in terms of the *discrete* orthonormal sequence $\{u_i(t_j)\}_{i,j=1}^{n}$. This is a finite dimensional version of a *sampling expansion* in signal processing.

Both forms, (1) and (2), are special cases of the expansion

$$f(t) = \sum_{i=1}^{n} l_i(f) u_i(t),$$

where the l_i's are (continuous) linear functionals. The extensions to infinite dimensional problems lead to Fourier series and Shannon's sampling theorem, as examples.

The problem of reconstruction (recovery) of a function in an infinite dimensional space (such as a finite energy signal) from *partial* or *indirect* information arises in many areas of applied mathematics, and can be viewed as an *inverse* problem. Examples of recovery problems from partial information include (i) *sampling problem*: given $f(t_n)$, $n \in N$, construct f; (ii) *extrapolation problem*: given $f(t)$ for $t \in T$, extrapolate f to a larger domain; (iii) given $f(t_n)$ and $f'(t_n)$ for a set of points $\{t_n\}$, construct f. Clearly these problems do not have a solution in general. These problems are classical problems in signal analysis. The field of signal processing is concerned with the representation, compression, and recovery of signals. Its principal tools are sampling expansions, filters, and transforms. Two or three decades ago, the common thread between inverse problems, signal processing, and moment problems was rather tenuous; these fields appeared dissimilar in the point of view, scope and intent. Researchers in one of these fields were often unfamiliar with the other two areas. This situation has changed drastically in recent years, and there is now significant interactions among these fields. This is partly due to some of the unifying common concepts and tools that are used in these fields (such as expansion theorems, wavelets, certain function spaces, etc.) and the impact of functional analysis and computational mathematics on recent developments.

In this paper, we consider the role of *reproducing kernel Hilbert spaces* (RKHS) in such problems. We recall that a Hilbert space of functions f on an interval T is said to be an RKHS if all the evaluation functionals $E_t(f) := f(t)$, $f \in H$, for each fixed $t \in T$, are continuous. Then, by the Reisz's representation theorem, for each $t \in T$ there exists a unique element, denote it by k_t, in H such that

$$(3) \qquad\qquad f(t) = \langle f, k_t \rangle, \; f \in H.$$

Let $k(t, s) := \langle k_s, k_t \rangle$ for $s, t \in T$. Then $k_s(t) = k(t, s)$, which is called the *reproducing kernel* (RK) of H. It is clear that every finite dimensional function space V is an RKHS with reproducing kernel $k(s, t) = \sum_{i=1}^{n} u_i(s)\overline{u_i(t)}$, where $\{u_i\}_{i=1}$ is any orthonormal basis for V.

If H is a separable Hilbert space with reproducing kernel $k(t, s)$ and orthonormal basis $\{u_n\}_{n=1}^{\infty}$, then

$$k(t, s) = \sum_{i=1}^{n} u_i(t)\overline{u}_i(s),$$

the series converging absolutely. Note that $L_2(T)$ is not an RKHS, while the subspace of all functions in $L_2(T)$ which are absolutely continuous and whose derivative is $L_2(T)$ is an RKHS under the appropriate inner product. $L_2(T)$ contains many interesting subspaces which are RKHS. For basic properties of RKHS, see [1], [10], [15] and [20].

2   Sampling Expansions and Signal Extrapolation

Let $f(t)$ be a (finite energy) signal defined on the real line. Consider its expansion $f(t) = \sum_{n=1}^{\infty} c_n u_n(t)$ for some orthonormal basis $\{u_n\}_{n=1}^{\infty}$. Suppose further that there exists an RK $k(t, s)$ such that for some real numbers t_n, one has $u_n(s) = k(t_n, s)$. Then $c_n = \langle f, u_n \rangle = f(t_n)$, and the series expansion becomes a *sampling theorem* which recovers $f(t)$ from its sampled values $f(t_n)$:

$$(4) \qquad\qquad f(t) = \sum_{n=1}^{\infty} f(t_n)\, u_n(t)$$

The sequence $\{u_n\}$ is called a *sampling sequence*. For example, for a π-band limited signal (i.e., a signal whose Fourier transform vanishes outside the interval $[-\pi, \pi]$), the reproducing kernel is $k(t,s) = \frac{\sin \pi(t-s)}{\pi(t-s)}$, and $u_n(t) = k(t,n)$ is an orthonormal basis in $L^2(\mathbf{R})$. In this case the above expansion reduces to the classical Shannon-Whittaker sampling theorem. This idea is generalized by Nashed and Walter [18] to the case of a general RKHS H_k with reproducing kernel $k(t,s)$, under the assumption that H_k is a closed subspace of the Sobolev space H^{-1}, the space of tempered distributions $f(t)$ whose Fourier transform $F(\omega)$ satisfies $\int_{-\infty}^{\infty} |F(\omega)|^2(|\omega|^2 + 1)^{-1}d\omega < \infty$. In general, the sampling sequence $\{u_n\}$ appearing in $f(t) = \sum f(t_n)u_n(t)$ need not be an orthogonal system. In this case the system $\{u_n, k(t_n, \cdot)\}$ has to satisfy the *biorthogonality condition*: $u_n(t_m) = \delta_{mn}$.

The general setup described above, in terms of reproducing kernel Hilbert spaces, allows one to include sampling theorems related to other transforms than the Fourier transform (as in the classical Shannon-Whittaker sampling theory), such as the Sturm-Louisville, Jacobi and Laguerre transforms, and sampling theorems using frames and wavelets. See Nashed and Walter [18], Zayed [30]. For delightful research tutorials on sampling theorems see also Butzer [5], Butzer, Splettstößer and Stens [6], and Higgins [9].

The Paley-Wiener theorem provides a direct characterization of the space of bandlimited functions. A π-bandlimited function f is the restriction to $\mathbf{R}$ of an entire function of exponential order one, i.e., $|f(t)| \leq Me^{a|t|}$. Thus, in theory, a bandlimited function f is determined by its value on any interval (actually much less information suffices for determining f). This motivates the following band-limited signal extrapolation problem: **How does one practically extrapolate a bandlimited signal f outside an interval $[-T,T]$ when $f(t)$ is given for $t \in [-T,T]$?** Since f is analytic, one obvious approach would be to compute the derivatives $f^{(n)}$ at $t = 0$ by using the values of f in $[-T,T]$ and then use the Taylor expansion to reconstruct $f(t)$. However this method is extremely unstable due to the instability in computing derivatives of high orders. Numerical differentiation is an ill-posed problem and the "*degree of ill-posedness*" increases by one with each order of differentiation. This is an example that dramatizes the unsuitability of certain expansion methods that appear to be analytically viable.

This signal extrapolation problem has many applications in signal processing, spectral estimation and limited-angle reconstruction in medical tomography. Since the early 70's there has been considerable interest in this problem, which is an ill-posed inverse problem. The extrapolated values change drastically when the given data in the interval change slightly. There are several modified algorithms to improve the extrapolation performance. However, to our knowledge, there is no extrapolation algorithm that estimates the error between the extrapolated and the true values outside the given interval for any nontrivial class of bandlimited signals, when the data are inaccurate. X-G. Xia and the author [28] have recently proposed a modified minimum norm solution method which estimates the error between the extrapolation error for some nontrivial classes of bandlimited signals, when the maximum magnitude of the data error is known. Some of these subspaces are again reproducing kernel Hilbert spaces of the Paley-Wiener space.

3 Moment Problems in RKHS

In this section we illustrate the role of RKHS in the recovery of a signal from its moments with respect a sequence of functions. The *a priori* knowledge that the signal belongs to a family of RKHS (as for example in the case of a bandlimited signal) generates a new family of functions which are linearly independent of the given sequence, and whose inner products with the unknown signal are known. This provides additional terms that are useful in the reconstruction processes, such as the Backus-Gilbert method.

Let H be a Hilbert space of functions defined on an interval T.

The moment problem. Find a function $f \in H$ such that

$$(5) \qquad \langle f, g_k \rangle = \mu_k \qquad k \in \mathcal{K}$$

where $\mathcal{K}$ is a subset of the set $\mathbf{Z}$ of all integers, $\{\mu_k\}_{k \in \mathcal{K}}$ belongs to l_2, and $\{g_k\}_{k \in \mathcal{K}}$ is a given system in $\boldsymbol{H}$.

The following results, (i) and (ii), about the existence and uniqueness of the solutions of this moment problem are known (see, for example, Young [29]).

(i) Uniqueness. The solution f of the moment problem is unique if and only if the system $\{g_k\}$ is complete.

Let $\boldsymbol{H}$ be a separable Hilbert space with an orthonormal basis $\{u_k\}$. Recall that $\{g_k\}$ is said to be a *Riesz-Fischer system* if there exists a bounded linear operator L on $\boldsymbol{H}$ into $\boldsymbol{H}$ such that $Lg_k = u_k$. A system $\{g_k\}$ is said to be a *Riesz basis* if there exists a bounded linear invertible (one-to-one and onto) operator L such that $Lg_k = u_k$.

(ii) Existence. A necessary and sufficient condition for the problem (5) to have a solution is that $\{g_k\}$ is a Riesz-Fischer system.

We can rewrite the moment problem (5) as an operator equation

(6) $$Af = \mu$$

where $A : H \rightarrow l^2$ is given by

(7) $$Af = \{\langle f, g_k \rangle\}_{k \in \mathcal{K}}.$$

Since, for any system $\{g_k\}_{k \in \mathcal{K}}$ in H, we have the orthogonal direct sum decomposition

$$H = \overline{\text{span}\{g_k\}} \oplus (\text{span}\{g_k\})^{\perp},$$

we immediately have the following results:

(iii) If the moment problem has a solution. then there is a unique *minimal norm* solution, which is characterized as the solution that belongs to the subspace $\overline{\text{span}\{g_k\}}$.

(iv) If the system $\{g_k\}$ is not complete, then any solution of the moment problem can be obtained by adding to the minimal norm solution a function in $(\text{span}\{g_k\})^{\perp}$.

(v) If $\{g_k\}$ is a Riesz basis, then the operator A defined in (7) is invertible and A^{-1} is given by

(8) $$A^{-1}\mu = \sum_i \sum_j (G^{-1})_{ji} \mu_i g_j$$

where G is the Gram matrix of the system $\{g_k\}$. In this case, the unique solution of the moment problem has the representation

$$f = \sum_{k \in \mathcal{K}} \mu_k \psi_k$$

where $\{\psi_k\}$ is the biorthogonal system of $\{g_k\}$ which can be constructed via

$$\psi_k = \sum_j (G^{-1})_{kj} g_j.$$

The above expansions for the solution of the moment problem are not useful in practice since A^{-1} is an unbounded operator and these expansions are highly sensitive to errors (the problem is ill-posed). A more useful approximating expansion is obtained via the Backus-Gilbert (BG) method. Let $H = L^2(T)$. The BG method constructs an approximation $\tilde{f}$ to f in the form

(9) $$\tilde{f}(t) = \sum_{k \in \mathcal{K}} \alpha_k(t)\mu_k \qquad t \in T$$

where $\alpha_k(t)$ are to be determined according to a certain criterion described below. Let

(10) $$A(t,t') = \sum_{k \in \mathcal{K}} \alpha_k(t)g_k(t').$$

Then we have the following averaging formula for $\tilde{f}$:

(11) $$\tilde{f}(t) = \int_T A(t,t')f(t')dt'.$$

A natural condition on the averaging kernel $A(t,t')$ is the following normalization condition for each fixed t

$$\int_T A(t,t')dt' = 1.$$

From (11) we can see that if $A(t,t')$ is the delta "function" $\delta(t-t')$, then $\tilde{f} = f$ for a continuous function f. This suggests that we choose $\alpha_k(t)$ in (9) such that the kernel function $A(t,t')$ in (10) is as 'close' to $\delta(t-t')$ as possible. Three useful criteria for measuring this closeness are analyzed in [27].

Up to now no additional a priori information about the signal f is assumed.

Now let $\boldsymbol{H}_\sigma, \sigma \in \mathcal{J}$, be a family of RKHS where $\mathcal{J}$ is an index set with a certain measure $m(\sigma)$, such as, $\mathcal{J} = (a,b), \mathcal{J} = \{1,2,\cdots\}$, $\mathcal{J} = \{1\}$, etc. Let $Q_\sigma(s,t)$ be the RK corresponding to $\boldsymbol{H}_\sigma, \sigma \in \mathcal{J}$. Assume that the original signal $f \in \boldsymbol{H}_\sigma$ for all $\sigma \in \mathcal{J}$. From the reproducing property of each $Q_\sigma(s,t)$ we have

(12) $$f(t) = \int_T f(s)Q_\sigma(s,t)ds \qquad \sigma \in \mathcal{J}, t \in T.$$

We now formulate the following *new moment problem:*

(13) $$\begin{cases} Q_\sigma(s,t), \ \sigma \in \mathcal{J}, \text{ are known.} \\[1ex] f \in \boldsymbol{H}_\sigma \text{ for every } \sigma \in \mathcal{J}. \\[1ex] \mu_k = \int_T f(t)g_k(t)dt, \ k \in \mathcal{K} \text{ are known.} \\[1ex] g_k, \ k \in \mathcal{K}, \text{ are known.} \\[1ex] f \text{ is the unknown function to be recovered.} \end{cases}$$

The families of the reproducing kernels and moment kernels $\{Q_\sigma, g_k\}$ generate a new family of kernel functions via

(14) $$\tilde{g}_{k,\sigma}(s) = \int_T g_k(t)Q_\sigma(t,s)dt.$$

For $\sigma \in \mathcal{J}$, let

(15) $$\mu_{k,\sigma} = \int_T f(s)\tilde{g}_{k,\sigma}(s)ds.$$

Then, from (12) we have

(16) $$\mu_{k,\sigma} = \mu_k \qquad \text{for } \sigma \in \mathcal{J}$$

where the μ_k are the original moments in (5). Thus, for the new moment problem, besides the moments μ_k and kernels g_k in (13) we have additional moments $\mu_{k,\sigma}$ and kernels $\tilde{g}_{k,\sigma}$ given in (14)– (16).

If there exist $k_0 \in \mathcal{K}$ and $\sigma_0 \in \mathcal{J}$ such that $g_{k_0} \notin H_{\sigma_0}$ and $\tilde{g}_{k,\sigma_0}, k \in \mathcal{K}$, are linearly independent, then $\tilde{g}_{k_0,\sigma_0}$ is linearly independent of $g_k, k \in \mathcal{K}$, and thus we have more functions that generate known moments for the new moment problem than the original one, which can be effectively utilized in inversion procedures, such as the BG method. The shape of the averaging kernel $A(t,t')$ in $\tilde{f}$ from the BG method may be better controlled with the above information for a general moment problem.

An interesting application of moment problems to magnetic resonance imaging is given by Zwaan [30], [32].

4 Sampling and Wavelet Sampling Solutions of Ill–Posed Problems

First kind integral equations, in which one attempts to find a solution $x(t)$ of the equation

(17) $$y(t) = \int_a^b K(t,s)x(s)ds$$

are notoriously difficult to solve. The reason is that in most cases, solving the equation (17) is an *ill–posed* problem. It may not have a solution and even if it does, the solution does not depend continuously on the data.

The difficulty is compounded if the data $y(t)$ are not known exactly but only approximately for a discrete set of t, and moreover are contaminated by noise. That is, instead of (17) we have to solve

(18) $$y_i = \int_a^b K(t_i,s)x(s)ds + \epsilon_i, \quad i = 1,2,\cdots,n$$

where ϵ_i is a random variable of noise.

A number of approaches are available (see e.g.. [3], [12], [14]), the most popular of which is regularization ([8], [16], [25]). However, most of them seem to suffer from either slow convergence or a tedious asymptotic analysis to determine an optimal value for a regularization (or related) parameter. There is a technique, due principally to Stenger [23], [24] which is used to solve certain integral equations numerically and whose rate of convergence is exponential. This technique uses cardinal series and has been applied to a number of different integral operators most of which, however, are not compact.

Fortunately a number of new results have appeared recently (such as [18], [2], [26], [30]) which will enable us to use signal analysis techniques for a wider class of problems. These may be coupled with an older approach involving problems such as (17) and (18) in a reproducing kernel Hilbert space setting ([14], [15], [16], [17].

In this section we describe a new approach to the solutions of inverse and ill-posed problems which is being studied by G. G. Walter and the author. The approach uses general sampling theorems in RKHS developed by the authors in [18].

Rather than attempting to solve (18) we shall use the usual engineering approach and replace the discrete values $\{y_i\}$ by the impulse train

$$(19) \qquad y^*(t) = \sum_{i=1}^{n} y_i \delta(t - i),$$

where $\delta(t - i)$ is the Dirac delta "function" (impulse) at $t = i$. The domain and range of the integral operator in (17) will both be assumed to be reproducing kernel Hilbert spaces (RKHS) with reproducing kernels P and Q, respectively, which we denote as H_P and H_Q. (This assumption is satisfied in many problems of interest.) We also suppose that H_Q is a closed subspace of the Sobolev space H^{-1} (so we can relate y to y^* as in [18]).

Our proposed procedure is as follows:

(i) Project the data y^* into the range space H_Q.

(ii) Approximate the projected function $\mathbf{P}y^*$ by a *sampling series*.

(iii) Solve the problem for the *partial* sums of the sampling series with a sampling series in H_P.

(iv) Find the resulting error and establish its order.

That (i) and (ii) are possible is shown in [18]. Step (iii) is feasible if the reproducing kernel has isolated zeros. Then as was shown in [18], $\{Q(t_n, t)\}$ and $\{P(s_n, s)\}$ are orthogonal sequences.

The integral operator $\mathbf{K}$ in (17) maps $P(s_n, s)$ into $K(t, s_n)$. If we expand x with respect to $\{P(s_n, s)\}$ we obtain

$$x(s) = \sum_n x(s_n) \frac{P(s_n, s)}{P(s_n, s_n)}$$

and similarly

$$y(t) = \sum_n y(t_n) \frac{Q(t_n, t)}{Q(t_n, t_n)}.$$

Hence we must solve

$$\sum_n y(t_n) \frac{Q(t_n, t)}{Q(t_n, t_n)} = (\mathbf{K}x)(t) = \sum_n x(s_n) \frac{K(t, s_n)}{K(s_n, s_n)}.$$

But at $t = t_m$ we have

$$y(t_m) = \sum_n x(s_n) \frac{K(t_m, s_n)}{P(s_n, s_n)}$$

and so $x(s_n)$ may be obtained by solving a linear system whose matrix is

$$\left[\frac{K(t_m, s_n)}{P(s_n, s_n)} \right].$$

The error calculations may be quite complex. However some types of error have been considered in [18]. The procedure can also be extended to treat (more abstract) operator equations acting between two RKHS, and to determine numerically least–squares solutions when y is not in the range of the operator, using the general framework of [12], [15], [16].

In summary, the proposed procedure is to attack a semi-discrete (or moment discretization) version of (18) of (17) by using the associated impulse train (19). This is projected onto the reproducing kernel Hilbert space H_Q, and is then approximated by the partial sums of a sampling series. The problem is finally to be solved for these partial sums and the error found and compared to other methods.

We conclude this paper by considering wavelet sampling solutions of (17)– (18). Associated with each RKHS $V_0 \subset L^2(\mathbf{R})$ there is a sequence $\{V_n\}$ of *dilation spaces*

$$V_{i+1} = \{ f \in L^2(\mathbf{R}) : f(x/2) \in V_i \}, i = 0, 1, \ldots$$

Clearly each V_i is an RKHS. If $V_{i+1} \supset V_i$ we obtain a *multiresolution ladder*

$$V_0 \subset V_1 \subset \cdots \subset V_i \subset V_{i+1} \subset \cdots \subset L^2(\mathbf{R}).$$

This is the setting for *wavelet analysis* [6]. The initial space V_0 is chosen as the closure of the span of translates of the *scaling function* $\varphi(x)$. The reproducing kernel for this space is

$$k(x, y) = \sum_{n=-\infty}^{\infty} \phi(x - n)\phi(y - n)$$

where $\{\phi(\cdot - n)\}$ is an orthonormal system. These scaling functions satisfy a scaling equation

$$\phi(x) = \sum c_k \phi(2x - k), \ \{c_k\} \in \ell^2.$$

The *"mother wavelet"* associated with ϕ is given by

$$\psi(x) = \sum (-1)^{k-1} c_k \phi(2x + k - 1).$$

Its translates form an orthonormal basis of the orthagonal complement W_1 of V_0 in V_1; i.e. $V_1 = V_0 \oplus W_1$ and

$$\psi_{jk}(x) = 2^{j/2} \psi(2^j x - k)$$

form an orthogonal basis of $L^2(\mathbf{R})$. For the classical Paley–Wiener case, the space V_0 is the space of π–bandlimited functions and $\phi(x) = \operatorname{sinc} \pi x$. Then V_1 is the space of 2π–bandlimited functions and W_1 is the set of functions whose Fourier transform has support in $[-2\pi, -\pi) \cup [\pi, 2\pi)$.

We now apply the sampling solution procedure to (18) but restrict consideration to H_Q which are wavelet subspaces. These subspaces are usually associated with a multiresolution analysis (MRA), a nested sequence $\{V_m\}$ of RKHS contained in $L^2(\mathbf{R})$ such that

$$\cdots \subset V_{-1} \subset V_0 \subset V_1 \subset \cdots \subset V_m \subset \cdots \subset L^2(\mathbf{R}).$$

Each of these subspaces has an associated sampling theorem under very broad hypotheses (see [18] [26])and fits the theory developed in our previous work. A similar approach in a wavelet context has been considered by Donoho and Johnstone who in a series of papers (see e.g [7]) have attacked the problem (18) by first projecting (19) onto V_m, but then using a "shrinking" procedure to change the wavelet coefficients. Their method is based on a statistical optimization technique and works best if $\{\epsilon_i\}$ is white noise. It uses the sample values $\{y_i\}$ as an approximation to the wavelet coefficients. Our method, because it uses the sampling series rather than the wavelet series, is exact when the noise is zero and $y \in V_m$. It also works when the $\{t_i\}$ are not uniformly distributed since nonuniform sampling exists in some wavelet subspaces ([2], [26], [30]).

Reproducing kernel Hilbert space methods have been effectively developed for the simultaneous regularization and approximation of ill-posed integral and operator equations (see, e.g., [16]). RKHS methods can also be coupled with other approaches to improperly posed problems, as described in the classification given by Payne [19]. The proposed approach should increase the effectiveness of the RKHS approach since *sampling expansions* (with derived error bounds) in wavelet and reproducing kernel subspaces will replace the integral representations. The interlocking in the areas of inverse/ill-posed problems, signal analysis and moment problems will advance all these areas.

References

[1] N. Aronszajn, Theory of reporducing kernels, *Trans. Amer. Math. Soc.*, **68** (1950), pp. 337–404.

[2] J.J. Benedetto and M. W. Frazier, editors, *Wavelets: Matehmatics and Applications*, CRC Press, Boca Raton, 1993.

[3] M. Bertero, C. De Mol, and E.R. Pike, Linear inverse problems with discrete data, I, II, *Inverse Problems*, **1** (1985), 301–330; **4** (1988), pp. 573–594.

[4] P. Butzer, A survey of the Whittaker–Shannon sampling theorem and some of its extensions, *J. Math. Res. Exposition*, **3** (1983), pp. 185–212.

[5] P. Butzer, W. Splettstößer and R. Stens, The sampling theorem and linear predictions in signal analysis, *Jahresber, Deutsch. Math.–Verein*, **90** (1988), pp. 1–60.

[6] I. Daubechies, *Ten Lectures on Wavelets*, SIAM, Philadelphia, 1992.

[7] D.L. Donoho and I.M. Johnstone, Minimax estimation via wavelet shrinkage, Tech. Rept., Dept. of Statistics, Stanford University, 1992.

[8] C.W. Groetsch, *the Theory of Tikhonov Regularization for Fredholm Equations of the First Kind*, Pitman, London–Boston, 1984.

[9] J. Higgins, Five short stories about the cardinal series, *Bull. Amer. Math. Soc.* **12** (1985), pp. 45–89.

[10] E. Hille, Introduction to the general theory of reproducing kernels, *The Rocky Mountain J. Math.*, **2** (1972), pp. 321–368.

[11] M.E.H. Ismail, M.Z. Nashed, A.E. Zayed and A.F. Ghaleb, editors, *Mathematical Analysis, Wavelets, and Signal Processing*, Contemporary Mathematics, Vol. 190, American Mathematical Society, Providence, R.I., 1995.

[12] M.Z. Nashed, Approximate regularized solutions to improperly posed linear integral and operator equations, in *Constructive and Computational Methods for Differential and Integral Equations* (D. Colton and R.P. Gilbert, Eds.), pp. 289–332. Springer–Verlag, Berlin–Heidelberg–New York, 1974.

[13] M.Z. Nashed, On moment–discretization and least–squares solutions of linear intergral equations of the first kind, *J. Math. Anal. Appl.*, **53** (1976), pp. 359–366.

[14] M.Z. Nashed, Operator–theoretic and computational approaches to ill–posed problems with applications to antenna theory, *IEEE Trans. Antennas and Propagation*, AP–29 (1981), pp. 220–231.

[15] M.Z. Nashed and G. Wahba, Convergence rates of approximate least–squares solutions of linear integral and operator equations of the first kind, *Math. Comp.*, **28** (1974), pp. 69–80.

[16] M.Z. Nashed and G. Wahba, Regularization and approximation of linear operator equations in reproducing kernel spaces, *Bull. Amer. Math. Soc.*, **80** (1974), pp. 1273–1218.

[17] M.Z. Nashed and G. Wahba, Generalized inverses in reproducing kernel spaces: an approach to regularization of linear operator equations. *SIAM J. Math. Anal.*, **5** (1974), pp. 974–987.

[18] M.Z. Nashed and G.G. Walter, General sampling theorems for functions in reproducing kernel Hilbert spaces, *Math. Control, Signals & Systems*, **4** (1991), pp. 363–390.

[19] L.E. Payne, *Improperly Posed Problems in Partial Differential Equations*, Regional Conference Series in Applied Mathematics, Vol. 22, SIAM, Philadelphia, 1975.

[20] H.S. Shapiro, *Topics in Approximation Theory*, Springer–Verlag, Berlin–Heidelberg, New York, 1981.

[21] I. Stakgold, *Boundary Value Problems of Mathematical Physics*, Vols. I and II, MacMillan, New York, 1967, 1968.

[22] I. Stakgold, *Green's Functions and Boundary Value Problems*, Wiley, New York, 1980.

[23] F. Stenger, Numerical methods based on Whittaker cardinal, or sinc functions, *SIAM Review*, **23** (1981), 165–224.

[24] F. Stenger, *Numerical Methods Based on Sinc and Analytic Functions*, Springer–Verlag, New York, 1993.

[25] A.N. Tikhonov and V.Y. Arsenin, *Solution of Ill-Posed Problems*, Winston, Wiley, New York, 1977.

[26] G.G. Walter, *Wavelets and Other Orthongonal Systems with Applications*, CRC Press, Boca Raton, 1994.

[27] X-G. Xia and M.Z. Nashed, The Backus-Bilgert method for signals in reproducing kernel Hilbert spaces and wavelet subspaces, *Inverse Problems*, **10** (1994), pp. 785–804.

[28] X-G. Xia and M.Z. Nashed, A modified minimum norm solution method for band-limited signal extrapolation with inaccurate data, to appear.

[29] R.M. Young, *An Introduction to Nonharmonic Fourier Series*, Academic Press, New York, 1980.

[30] A.I. Zayed, *Advances in Shannon's Sampling Theory*, CRC Press, Boca Raton, 1993.

[31] M. Zwaan, Moment Problems in Hilbert Space with Applications to Magnetic Resonance Imaging, (CWI/Tract **89** P.O. Box 8079, Amsterdam, The Netherlands), 1993.

[32] M. Zwaan, Approximation of the solution to the moment problem in a Hilbert space, *Numer. Funct. Anal. Optimiz.*, **11** (1990), pp. 601–612.

[33] A. Zwaan, MRI reconstruction as a moment problem, *Math. Meth. in Appl. Sci.*, **15** (1992), 661–675.

Multiple Steady States in Coaxial Discs Flow

David O. Olagunju[*]

Abstract

Consider the creeping flow of a viscoelastic fluid between two coaxial disks. We show that for Deborah, De in a neighborhood of a critical value $\mathrm{De}_c = \pi/[|1 - \delta|\sqrt{\beta(3 + 2\beta)}]$, $\delta \neq 1$, there is a bifurcation from the base torsional flow. The bifurcation is supercritical if the retardation parameter $\beta < \beta_c$ and subcritical if $\beta > \beta_c$ where $\beta_c \simeq 0.592$.

1 Introduction

The torsional flow of a viscoelastic fluid confined to the space between two coaxial discs has many important applications. In 1983, Phan–Thien [10] showed that the base torsional flow is unstable if the Deborah number, De, a measure of the relaxation time of the fluid, exceeds the critical value $\mathrm{De}_c = \pi/\sqrt{\beta(3 + 2\beta)}$. Bifurcation from this base flow was studied numerically by Huilgol and Keller [5], Walsh [14], Crewther et al. [3], and Ji et al. [6]. This paper is concerned with analytical study of the bifurcation problem. We will show that in a neighborhood of the critical Deborah number, which depends on the retardation parameter and the ratio of angular speeds, there are multiple steady state solutions. Specifically, we will show that in a neighborhood of the critical Deborah number there is a bifurcation from the base torsional flow. In addition we will show that the nature of the bifurcation depends on the retardation parameter β, the ratio of the polymer viscosity to the total viscosity of the fluid.

2 Problem formulation

Consider the flow of a non–Newtonian fluid in the gap between two coaxial discs in which the top and bottom discs rotate with constant angular speeds Ω_T and Ω_B respectively. We will assume that the ratio of the plate separation to radius h/a is very small (Figure 1).

---

[*]Department of Mathematical Sciences, University of Delaware, Newark, DE 19716.

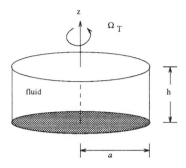

FIG. 1. *Coaxial discs flow*

In the following analysis we will consider only steady state, axially symmetric, creeping flows. For $(h/a) << 1$, one can show [8, 10] that for an Oldroyd–B fluid, the dimensionless equations (after eliminating pressure) in cylindrical coordinates are as follows:

Continuity equation

$$\text{(1)} \qquad \frac{\partial u}{\partial r} + \frac{u}{r} + \frac{\partial w}{\partial z} = 0.$$

Momentum equations

$$\text{(2)} \qquad (1-\beta)\frac{\partial^3 u}{\partial z^3} + \frac{\partial^2 \Sigma}{\partial r \partial z} + \frac{\partial^2 \gamma}{\partial z^2} + \frac{1}{r}(\frac{\partial \Sigma}{\partial z} - \frac{\partial \Delta}{\partial z}) = 0,$$

$$\text{(3)} \qquad (1-\beta)\frac{\partial^2 v}{\partial z^2} + \frac{\partial \zeta}{\partial r} + \frac{2}{r}\zeta + \frac{\partial \Pi}{\partial z} = 0.$$

Constitutive equations

$$\text{(4)} \qquad \Sigma = -\text{De}[u\frac{\partial \Sigma}{\partial r} + w\frac{\partial \Sigma}{\partial z} - 2\Sigma\frac{\partial u}{\partial r} - 2\gamma\frac{\partial u}{\partial z}],$$

$$\text{(5)} \qquad \zeta = -\text{De}[u\frac{\partial \zeta}{\partial r} + w\frac{\partial \zeta}{\partial z} + \Sigma(\frac{v}{r} - \frac{\partial v}{\partial r})$$
$$-\zeta(\frac{u}{r} + \frac{\partial u}{\partial r}) - \Pi\frac{\partial u}{\partial z} - \gamma\frac{\partial v}{\partial z}],$$

$$\text{(6)} \qquad \gamma = -\text{De}[u\frac{\partial \gamma}{\partial r} + w\frac{\partial \gamma}{\partial z} - \gamma(\frac{\partial u}{\partial r} + \frac{\partial w}{\partial z})$$
$$-\Gamma\frac{\partial u}{\partial z} - \Sigma\frac{\partial w}{\partial r}] + \beta\frac{\partial u}{\partial z},$$

$$(7) \qquad \Delta = -\text{De}[u\frac{\partial \Delta}{\partial r} + w\frac{\partial \Delta}{\partial z} + 2\zeta(\frac{v}{r} - \frac{\partial v}{\partial r})$$
$$-2\frac{u}{r}\Delta - 2\Pi\frac{\partial v}{\partial z}],$$

$$(8) \qquad \Pi = -\text{De}[u\frac{\partial \Pi}{\partial r} + w\frac{\partial \Pi}{\partial z} + \gamma(\frac{v}{r} - \frac{\partial v}{\partial r})$$
$$-\Pi(\frac{u}{r} + \frac{\partial w}{\partial z}) - \Gamma\frac{\partial v}{\partial z} - \zeta\frac{\partial w}{\partial r}] + \beta\frac{\partial v}{\partial z},$$

$$(9) \qquad \Gamma = -\text{De}[u\frac{\partial \Gamma}{\partial r} + w\frac{\partial \Gamma}{\partial z} - 2\gamma\frac{\partial w}{\partial r} - 2\Gamma\frac{\partial w}{\partial z}]$$
$$+2\beta\frac{\partial w}{\partial z}.$$

Boundary conditions

$$(10) \qquad u = w = 0, \quad v = \delta r \quad \text{on} \quad z = 0,$$

$$(11) \qquad u = w = 0, \quad v = r, \quad \text{on } z = 1.$$

Here (u, v, w) is the velocity vector and

$$\begin{pmatrix} \Sigma & \zeta & \gamma \\ \zeta & \Delta & \Pi \\ \gamma & \Pi & \Gamma \end{pmatrix}$$

is the stress tensor. The dimensionless parameters appearing above are the Deborah number De, the retardation parameter β, and the rotation ratio $\delta = \Omega_B/\Omega_T$. Note that $0 \le \beta \le 1$.

3 Similarity solutions

The complete formulation of the problem, defined by (1)– (11) requires an additional boundary condition at the interface between the fluid and air. In the present analysis we will ignore this extra condition and instead assume that the system is infinite in the radial direction. For an analysis incorporating free surface effects see for example Joseph [7], Olagunju [8] and Shipman *et al.* [12].

It is known that this problem admits a similarity solution of von–Kármán type [5, 8, 10, 14]. Introduce the following similarity variables

$$u = r\frac{df}{dz}, \quad w = -2f(z), \quad v - rg_0 = rg(z),$$

$$\Sigma = r^2\hat{\Sigma}(z), \quad \zeta = r^2\hat{\zeta}(z), \quad \gamma = r\hat{\gamma}(z),$$

$$\Delta - r^2\Delta_0 = r^2\hat{\Delta}(z), \quad \Pi - r\Pi_0 = r\hat{\Pi}(z), \quad \Gamma = \hat{\Gamma}(z),$$

where $g_0 = (1-\delta)z + \delta$, $\Pi_0 = (1-\delta)\beta$, and $\Delta_0 = 2\beta(1-\delta)^2\text{De}$. Note that $\delta = 1$ corresponds to rigid body motions. Substituting in the governing equations and dropping the hats we obtain

(12) $$\mathcal{L}_1\mathbf{q} - \text{De}\mathcal{L}_2\mathbf{q} + \mathcal{N}(\text{De}, \mathbf{q}) = 0,$$

and the boundary conditions

(13) $$f = f' = g = 0 \quad \text{at} \quad z = 0, 1$$

where

(14) $$\mathbf{q} = (f, g, \Sigma, \zeta, \gamma, \Delta, \Pi, \Gamma)^T.$$

Let $D \equiv \frac{d}{dz}$, then

(15) $$\mathcal{L}_1 = \begin{pmatrix} (1-\beta)D^4 & 0 & 3D & 0 & D^2 & -D & 0 & 0 \\ 0 & (1-\beta)D^2 & 0 & 4 & 0 & 0 & D & 0 \\ 0 & 0 & -1 & 0 & 0 & 0 & 0 & 0 \\ 0 & 0 & 0 & -1 & 0 & 0 & 0 & 0 \\ \beta D^2 & 0 & 0 & 0 & -1 & 0 & 0 & 0 \\ 0 & 0 & 0 & 0 & 0 & -1 & 0 & 0 \\ 0 & \beta D & 0 & 0 & 0 & 0 & -1 & 0 \\ -4\beta D & 0 & 0 & 0 & 0 & 0 & 0 & -1 \end{pmatrix},$$

(16) $$\mathcal{L}_2 = \begin{pmatrix} 0 & 0 & 0 & 0 & 0 & 0 & 0 & 0 \\ 0 & 0 & 0 & 0 & 0 & 0 & 0 & 0 \\ 0 & 0 & 0 & 0 & 0 & 0 & 0 & 0 \\ -\Pi_0 D^2 & 0 & 0 & 0 & -g_0' & 0 & 0 & 0 \\ 0 & 0 & 0 & 0 & 0 & 0 & 0 & 0 \\ 0 & -2\Pi_0 D & 0 & 0 & 0 & 0 & -2g_0' & 0 \\ 2\Pi_0 D & 0 & 0 & 0 & 0 & 0 & 0 & -g_0' \\ 0 & 0 & 0 & 0 & 0 & 0 & 0 & 0 \end{pmatrix},$$

$$\mathcal{N}(\text{De}, \mathbf{q}) = \text{De} \begin{pmatrix} 0 \\ 0 \\ 2(fD\Sigma + \gamma D^2 f) \\ 2fD\zeta + \Pi D^2 f + \gamma Dg \\ 2fD\gamma - 2\gamma Df + \Gamma D^2 f \\ 2(fD\Delta + \Pi Dg) \\ 2fD\Pi - 2\Pi Df + \Gamma Dg \\ 2(fD\Gamma - 2\Gamma Df) \end{pmatrix}.$$

(17)

4 Steady states

One solution of the problem defined by (12) and (13) is $\mathbf{q} = 0$ for all values of the parameters β and De. The linearization of this problem about the trivial solution is

(18) $$\mathcal{L}_1 \mathbf{q} - \text{De}\mathcal{L}_2 \mathbf{q} = 0.$$

Equation (18) and (13) define an eigenvalue problem. It is straight forward to show that there is an infinite sequence of simple eigenvalues $\text{De}_n(\beta)$ [3, 8, 10, 14], the smallest of which is

(19) $$\text{De}_c = \frac{\pi}{|1 - \delta|\sqrt{\beta(3 + 2\beta)}}, \qquad \delta \neq 1.$$

This critical value (for the case $\delta = 0$) was first obtained by Phan–Thien[10]. Let $\lambda = \text{De} - \text{De}_c$, then the nonlinear problem (12) and (13) can be written in abstract form

(20) $$\mathcal{F}(\lambda, \mathbf{q}) = 0$$

where

$$\mathcal{F} : R \times \mathcal{X} \to \mathcal{Y},$$

and $\mathcal{X}$, $\mathcal{Y}$ are Banach spaces [9]. We will consider λ the bifurcation parameter. Note that we have suppressed the dependence on β. By definition it follows that

$$\mathcal{F}(\lambda, 0) = 0$$

The main result is the following.

THEOREM 4.1.

1. $\lambda = 0$ *is a bifurcation point of (20).*

2. *For $\beta < \beta_c$ bifurcation is supercritical and for $\beta > \beta_c$, bifurcation is subcritical, where β_c is the positive root of*

$$(21) \qquad 228\beta^4 + 55\beta^3 - 764\beta^2 - 273\beta + 390 = 0.$$

That $\lambda = 0$ is bifurcation point follows from the fact that $\lambda = 0$ is a simple eigenvalue of the pair $(\mathcal{L}_1, \mathcal{L}_2)$ [9]. For the definition of a simple eigenvalue in this context see Crandall and Rabinowitz [2] and Chow and Hale [1]. In order to determine the nature of the bifurcation we apply the Lyapunov– Schmidt reduction [1, 4, 11, 13]. Let φ be the normalized eigenfunction of $\mathcal{L} \equiv \mathcal{L}_1 - \mathrm{De}_c\mathcal{L}_2$, i.e. such that $< \mathcal{L}_2\varphi, \varphi^* >= 1$, where φ^* is the eigenfunction of the adjoint $\mathcal{L}^*$, and $<, >$ is the L^2 inner product. We assume that the spaces can be decomposed as follows

$$\mathcal{X} = \mathcal{X}_0 \oplus \mathcal{X}_1, \quad \mathcal{Y} = \mathcal{Y}_0 \oplus \mathcal{Y}_1$$

where $\mathcal{X}_0 = [\varphi]$, $\mathcal{Y}_0 = [\mathcal{L}_2\varphi]$ and $[,]$ denotes span. The solution can then be written as $\mathbf{q} = A\varphi + \mathbf{w}$, where A is a constant and $\mathbf{w} \in \mathcal{X}_1$. It can then be shown that equations (12)–(13) are equivalent to the pair

$$(22) \qquad \mathcal{L}\mathbf{w} + N(\lambda, \mathbf{w})- < N, \varphi^* > \mathcal{L}_2\varphi = 0,$$

and

$$(23) \qquad < \mathcal{L}\mathbf{w} + N(\lambda, \mathbf{w}), \varphi^* >= 0.$$

Equation (22) can be solved uniquely for $\mathbf{w} = \mathbf{w}(A, \lambda)$. Substituting this solution into (23) gives the bifurcation equation which we write as

$$(24) \qquad g(A, \lambda) = 0.$$

This equation cannot be written down explicitly. However, since we are mainly interested in the local behavior of g in the neighborhood of $(0,0)$ we use power series to solve (24). A lengthy calculation with Maple gives

(25)
$$\lambda = \lambda_2 A^2 + o(A^2),$$

or

(26)
$$De = De_c + \lambda_2 A^2 + o(A^2),$$

where

(27)
$$\lambda_2 = \frac{\pi^3}{2} \frac{(228\beta^4 + 55\beta^3 - 764\beta^2 - 273\beta + 390)}{|1 - \delta|^3 [\beta(3 + 2\beta)]^{5/2}}, \qquad \delta \neq 1.$$

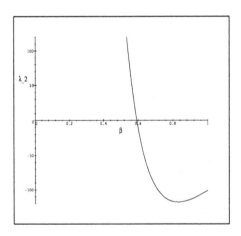

FIG. 2. *Plot of λ_2 vs. β for $\delta = 0$*

The direction of the bifurcation is determined by the sign of λ_2. From Figure 2, we see that there is a value β_c of β such that λ_2 is negative if $\beta > \beta_c$ and positive if $\beta < \beta_c$, where $\beta_c \simeq 0.592$ is the positive solution of the equation $\lambda_2 = 0$ which is (21). Thus for $\beta < \beta_c$ bifurcation is supercritical and for $\beta > \beta_c$ it is subcritical (Figure 3). At $\beta = \beta_c$

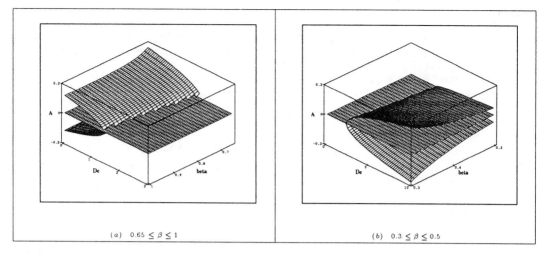

(a) $0.65 \leq \beta \leq 1$ (b) $0.3 \leq \beta \leq 0.5$

FIG. 3. *Bifurcation surfaces for $\delta = 0$, showing (a)subcritical and (b) supercritical pitchfork bifurcations*

there is a degeneracy and bifurcation cannot be determined at this point. For relevant parameter values our results agree with those obtained from numerical computations. The calculations of Walsh [14] for the Maxwell fluid ($\beta = 1$) show a subcritical bifurcation while those of Crewther et al., for the case $\delta = 1/2$ and $\beta = 1/2$, give a supercritical bifurcation at De_c.

5   Conclusion

An analysis of the steady state solutions of a viscoelastic fluid of Oldroyd–B type undergoing steady shear between two coaxial discs has been carried out. For creeping flows (i. e. in the absence of inertia), we found that the problem has a unique similarity solution of small norm on one side of the critical Deborah number $\mathrm{De}_c = \pi/[|1 - \delta|\sqrt{\beta(3 + 2\beta)}]$ and at least three such solutions on the other side. This was accomplished via a bifurcation analysis using the method of Lyapunov–Schmidt. The direction of the bifurcation was shown to be determined by the retardation parameter, β. Bifurcation is supercritical if $\beta < \beta_c$ and subcritical if $\beta > \beta_c$. The value of $\beta_c \simeq 0.592$ is given by equation (21). These results agree with numerical results obtained by Crewther et al. [3] and Walsh [14] for the relevant parameter values.

References

[1] S. Chow, and J. K. Hale, *Methods of Bifurcation Theory*, (1982), Springer– Verlag, New York.

[2] M. G. Crandall, and P. H. Rabinowitz, *Bifurcation from simple eigenvalues*, J. Fun. Anal., 8 (1971) pp. 321–340.

[3] I. Crewther, R. R. Huilgol, and R. Jozsa, *Axisymmetric and non-axisymmetric flows of a non-newtonian fluid between coaxial rotating discs*, Phil Trans R. Soc. Lond. A 337 (1991) pp. 467–495.

[4] M. Golubitsky, and D. G. Schaeffer, *Singularities and Groups in Bifurcation Theory*, vol 1 (1985). Springer–Verlag, N. Y.

[5] R. R. Huilgol, and H. B. Keller, J. non–Newt. Fluid Mech., 18 (1985), pp. 101–110.

[6] Z. Ji, K. R. Rajagoppal, and A. Z. Szeri, *Multiplicity of solutions in von Karman flows of viscoelastic fluids*, J non–Newt. Fluid Mech., 36 (1990) pp. 1–25.

[7] D. D. Joseph, *Slow motion and viscometric motion; stability and bifurcation of the rest state of a simple fluid*, Arch. Rat. Mech., 56 (1974) pp. 100–157.

[8] D. O. Olagunju, *Effect of free surface and inertia on viscoelastic parallel plate flow*, J Rheol., 38(1) (1994) pp. 151–167.

[9] D. O. Olagunju, *Instabilities and bifurcations of von Kármán similarity solutions in swirling viscoelastic flow*, J. Appl. Math. Phys. (ZAMP), 46 (1995) pp. 224–238.

[10] N. Phan-Thien, *Coaxial-disk flow of an Oldroyd–B fluid: Exact solution and stability*, J. Non–Newt. Fluid Mech., 13 (1983) pp. 325–340.

[11] D. H. Sattinger, *Group Theoretic Methods in Bifurcation Theory*, Lecture Notes in Math., 762 (1979). Springer–Verlag, N. Y.

[12] R. W. G. Shipman, M. M. Denn, and R. Keunings, *Free surface effects in torsional parallel-plate rheometry*, Ind. Engng. Chem. Res. 30 (1991) pp. 918–922.

[13] I. Stakgold, *Branching of solutions of nonlinear equations*, SIAM Review, 13 (3) (1971) pp. 289–332.

[14] W. P. Walsh, *On a flow of a non-Newtonian fluid between rotating, coaxial discs*, J. Appl. Math. Phys. (ZAMP) 38 (1987) pp. 495–511.

Quenching for the Heat Equation with a Nonlocal Nonlinearity*

W. E. Olmstead[†]   Catherine A. Roberts[‡]

Abstract

The quenching problem is examined for a one-dimensional heat equation with a nonlinear source term whose singular part is governed by a nonlocal effect. Sufficient conditions are derived for both quenching and non-quenching behavior. For the quenching case, the growth rate of the norm is determined.

This paper is dedicated to Ivar Stakgold on the occasion of his 70th birthday. Best wishes to a mentor and a friend.

1 Introduction

We examine a quenching problem for the heat equation in a one-dimensional strip of finite width that contains a nonlinear heat source of a nonlocal nature. In particular, we consider

$$
(1) \qquad \frac{\partial v}{\partial t}(x,t) = \frac{\partial^2 v}{\partial x^2}(x,t) + \alpha[1 - u(t)]^{-p} \,, \quad 0 < x < \ell \,, \ t > 0 \,,
$$

where $\alpha > 0$, $p > 0$ and

$$
(2) \qquad u(t) \equiv \frac{1}{\ell} \int_0^\ell |v(x,t)|\, dx \,\,.
$$

The temperature $v(x,t)$ that satisfies (1) is also required to fulfill the initial condition

$$
(3) \qquad v(x,0) = v_0(x) \,, \ \ 0 \leq v_0(x) \leq 1 - \delta \,, \ \ 0 < \delta < 1 \,,
$$

and the Dirichlet boundary conditions

$$
(4) \qquad v(0,t) = 0 \,, \ \ v(\ell,t) = 0 \,, \ \ t > 0 \,\,.
$$

While some consideration of nonlocal effects in quenching has been made in [1], we are unaware of any prior attempt to include this effect in the singular part of the nonlinear source term, as in (1). This novel form suggests a slight modification for the notion of the

*The research of the first author was supported by NSF Grant DMS-9401016 and that of the second author by NSF Grant DMS-9510082.

†Department of Engineering Sciences and Applied Mathematics, Northwestern University, Evanston, IL 60208.

‡Department of Mathematics, Northern Arizona University, Flagstaff, AZ 86011.

quenching. Thus, within the context of (1)-(4), we define the quenching effect as follows: *The non-negative solution $v(x,t)$ quenches whenever there exists a $\hat{t} < \infty$ such that*

$$(5) \qquad \lim_{t \to \hat{t}-} v(x,t) = \hat{v}(x) < \infty \ , \quad \lim_{t \to \hat{t}-} \frac{\partial v}{\partial t}(x,t) = \infty \ , \quad 0 < x < \ell \ .$$

It is evident that the occurrence of (5) is associated with $u(t) \to 1$ as $t \to \hat{t}^-$. Indeed, we will analyze this quenching problem from that viewpoint by converting (1)-(4) to a nonlinear Volterra equation for $u(t)$, and then implement techniques developed in [2],[3],[4] to analyze such integral equations. As a first step, we express (1)-(4) in the form

$$(6) \qquad v(x,t) = \alpha \int_0^t \left[\int_0^\ell G(x,t \mid \xi,s)d\xi \right] [1 - u(s)]^{-p}ds + \tilde{h}(x,t) \ ,$$

where

$$(7) \qquad \tilde{h}(x,t) \equiv \int_0^\ell G(x,t \mid \xi,0)v_0(\xi)d\xi \ ,$$

and the Green's function can be expressed (c.f. [6]) in the form,

$$(8) \qquad G(x,t \mid \xi,s) = H(t-s)\left[\frac{2}{\ell} \sum_{n=1}^\infty \sin\left(\frac{n\pi\xi}{\ell} \right) \sin\left(\frac{n\pi x}{\ell} \right) e^{-\frac{n^2\pi^2}{\ell^2}(t-s)} \right] \ .$$

Since $G(x,t \mid \xi,s) \geq 0$, it is clear from (3), (6) and (7) that $v(x,t) \geq 0$ whenever it exists. Thus, integration of (6) yields

$$(9) \qquad u(t) = Tu(t) \equiv \alpha \int_0^t k(t-s)[1 - u(s)]^{-p}ds + h(t) \ , \quad t \geq 0 \ .$$

Here the kernel is given by

$$(10) \qquad \begin{aligned} k(t-s) \ &\equiv \frac{1}{\ell} \int_0^\ell \int_0^\ell G(x,t \mid \xi,s)d\xi \, dx \\ &= 1 - \frac{4}{\ell} \int_0^{t-s} G\left(\frac{\ell}{2}, t-s \, \Big| \, \frac{\ell}{2}, \tau \right) d\tau = \frac{8}{\pi^2} \sum_{n=1}^\infty \frac{1}{(2n-1)^2} e^{-\frac{(2n-1)^2\pi^2}{\ell^2}(t-s)} \ , \end{aligned}$$

where the latter two equalities follow from direct computation. It is further seen from (10) that

$$(11) \qquad 0 \leq k(t) \leq 1 \ , \quad k'(t) \leq 0 \ , \quad t \geq 0 \ .$$

The function $h(t)$ depends only upon the initial data. It is given by

$$(12) \qquad h(t) \equiv \frac{1}{\ell} \int_0^\ell \tilde{h}(x,t)dx = \frac{1}{\ell} \int_0^\ell \int_0^\ell G(x,t \mid \xi,0)v_0(\xi)d\xi \, dx \ ,$$

and has the property that

$$(13) \qquad 0 < h(t) \leq 1 - \delta \ , \quad t \geq 0 \ .$$

In order to facilitate the analysis to follow, we will restrict the initial data to be such that $h(t)$ satisfies

$$(14) \qquad h'(t) + \alpha k(t)[1 - h(0)]^{-p} > 0 \ , \quad t \geq 0 \ .$$

Our examination of the quenching problem for (1)-(4) will focus on the nonlinear Volterra equation (9). In the context of the integral equation, we will refer to quenching of the solution as the situation in which there exists a $\hat{t} < \infty$ such that

(15) $$\lim_{t \to \hat{t}^-} u(t) = 1 \ , \ \ \lim_{t \to \hat{t}^-} u'(t) = \infty \ .$$

We will be able to demonstrate that (15) implies (5), and provide a sufficient condition for that occurrence.

2 Existence and Nonexistence

We will investigate the possibility of quenching by examining (9) for a continuously differentiable solution $u(t)$ such that

(16) $$0 \le u(t) \le m < 1 , \ 0 \le u'(t) < \infty , \ 0 \le t < t^* \ .$$

It is clear that whenever (9) has a solution, bounded above by unity, then it must be positive in view of the positivity of $k(t-s)$ and $h(t)$. To see that it must also be increasing, differentiate (8) to obtain

(17) $$u'(t) = h'(t) + \alpha k(t)[1 - h(0)]^{-p} + \alpha p \int_0^t k(t-s)[1 - u(s)]^{-p-1} u'(s) ds \ .$$

From (14), $u'(0) > 0$ and hence $u'(t) > 0$ at least on some small interval $0 \le t < \bar{t} < \infty$. Moreover, it is impossible that $u'(\bar{t}) = 0$ since the right side of (17) would be positive at $t = \bar{t}$. Thus $u'(t) > 0$ whenever it exists.

In order to establish the existence of a continuously differentiable solution of (9) that satisfies (16), it will suffice to demonstrate that (9) has a continuous solution bounded above by unity. Once that has been shown, then (17) can be interpreted as a linear Volterra equation of the second kind for $u'(t)$. Well known results (c.f. [5]) for such linear problems then insure the existence of $u'(t)$.

To prove the existence and uniqueness of a continuous solution of (9), we will employ contraction mapping arguments for the operator T. The properties of T are such that we can restrict ourselves to the closed subset of continuous functions having the property that $0 \le u(t) \le m < 1$. We use the norm

(18) $$\|u\| \equiv \sup_{0 \le t < t^*} |u(t)| \ .$$

The contraction mapping theorem requires that T map the closed subset into itself. Thus, it follows from (9) that we should impose the constraint

(19) $$Tu(t) \le 1 - \delta + \alpha[1 - m]^{-p} I(t) \le m , \ 0 \le t < t^* ,$$

where

(20) $$I(t) \equiv \int_0^t k(t-s) ds = \frac{8\ell^2}{\pi^4} \sum_{n=1}^{\infty} \frac{1}{(2n-1)^4} \left[1 - e^{-\frac{(2n-1)^2 \pi^2}{\ell^2} t} \right] \ .$$

It is easily verified that $I(t)$ has the properties

(21) $$I'(t) > 0 \ , \ 0 \le I(t) < \frac{\ell^2}{12} , \ 0 \le t < \infty \ ,$$

where the upper bound is attained as $t \to \infty$.

To establish the contraction property of T, we consider some arbitrary $u_1(t)$ and $u_2(t)$ from the closed subset of continuous functions, which are non-negative and bounded above by m. It follows from (9) that

$$(22) \qquad \|Tu_1 - Tu_2\| \leq p\alpha[1 - m]^{-p-1}I(t)\,\|u_1 - u_2\| \quad .$$

Thus T is a contraction mapping provided that

$$(23) \qquad p\alpha[1 - m]^{-p-1}I(t) < 1 \ , \quad 0 \leq t < t^* \ .$$

Upon satisfying both (19) and (23), the contraction mapping theorem guarantees the existence and uniqueness of a continuous solution to (1), which satisfies $0 \leq u(t) \leq m$, for $0 \leq t < t^*$. In order to have the upper bound on $u(t)$ hold for the largest possible value of t^*, an optimal choice of m is needed. An examination of (19) and (23) reveals this to be

$$(24) \qquad m = 1 - \frac{p\delta}{p+1} \quad .$$

This then implies that the limiting value of t^* for the contraction property is given by

$$(25) \qquad I(t^*) = \frac{1}{p\alpha}\left(\frac{p\delta}{p+1}\right)^{p+1} \quad .$$

Thus, we have shown that (9) has a unique solution, which is continuously differentiable and satisfies (16), for $0 \leq t < t^*$ where t^* is determined by (25). It is possible that $t^* = \infty$, whereupon (9) has a continuously differentiable solution for all $t \geq 0$ and, consequently, there is no quenching of the solution. We can characterize this case of non-quenching by appealing to (21) and (25). This is, if

$$(26) \qquad \alpha < \frac{12}{p\ell^2}\left(\frac{p\delta}{p+1}\right)^{p+1} \quad ,$$

then quenching does not occur.

When (25) implies a $t^* < \infty$, there is the possibility of quenching. Moreover, if quenching does in fact occur as $t \to \hat{t}^-$, then

$$(27) \qquad \hat{t} \geq t^* \ .$$

Next, we examine the possibility that quenching does occur. We will seek a t^{**} such that (9) can not possess a continuous solution for $t \geq t^{**}$, and show that this corresponds to quenching. Under the assumption that (9) has a continuous solution, $0 \leq u(t) \leq m < 1$ for $0 \leq t \leq t^{**}$, we have that

$$(28) \qquad u(t) \geq J(t) \equiv \alpha\int_0^t k(t^{**} - s)[1 - u(s)]^{-p}ds \ , \quad 0 \leq t \leq t^{**} \ ,$$

since $k(t^{**} - s) \geq k(t - s)$. It then follows that

$$(29) \qquad J'(t) = \alpha k(t^{**} - t)[1 - u(t)]^{-p} \geq \alpha k(t^{**} - t)[1 - J(t)]^{-p} \ .$$

Integration of (29) yields

$$(30) \qquad \int_0^{J(t^{**})}(1 - J)^p dJ = \frac{1 - [1 - J(t^{**})]^{p+1}}{p+1} = \alpha\int_0^{t^{**}} k(t^{**} - t)dt = \alpha I(t^{**}) \ .$$

The form of (30) implies that the limiting value of t^{**} is detemined by

$$(31) \qquad I(t^{**}) = \frac{1}{\alpha(p+1)} \ .$$

If (31) is satisfied, then $J(t^{**}) = 1 \leq u(t^{**})$, which contradicts the assumed bound on $u(t)$. Thus we conclude that if (31) is satisfied, then there exists a $\hat{t}$ such that

$$(32) \qquad \lim_{t \to \hat{t}-} u(t) = 1 \ ,$$

where
$$(33) \qquad \hat{t} \leq t^{**} \ .$$

Moreover, from (17) we have

$$
\begin{aligned}
(34) \qquad u'(t) \ &\geq \ h'(t) + \alpha k(t)[1 - h(0)]^{-p} + \alpha p k(t) \int_0^t [1 - u(s)]^{-p-1} u'(s) ds \\
&= \ h'(t) + \alpha k(t)[1 - u(t)]^{-p} \ ,
\end{aligned}
$$

from which follows that
$$(35) \qquad \lim_{t \to \hat{t}-} u'(t) = \infty \ .$$

We note that (32) and (35) constitute our definition (15) of the quenching of the solution to the integral equation. To obtain a sufficient condition for this quenching of $u(t)$, we compare (31) with the upper bound on $I(t)$ given in (21). Thus, if

$$(36) \qquad \alpha > \frac{12}{\ell^2(p+1)} \ ,$$

then (31) can be satisfied for some $t^{**} < \infty$, which then implies that (15) is satisfied.

It remains to be shown in the case of quenching that (15) implies (5). That requires some information about the growth of $u(t)$ as $t \to \hat{t}-$.

3 Asymptotic Growth near Quenching

It is assumed here that the case in which $u(t) \to 1$ as $t \to \hat{t}$ does apply. Based upon that assumption, we will develop a self-consistent asymptotic analysis of (9) that determines the behavior of $u(t)$ as $t \to \hat{t}$. For this analysis, we begin by expressing (9) as

$$(37) \qquad u(t) = \alpha \int_{0_l}^t [1 - K(t - s)][1 - u(s)]^{-p} ds + h(t) \ , \quad t \geq 0 \ ,$$

where from (10),
$$(38) \qquad K(t) = \frac{4}{\ell} \int_0^t G\left(\frac{\ell}{2}, t \,\Big|\, \frac{\ell}{2}, \tau\right) d\tau \ .$$

To put $K(t)$ into a more useful form for the asymptotic analysis, we use an equivalent form of the Green's function given as

$$
\begin{aligned}
(39) \qquad G(x, t \mid \xi, s) \ = \ &\frac{H(t - s)}{2[\pi(t - s)]^{1/2}} \sum_{n=-\infty}^{\infty} \left\{ \exp\left[-\frac{(x - \xi - 2n\ell)^2}{4(t - s)} \right] \right. \\
&\left. - \exp\left[-\frac{(x + \xi - 2n\ell)^2}{4(t - s)} \right] \right\}
\end{aligned}
$$

It then follows that (38) can be expressed as

$$(40) \qquad K(t) = \frac{2}{\ell\pi^{1/2}} \int_0^t \frac{1 + R(t-\tau)}{(t-\tau)^{1/2}} d\tau = \frac{2}{\ell\pi^{1/2}} \int_0^t \frac{1 + R(\tau)}{\tau^{1/2}} d\tau ,$$

where $R(\tau)$ denotes all the exponential terms arising from (39) with $x = \xi = \ell/2$.

Upon substitution of (40) into (37), and then differentiating, we obtain

$$(41) \quad u'(t) = \alpha[1 - u(t)]^{-p} - \frac{2\alpha}{\ell\pi^{1/2}} \int_0^t \frac{1 + R(t-s)}{(t-s)^{1/2}}[1 - u(s)]^{-p} ds + h'(t) , \quad t \ge 0 .$$

We will consider (41) as an alternate form of (37) or (9), which facilitates the asymptotic analysis near quenching.

To investigate (41) as $t \to \hat{t}^-$, it is convenient to introduce the transformation

$$(42) \qquad \eta = (\hat{t} - t)^{-1} - \eta_0 , \quad \eta_0 = (\hat{t})^{-1} , \quad w(\eta) = u(t) .$$

Thus, quenching occurs as $\eta \to \infty$, so that

$$(43) \qquad \lim_{\eta \to \infty} w(\eta) = 1 , \quad \lim_{\eta \to \infty} w'(\eta) = \infty .$$

Using (42), we follow [2],[4] to convert (41) into the asymptotic equality as $\eta \to \infty$,

$$(44) \qquad \eta^2 w'(\eta) \sim \alpha[1 - w(\eta)]^{-p} - \frac{\alpha}{2\pi i} \int_{c-i\infty}^{c+i\infty} \eta^{1-z} \frac{\Gamma(1-z)}{\Gamma\left(\frac{3}{2} - z\right)} M[\Phi; z] dz + h(\hat{t}) ,$$

where $M[\Phi; z]$ is the Mellin transform

$$(45) \qquad M[\Phi; z] \equiv \int_0^\infty \tau^{z-1} \Phi(\tau) d\tau , \quad \Phi(\tau) \equiv (\tau + \eta_0)^{-3/2}[1 - w(\tau)]^{-p} .$$

The complex path integral in (44) is taken along a vertical path lying within the strip of analyticity of the integrand. Also, in deriving (44), a result from [4] was used to show that contributions from the exponential terms $R(\tau)$ can be neglected at leading order.

We will seek an asymptotic balance in (44) by considering

$$(46) \qquad w(\eta) \sim 1 - C\eta^{-q} \text{ as } \eta \to \infty , \quad q > 0 , \quad C > 0 .$$

It then follows from (44) that to leading order

$$(47) \qquad Cq\eta^{1-q} \sim \alpha C^{-p}\eta^{pq} , \quad \text{as } \eta \to \infty ,$$

where the integral term makes only a lower order contribution. The implication of (47) is that

$$(48) \qquad q = \frac{1}{p+1} , \quad C = [\alpha(p+1)]^{\frac{1}{p+1}} .$$

Thus we have determined that the asymptotic growth near quenching is given by

$$(49) \qquad \begin{aligned} &u(t) \sim 1 - [\alpha(p+1)(\hat{t} - t)]^{\frac{1}{p+1}} , \\ &u'(t) \sim \frac{[\alpha(p+1)]^{\frac{1}{p+1}}}{p+1}(\hat{t} - t)^{-\frac{p}{p+1}} , \text{as } t \to \hat{t}^- . \end{aligned}$$

4 Conclusions

It remains to relate the results of sections 2 and 3 on the quenching of $u(t)$ to the quenching of $v(x,t)$ as defined by (5). In section 2, we derived a sufficient condition (26) for the non-quenching of $u(t)$. It is clear that this likewise implies the non-quenching of $v(x,t)$. Also in section 2, we derived a sufficient condition (36) for the quenching of $u(t)$, with its asymptotic behavior determined in section 3. From (49), we find that

$$(50) \qquad [1 - u(t)]^{-p} \sim [\alpha(p+1)(\hat{t} - t)]^{-\frac{p}{p+1}} , \quad \text{as } t \to \hat{t}^- ,$$

which implies an integrable singularity in (6). Thus, it follows that

$$(51) \qquad \lim_{t \to \hat{t}^-} v(x,t) = \hat{v}(x) < \infty .$$

To investigate $\dfrac{\partial v}{\partial t}(x,t)$ as $t \to \hat{t}^-$, it is useful to express (1), subject to the Dirichlet conditions (4), in the equivalent form

$$(52) \qquad v(x,t) = \frac{1}{\ell} \int_0^\ell [x(\ell - \xi) - \ell(x - \xi)H(x - \xi)] \left\{ \alpha[1 - u(t)]^{-p} - \frac{\partial v}{\partial t}(\xi,t) \right\} d\xi ,$$

where $H(x - \xi)$ is the Heaviside function. Since $v(x,t)$ remains bounded as $t \to \hat{t}^-$, it is clear from (52) that

$$(53) \qquad \frac{\partial v}{\partial t}(x,t) \sim \alpha[1 - u(t)]^{-p} \sim \alpha[\alpha(p+1)(\hat{t} - t)]^{-\frac{p}{p+1}} , \quad \text{as } t \to \hat{t}^- , \ 0 < x < \ell .$$

Having thus established that

$$(54) \qquad \lim_{t \to \hat{t}^-} \frac{\partial v}{\partial t}(x,t) = \infty , \quad 0 < x < \ell ,$$

together with (51), then our definition of quenching (5) is fulfilled whenever (36) is satisfied.

References

[1] K. Deng, *Dynamical behavior of solutions of a semilinear heat equation with nonlocal singularity*, SIAM J. Math. Anal. 26 (1995), pp. 98-111.

[2] W. E. Olmstead and C. A. Roberts, *Explosion in a diffusive strip due to a concentrated nonlinear source*, Meth. and Appl. Anal. 1 (1994), pp. 434-445.

[3] C. A. Roberts, D. G. Lasseigne and W. E. Olmstead, *Volterra equations which model explosion in a diffusive medium*, J. Integral Eqns and Appl. 5 (1993), pp. 531-545.

[4] C. A. Roberts, and W. E. Olmstead, *Growth rates for blow-up solutions of nonlinear Volterra equations*, Q. Appl. Math. (to appear).

[5] I. Stakgold, *Boundary Value Problems of Mathematical Physics, Vol. I.*, Macmillan, New York, 1967.

[6] _____, *Boundary Value Problems of Mathematical Physics, Vol. II.*, Macmillan, New York, 1968.

DECAY BOUNDS IN QUASILINEAR PARABOLIC PROBLEMS

L. E. Payne
Department of Mathematics
Cornell University
Ithaca, N.Y. 14853
and
G. A. Philippin
Département de Mathématiques
Université Laval
Québec, Canada G1K 7P4

1. Introduction.

In a recent paper [4], the authors derived decay bounds, in the form of a maximum principle, for solutions of the semilinear parabolic equation

$$\Delta u + f(u) = \frac{\partial u}{\partial t}, \ x \in \Omega, \ t > 0 \tag{1.1}$$

under appropriate hypotheses on the form of f. Here Δ is the Laplace operator and Ω is a bounded region in $\mathbb{R}^N$. In fact their maximum principle involved combinations of u and its spatial derivatives. For related results on maximum principles for parabolic equations see e.g. Serrin [6] and the references cited in Protter–Weinberger [5], Sperb [7], and Walter [8].

In this paper we derive similar maximum principles for classical solutions of the equation

$$\sum_{i=1}^{N} \frac{\partial}{\partial x_i} \left[g(|\nabla u|^2) \frac{\partial u}{\partial x_i} \right] - \frac{\partial u}{\partial t} = 0 \quad x \in \Omega, \ t > 0 \tag{1.2}$$

where g is assumed to be a positive C^2 function of its argument and to satisfy the condition

$$G(s) = g(s) + 2sg'(s) > 0, \ \forall \ s \geq 0. \tag{1.3}$$

Condition (1.3) implies that the summation operator in (1.2) is elliptic.

We will be primarily concerned with the initial–boundary value problem in which u satisfies

$$u(x,t) = 0, \ x \in \partial\Omega, \ t > 0 \tag{1.4}$$

where $\partial\Omega$ is the boundary of Ω, and

$$u(x,0) = f(x), \tag{1.5}$$

although our maximum principle will apply to arbitrary solutions of (1.2), (1.3).

It is well known that solutions of (1.2), (1.3) satisfy the parabolic maximum principle (see Protter–Weinberger [5] or Nirenberg [2]) which implies that

$$|u(x,t)| \leq \max_{x \in \Omega} |u(x,0)|. \tag{1.6}$$

In fact it can be easily shown that q^2 satisfies an inequality of the form

$$g\Delta(q^2) + 2g'u_{,i}\, u_{,j}\, (q^2)_{,ij} + w_k(q^2)_{,k} \geq (q^2)_{,t}, \tag{1.7}$$

which implies that q^2 takes its maximum value either at $t = 0$ or on $\partial\Omega$, $t > 0$.

For the special case in which u satisfies (1.2)–(1.5) and the boundary $\partial\Omega$ is a $C^{2+\epsilon}$ surface whose average curvature at each point is nowhere negative, explicit decay bounds for a solution and its gradient in terms of initial data are derived. Again these bounds are in the form of a maximum principle involving the solution and its derivatives.

In this paper we adopt the summation convention of summing over repeated indices (in any term) from 1 to N, and a comma denotes partial differentiation. Also for simplicity we set

$$q^2 = |\nabla u|^2. \tag{1.8}$$

Using this notation, (1.2) is rewritten as

$$[g(q^2)u_{,i}]_{,i} - u_{,t} = 0, \ x \in \Omega, \ t > 0. \tag{1.9}$$

In the next section we derive a maximum principle for a function $\Phi(x,t)$ defined as

$$\Phi(x,t) = \left[\int_0^{q^2} G(s)ds + \alpha u^2 \right] e^{2\alpha\beta t} \tag{1.10}$$

for an arbitrary positive constant α and a positive constant β to be chosen. The specific choice for β will depend on properties of the function g.

2. Maximum Principle for $\Phi(x,t)$.

For solutions of (1.2) we establish the following:

Theorem 1. *Let $u(x,t)$ be a classical solution of (1.2), (1.3) in $\Omega \times \mathbb{R}^+$ with $N \geq 2$. Suppose that either*

$$a) \qquad g' > 0$$

and that positive constants λ and $\beta \in (0,1]$ can be found such that

$$g - A(N,\lambda,\beta)q^2 g' \geq 0 \tag{2.1}$$

with

$$A(N,\lambda,\beta) := \max\left\{N\lambda, \frac{N\lambda + \lambda^{-1} + 2\beta - 4}{1 - \beta}\right\}, \tag{2.2}$$

and

$$q^2 G(q^2) - \beta \int_0^{q^2} G(s)ds \geq 0, \tag{2.3}$$

or

$$b) \qquad g' < 0$$

and (2.3) is satisfied for some $\beta \in (0,1)$ together with the condition

$$g + B(N,\beta)g'q^2 \geq 0, \tag{2.4}$$

where

$$B(N,\beta) = \max \left\{ N - 1, \frac{3 - 2\beta}{1 - \beta} \right\}. \tag{2.5}$$

Then $\Phi(x,t)$ defined by (1.10) satisfies a parabolic inequality of the form

$$L\Phi := g\Delta\Phi + 2g'\Phi_{,ik}\, u_{,i}\, u_{,k} + q^{-2}W_k\Phi_{,k} - \Phi_{,t} \geq 0 \text{ in } \Omega \times \mathbb{R}^+, \tag{2.6}$$

where $W := (w_1, w_2, \ldots, w_N)$ is a bounded vector field in $\Omega \times \mathbb{R}^+$.

It is worth noting that as long as $(N\lambda + \lambda^{-1} + 2\beta - 4)/(1 - \beta) \geq N\lambda$ then the optimal choice for λ in (2.2) is

$$\lambda = N^{-1/2}. \tag{2.7}$$

On the other hand if we choose a λ to make the two bracketed terms on the right of (2.2) equal, we may be imposing an unnecessary restriction on the size of β even for $N = 2,3$. It is also clear that in (2.5)

$$(3 - 2\beta)/(1 - \beta) > 3 \tag{2.8}$$

for $\beta \in (0,1)$, so that for $N \leq 4$

$$B(N,\beta) = (3 - 2\beta)/(1 - \beta). \tag{2.9}$$

Also if the strict inequality is imposed in (2.4) then (2.4) implies (1.3).

For completeness we state without proof the equivalent theorem for $N = 1$. The proof is straightforward.

Theorem 1. *Let g be such that*

$$u_{,x}^2\, G(u_{,x}^2) - \int_0^{u_{,x}^2} G(s)ds \geq 0. \tag{2.10}$$

Then the function $\Phi(x,t)$ defined as

$$\Phi(x,t) = e^{2\alpha t}\left\{ \int_0^{u_x^2} G(s)ds + \alpha u^2 \right\} \tag{2.11}$$

satisfies a parabolic inequality of the form

$$G(u_{,x}^2)\Phi_{,xx} - u_{,x}^{-2}\, w\Phi_{,x} - \Phi_{,t} \geq 0 \tag{2.12}$$

where $w(x,t)$ is a bounded function.

Proof of Theorem 1

From (1.10) we compute

$$\Phi_{,i} = 2e^{2\alpha\beta t}\{Gu_{,k}\,u_{,ik} + \alpha uu_{,i}\},\tag{2.13}$$

$$\Phi_{,ij} = 2e^{2\alpha\beta t}\{2G'u_{,k}u_{,ik}\,u_{,\ell}\,u_{,j\ell} + G[u_{,k}\,u_{,ijk} + u_{,ik}\,u_{,ij}]$$
$$+ \alpha[u_{,i}\,u_{,j} + uu_{,ij}]\},\tag{2.14}$$

$$\Delta\Phi = 2e^{2\alpha\beta t}\{2G'u_{,k}u_{,ik}\,u_{,\ell}\,u_{,i\ell} + G[u_{,k}\,\Delta u_{,k} + u_{,ij}\,u_{,ij}]$$
$$+ \alpha[q^2 + u\Delta u]\},\tag{2.15}$$

and

$$\Phi_{,t} = 2e^{2\alpha\beta t}\Big\{\alpha\beta\Big[\int_0^{q^2} G(s)ds + \alpha u^2\Big] + Gu_{,i}\,u_{,it} + \alpha uu_{,t}\Big\}.\tag{2.16}$$

From (1.9) rewritten as

$$\Delta u = g^{-1}u_{,t} - 2[\log g]'u_{,i}\,u_{,k}\,u_{,ik}\tag{2.17}$$

we obtain

$$u_{,k}\,\Delta u_{,k} = -2g'g^{-2}u_{,i}\,u_{,k}\,u_{,ik}\,u_{,t} + g^{-1}u_{,k}\,u_{,kt} - 4[\log g]''(u_{,i}\,u_{,k}\,u_{,ik}\,)^2$$
$$- 2[\log g]'[u_{,i}\,u_{,j}\,u_{,k}\,u_{,ijk} + 2u_{,k}\,u_{,ik}\,u_{,j}\,u_{,ij}].\tag{2.18}$$

Inserting (2.17) and (2.18) into (2.15) and using (2.14) and (2.16) we have after some reduction

$$g\Delta\Phi + 2g'\Phi_{,ik}\,u_{,i}\,u_{,k} - \Phi_{,t}$$
$$= e^{2\alpha\beta t}\Big\{8[gg' + q^2gg'' - q^2(g')^2]u_{,k}\,u_{,ik}\,u_{,\ell}\,u_{,i\ell}$$
$$- 4Gg'\Delta uu_{,i}\,u_{,k}\,u_{,ik} + 8[3(g')^2 - gg''](u_{,i}\,u_{,k}\,u_{,ik}\,)^2$$
$$+ 2gGu_{,ik}\,u_{,ik} - 2\alpha^2\beta u^2 + 2\alpha\Big(q^2G - \beta\int_0^{q^2} G(s)ds\Big)\Big\}\tag{2.19}$$

Consider the first case: $g' \geq 0$. Using the arithmetic–geometric mean inequality and the fact that $(\Delta u)^2 \leq Nu_{,ij}\,u_{,ij}$, it follows that

$$2\Delta uu_{,i}\,u_{,k}\,u_{,ik} \leq \lambda q^2(\Delta u)^2 + \lambda^{-1}q^{-2}(u_{,i}\,u_{,k}\,u_{,ik}\,)^2$$
$$\leq \lambda Nq^2u_{,ij}\,u_{,ij} + \lambda^{-1}q^{-2}(u_{,i}\,u_{,k}\,u_{,ik}\,)^2\tag{2.20}$$

where the constant $\lambda > 0$ is arbitrary. Combining (2.19), (2.20) and using (2.3) we arrive at the inequality

$$g\Delta\Phi + 2g'\Phi_{,ik}\,u_{,i}\,u_{,k} - \Phi_{,t}$$
$$\geq e^{2\alpha\beta t}\{2G(g - N\lambda q^2g')u_{,ik}\,u_{,ik}$$
$$+ 2[12(g')^2 - 4gg'' - Gg'\lambda^{-1}q^{-2}](u_{,i}\,u_{,k}\,u_{,ik}\,)^2$$
$$+ 8[gg' + q^2gg'' - q^2(g')^2]u_{,k}\,u_{,ik}\,u_{,j}\,u_{,ij} - 2\alpha^2\beta u^2\}.\tag{2.21}$$

We now use (2.1) and Schwarz's inequality

$$q^2 u_{,ik} u_{,ik} \geq u_{,k} u_{,ik} u_{,j} u_{,ij} \tag{2.22}$$

in (2.21) to obtain

$$
\begin{aligned}
g\Delta\Phi &+ 2g'\Phi_{,ik} u_{,i} u_{,k} - \Phi_{,t} \\
&\geq e^{2\alpha\beta t}\{2[12(g')^2 - 4gg'' - Gg'\lambda^{-1}q^{-2}](u_{,i} u_{,k} u_{,ik})^2 \\
&\quad + (8[gg' + q^2 gg'' - q^2(g')^2] + 2Gq^{-2}[g - N\lambda q^2 g'])u_{,k} u_{,ik} u_{,\ell} u_{,i\ell} \\
&\quad - 2\alpha^2\beta u^2\}.
\end{aligned} \tag{2.23}
$$

But from (2.13)

$$u_{,k} u_{,ik} u_{,j} u_{,ij} = G^{-2}\alpha^2 u^2 q^2 + \cdots \tag{2.24}$$

and

$$(u_{,i} u_{,k} u_{,ik})^2 = G^{-2}\alpha^2 u^2 q^4 + \cdots, \tag{2.25}$$

where dots indicate terms containing first order spatial derivatives of Φ. The insertion of (2.24) and (2.25) into (2.23) gives

$$
\begin{aligned}
L\Phi &\geq 2e^{2\alpha\beta t}G^{-1}\alpha^2 u^2\{(4 - \lambda^{-1} - N\lambda)q^2 g' + g - \beta G\} \\
&= 2e^{2\alpha\beta t}G^{-1}\alpha^2 u^2(1 - \beta)\{g + (4 - \lambda^{-1} - N\lambda - 2\beta)(1 - \beta)^{-1}q^2 g'\}, \quad (2.26)
\end{aligned}
$$

which is nonnegative by (2.1). This completes the proof of the theorem when $g' > 0$.

Consider next the case $g' < 0$. From (2.19) we must now provide a lower bound for $\Delta u u_{,i} u_{,k} u_{,ik}$. To this end we make use of an inequality derived in [3], i.e.,

$$
\begin{aligned}
2\Delta u u_{,i} u_{,k} u_{,ik} &\geq -(N - 1)u_{,ij} u_{,ij} + q^{-4}(u_{,ik} u_{,i} u_{,k})^2 \\
&\quad + (N - 1)q^{-2}u_{,ik} u_{,k} u_{,ij} u_{,j},
\end{aligned} \tag{2.27}
$$

reducing (2.19) with the use of (2.3) to

$$
\begin{aligned}
g\Delta\Phi &+ 2g'\Phi_{,ik} u_{,i} u_{,k} - \Phi_{,t} \\
&\geq e^{2\alpha\beta t}\{(8[gg' + q^2 gg'' - q^2(g')^2] - 2(N - 1)Gg')u_{,ik} u_{,k} u_{,ij} u_{,j} \\
&\quad + (8[3(g')^2 - gg''] - 2q^{-2}g'G)(u_{,ik} u_{,i} u_{,k})^2 \\
&\quad + 2G[g + (N - 1)q^2 g']u_{,ik} u_{,ik} - 2\alpha^2\beta u^2\}.
\end{aligned} \tag{2.28}
$$

An application of (2.4) and (2.22) together with (2.24) and (2.25) leads after some reduction to

$$L\Phi \geq 2e^{2\alpha\beta t}G^{-1}\alpha^2 u^2(1 - \beta)\{g + (3 - 2\beta)(1 - \beta)^{-1}q^2 g'\} \geq 0 \tag{2.29}$$

by (2.4). This completes the proof of Theorem 1.

Making use of the standard maximum principle for parabolic inequalities we have as a consequence of Theorem 1, as in [4].

Theorem 2. *Let $u(x,t)$ be a classical solution of (1.2), (1.4) in which either g' is positive and constants λ, β are chosen to satisfy (2.1), (2.3) or g' is negative and (2.3), (2.4) are satisfied. Then $\Phi(x,t)$ defined by (1.10) takes its maximum value either at a boundary point $(\widehat{x},\widehat{t})$ with $\widehat{x} \in \partial\Omega$, $\widehat{t} > 0$, or at an interior critical point $(\overline{x},\overline{t})$ of u with $\overline{x} \in \Omega$, $\overline{t} > 0$ or initially at a point $(\widetilde{x},0)$ with $\widetilde{x} \in \Omega$, i.e., we have*

$$\Phi(x,t) \leq \max \begin{cases} \Phi(\widehat{x},\widehat{t}), & \widehat{x} \in \partial\Omega, \ \widehat{t} > 0 \\ \Phi(\overline{x},\overline{t}), & \text{with } \nabla u(\overline{x},\overline{t}) = 0 \\ \max_{x \in \overline{\Omega}} \Phi(x,0), & x \in \overline{\Omega}. \end{cases} \tag{2.30}$$

3. Elimination of the First Possibility in (2.30).

As in [4] we impose a condition on the geometry of $\partial\Omega$ and obtain the following theorem:

Theorem 3. *Let $u(x,t)$ be a classical solution of (1.2), (1.4) satisfying the conditions of Theorem 1 with $\Phi(x,t)$ defined by (1.10). Let Ω be a bounded region in $\mathbb{R}^N$ with $C^{2+\epsilon}$ boundary $\partial\Omega$ at each point of which the average curvature, K, is non–negative. Then $u(x,t)$ cannot take its maximum value on $\partial\Omega$ unless $u(x,t) \equiv 0$.*

Before presenting the proof of Theorem 3, which is similar to the proof of Theorem 2 in [4], we point out that there is an error in the statement and proof of Theorem 2 in [4] for the case of homogeneous Neumann boundary conditions. This error is being corrected in a corrigendum.

Since the proof of Theorem 3 uses the arguments of [4] it will be sufficient to sketch the proof here. We assume that the maximum of Φ does occur at a point $\widehat{P}$ on $\partial\Omega$ and apply the maximum principle on the boundary (see e.g. Protter–Weinberger [5] or Friedman [1]) to establish a contradiction.

At any point on $\partial\Omega$ we have

$$\frac{\partial \Phi}{\partial n} = 2e^{2\alpha\beta t} G u_n u_{nn}, \tag{3.1}$$

where $\dfrac{\partial}{\partial n}$ is the outward normal derivative and u_n and u_{nn} are the first and second normal derivatives respectively. Since $\partial\Omega$ is a $C^{2+\epsilon}$ surface the differential equation (1.9) is satisfied on $\partial\Omega$ for $t > 0$. In normal coordinates (1.9) is written at points on $\partial\Omega$ as

$$G u_{nn} + g(N-1)K u_n = 0, \tag{3.2}$$

where K is the average curvature of $\partial\Omega$. The insertion of (3.2) into (3.1) gives

$$\frac{\partial \Phi}{\partial n} = -2(N-1)e^{2\alpha\beta t} g K u_n^2 \tag{3.3}$$

at each point of $\partial\Omega$. The maximum principle and (3.3) imply that if the maximum of Φ occurs at $\widehat{P}$ on $\partial\Omega$, then $\dfrac{\partial \Phi}{\partial n} = 0$ at $\widehat{P}$, which means that at $\widehat{P}$ either $K = 0$ or $u_n^2 = 0$. In addition, $\Phi \equiv$ constant in Ω, for $t > \widehat{t}$. Now if $u_n^2 = 0$ at $\widehat{P}$ it follows that Φ must vanish identically, which can happen only if $u \equiv 0$. Thus if $\Phi \equiv$ constant for $t \geq \hat{t}$ and $u \not\equiv 0$ we have

$$\Phi(x,t) = e^{2\alpha\beta t} \int_0^{q_M^2} G(s)ds = \Phi_{Max} > 0, \tag{3.4}$$

where q_M is the maximum value of q on $\partial\Omega$. This implies that q^2 is a positive constant at each point of $\partial\Omega$ for $t > \hat{t}$, and hence since $\dfrac{\partial\Phi}{\partial n}$ vanishes at each point of $\partial\Omega$, $t \geq \hat{t}$, K must vanish identically on $\partial\Omega$. But this is impossible since $\partial\Omega$ is a closed surface and Theorem 3 is proved.

4. Decay Estimates.

Up to now the choice of α has been arbitrary. We now show that with an appropriate choice of α the second possibility in (2.30) is also ruled out. Let d be the radius of the largest inscribed ball in Ω, then this result is achieved by the following theorem.

Theorem 4. *Let u be a classical solution of (1.2)–(1.5) in a bounded region Ω whose $C^{2+\epsilon}$ boundary has nonnegative average curvature at each point. Then if $g' > 0$, (2.1), (2.3) are satisfied, and*

$$\alpha < \frac{\pi^2}{4d^2}g(0) \tag{4.1}$$

or if $g' < 0$, (2.3), (2.4) are satisfied and

$$\alpha < \frac{\pi^2}{4d^2}\frac{(B-2)}{B}g(p_M^2), \tag{4.2}$$

then $\Phi(x,t)$, given by (1.10) must take its maximum value at $t = 0$. Here

$$p_M^2 = \sup_{x \in \Omega} |\nabla f|^2. \tag{4.3}$$

In proving Theorem 4 we assume to the contrary that Φ takes its maximum value at $(\bar{x}, \bar{t})$, an interior critical point of u. (We showed in the previous section that the maximum could not occur on $\partial\Omega$.) Thus we have

$$e^{2\alpha\beta t}\left[\int_0^{q^2} G(s)ds + \alpha u^2\right] \leq \alpha e^{2\alpha\beta\bar{t}}(u^2)_M$$

where

$$(u^2)_M = \max_{x \in \Omega} u^2(x,\bar{t}). \tag{4.4}$$

Evaluating the left hand side at $t = \bar{t}$ we obtain

$$\int_0^{q^2} G(s)ds \leq \alpha[(u^2)_M - u^2]. \tag{4.5}$$

But if $g' > 0$, it follows that

$$\int_0^{q^2} G(s)ds \geq q^2 g(q^2) \geq q^2 g(0),\tag{4.6}$$

so that

$$q^2 \leq \frac{\alpha}{g(0)}[(u^2)_M - u^2].\tag{4.7}$$

As in [4] this inequality is integrated from the point $(\bar{x}, \bar{t})$ to the boundary to give

$$\alpha \geq \frac{\pi^2}{4d^2} \, g(0).\tag{4.8}$$

Since (4.1) and (4.8) are in contradiction we conclude that in this case Φ cannot take its maximum value at an interior critical point.

If $g' < 0$ and (2.3), (2.4) are satisfied we again have (4.5). Use of (2.4) in (4.5) leads to

$$(B-2)B^{-1}\int_0^{q^2} g(s)ds \leq \alpha[(u^2)_M - u^2].\tag{4.9}$$

Since $g' < 0$ we have

$$g(s) \geq g(q^2) \geq g(q_M^2) \geq g(p_M^2)\tag{4.10}$$

the final inequality in (4.10) following from the standard maximum principle for q^2 and the fact that the maximum cannot occur on $\partial\Omega$. The arguments used in proving the last part of this statement are precisely those used in Section 3. Thus from (4.9), (4.10) it follows that

$$q^2 \leq \frac{\alpha B}{(B-2)g(p_M^2)}[(u^2)_M - u^2].\tag{4.11}$$

Recall that $B > 3$. An integration of (4.11) leads as before to

$$\alpha \geq \frac{(B-2)g(p_M^2)}{B}\left(\frac{\pi^2}{4d^2}\right),\tag{4.12}$$

which is in contradiction to (4.2). Thus Theorem 4 is established.

It follows then that under the conditions of Theorem 4

$$\int_0^{q^2} G(s)ds + \alpha u^2 \leq \max_{x \in \Omega}\left\{\int_0^{p^2} G(s)ds + \alpha f^2\right\}e^{-2\alpha\beta t}\tag{4.13}$$

where $p^2 = |\nabla f|^2$. This bound is explict, and it decays exponentially in time.

Clearly, some restriction on the curvature of $\partial\Omega$ is necessary for the establishment of (4.13). If the average curvature is sufficiently negative at some point we would expect q^2 to become very large there and to tend to infinity at a point as the curvature tends to $-\infty$. Thus one would expect that without the restriction that K be nonnegative, $\Phi(x, t)$ might well take its maximum value on $\partial\Omega$.

5. Examples and Applications.

In this section we give examples of functions $g(q^2)$ which satisfy the conditions of Theorem 1. For instance, for $N \leq 4$ the function

$$g = (1 + q^2)^{1/2} \tag{5.1}$$

clearly satisfies $g' > 0$. Choosing

$$\lambda = N^{-1/2} \tag{5.2}$$

we observe that

$$g - N\lambda q^2 g' = (1 + q^2)^{-1/2} \left\{ 1 + \left(1 - \frac{\sqrt{N}}{2} \right) q^2 \right\} \geq 0, \tag{5.3}$$

so that β may be chosen to satisfy

$$g - \frac{4 - 2\sqrt{N} - 2\beta}{1 - \beta} g' q^2 = (1 + q^2)^{-1/2} \left\{ 1 + \frac{2 - \sqrt{N} - \beta}{1 - \beta} q^2 \right\} \geq 0. \tag{5.4}$$

This will insure that (2.1) is satisfied. But (5.4) will be nonnegative provided $\beta \leq (3 - \sqrt{N})/2$. The suggested choice for β then is

$$\beta = (3 - \sqrt{N})/2. \tag{5.5}$$

Since $G'(s) > 0$, (2.3) is clearly satisfied.

As an example of a g for which $g' < 0$ we consider for $N \leq 4$

$$g = (1 + q^2)^{-\varepsilon} \tag{5.6}$$

for a range of ε to be determined. Clearly ε cannot be greater than $1/2$ unless we restrict the "size" of the initial data. Integrating we may rewrite (2.3) as

$$q^2 G - \beta q^2 g - \beta \int_0^{q^2} s g'(s) ds \geq 0. \tag{5.7}$$

Since the third term on the left is nonnegative, it is sufficient to choose β to satisfy (2.4) and

$$G - \beta g \geq . \tag{5.8}$$

It is easily checked that if we choose

$$0 < \beta \leq 1 - 2\varepsilon \tag{5.9}$$

(5.8) will be satisfied while (2.4) is satisfied if

$$\beta \leq \frac{1 - 3\varepsilon}{1 - 2\varepsilon}. \tag{5.10}$$

Thus if $\varepsilon < 1/3$ a positive β can be found such that Theorem 1 is satisfied for $g = (1+q^2)^{-\varepsilon}$.

We remark that since q^2 satisfies a maximum principle if $G(q^2) > 0$, a larger β depending on p_M can be used. For instance, since

$$g + \frac{3-2\beta}{1-\beta}q^2 g' = \frac{1 + [1 - (\frac{3-2\beta}{1-\beta})\varepsilon]q^2}{(1+q^2)^{1+\varepsilon}} \tag{5.11}$$

inequality (2.4) is satisfied for $\frac{1}{2} \le \varepsilon < \frac{1}{3}$ if

$$\beta \le \frac{1 + (1-3\varepsilon)p_M^2}{1 + (1-2\varepsilon)p_M^2}. \tag{5.12}$$

For both $g' > 0$ and $g' < 0$ we have derived a decay inequality of the form

$$\int_0^{q^2} G(s)ds + \alpha u^2 \le Q_0^2 e^{-2\alpha\beta t}, \tag{5.13}$$

where

$$Q_0^2 = \sup_\Omega \left\{ \int_0^{p^2} G(s)ds + \alpha f^2 \right\}, \tag{5.14}$$

with α satisfying either (4.1) if $g' > 0$ or (4.2) if $g' < 0$.

In the case $g' > 0$, (5.13) implies that

$$g(0)q^2 + \alpha u^2 \le Q_0^2 e^{-2\alpha\beta t} \tag{5.15}$$

or

$$q^2 \le g_0^{-1}\{Q_0^2 e^{-2\alpha\beta t} - \alpha u^2\}. \tag{5.16}$$

From an arbitrary point (x,t) in Ω we integrate to the boundary for fixed t obtaining

$$\left| \int_0^{u(x,t)} \frac{d\eta}{\sqrt{\frac{Q_0^2}{\alpha}e^{-2\alpha\beta t} - \eta^2}} \right| \le \alpha^{1/2}[g(0)]^{-1/2}\delta \tag{5.17}$$

or

$$\left| \sin^{-1} \frac{u\sqrt{\alpha}}{Q_0}e^{\alpha\beta t} \right| \le \alpha^{1/2}[g(0)]^{-1/2}\delta, \tag{5.18}$$

where δ is the distance from (x,t) to the boundary. But this implies

$$|u(x,t)| \le \sin(\alpha^{1/2}[g(0)]^{-1/2}\delta)Q_0\alpha^{-1/2}e^{-\alpha\beta t}. \tag{5.19}$$

This bound has some interesting features. It is explicit, it tends to zero as x tends to the boundary and it exhibits the temporal decay.

For $g' < 0$ the constant α is chosen to satisfy (4.2), and as in the derivation of (4.9)–(4.11) we have from (5.13)

$$\frac{(B-2)}{B}g(p_M^2)q^2 + \alpha u^2 \le Q_0^2 e^{-2\alpha\beta t}. \tag{5.20}$$

Integrating as in the case $g' > 0$ we now obtain

$$|u(x,t)| \leq \sin \left(\left[\frac{B}{(B-2)} \alpha g^{-1}(p_M^2) \right]^{1/2} \delta \right) Q_0 \alpha^{-1/2} e^{-\alpha\beta t}. \tag{5.21}$$

In physical contexts the quantity $qg(q^2)$ at a point on $\partial\Omega$ represents the magnitude of the flux at that point. To find a bound for this quantity we start with

$$\int_0^{q^2} G(s)ds \leq Q_0^2 e^{-2\alpha\beta t} \quad \text{on} \quad \partial\Omega. \tag{5.22}$$

If $g' > 0$ we may rewrite this inequality as

$$q^2 g(q^2) + \int_0^{q^2} sg'(s)ds \leq Q_0^2 e^{-2\alpha\beta t}, \tag{5.23}$$

which implies that on $\partial\Omega$

$$q^2 g(q^2) \leq Q_0^2 e^{-2\alpha\beta t}. \tag{5.24}$$

But

$$qg(q^2) = [q^2 g(q^2)]^{1/2} [g(q^2)]^{1/2} \leq Q_0 [g(p_M^2)]^{1/2} e^{-\alpha\beta t}, \tag{5.25}$$

the desired bound.

If $g' < 0$ then as in the derivation of (4.9) we have

$$\left(1 - \frac{2}{B}\right) \int_0^{q^2} g(s)ds \leq Q_0^2 e^{-2\alpha\beta t} \tag{5.26}$$

or since $g(s) \geq g(q^2)$,

$$q^2 g(q^2) \leq \frac{B}{(B-2)} Q_0^2 e^{-2\alpha\beta t}, \tag{5.27}$$

which implies that

$$qg(q^2) \leq \left[\frac{B}{B-2}\right]^{1/2} Q_0 \sqrt{g(0)} e^{-\alpha\beta t}. \tag{5.28}$$

REFERENCES

1. Friedman, A., *Remarks on the maximum principle for parabolic equations and its applications*, Pac. J. Math. **8** (1958), 201–211.
2. Nirenberg, L., *A strong maximum principle for parabolic equations*, Comm. Pure Appl. Math. **6** (1953), 167–177.
3. Payne, L. E. and Philippin, G. A., *On some maximum principles involving harmonic functions and their derivatives*, SIAM Journal of Math. Anal. **10** (1979), 96–104.
4. Payne, L. E. and Philippin, G. A., *Decay bounds for solutions of second order parabolic problems and their derivatives*, Math. Models and Methods in Appl. Sci. **5** (1995), 95–110.
5. Protter, M. H. and Weiberger, H. F., *Maximum Principles in Differential Equations*, Prentice–Hall, 1967.
6. Serrin, J., *Gradient estimates for solutions of nonlinear elliptic and parabolic equations*, Contributions to nonlinear functional analysis (Proc. Symp. Math. Res. Cent. Univ. Wisc. 1971) Academic Press, New York (1971).
7. Sperb, R. P., *Maximum Principles and Their Applications*, Math. in Sci. and Eng., Vol. 157, Academic Press (1981).
8. Walter, W., *Differential and Integral Inequalities*, Ergebnisse der Mathematik und ihrer Grenzgebiete, vol. 55, Springer (1964).

Some complementary estimates in the Dead Core problem

by

René Sperb
Seminar für Angewandte Mathematik
ETH Zürich
8092 Zürich, Switzerlannd

email: sperb@sam.math.ethz.ch

Abstract

This article is concerned with some estimates in the problem $\Delta u = c^2 u^p$ in Ω, $u = 1$ on $\partial\Omega$. The quantities of interest here are the critical value of c for which a dead core Ω_0 exists, the location of Ω_0 and the effectiveness factor.

1 Introduction

This article is concerned with the problem

$$(1.1) \qquad \begin{cases} \Delta u = c^2 u^p & \text{in } \Omega \subset \mathbb{R}^N, \\ u = 1 & \text{on } \partial\Omega, \end{cases}$$

with $p \in (0,1)$. This precise form of the nonlinearity was chosen for simplicity only and in fact one can replace u^p by a more general function $f(u)$ satisfying

$$f(0) = 0, \; f'(s) \geq 0 \; \text{ and } \; \int_0^1 \frac{ds}{f(s)} < \infty .$$

It was shown in [3] that for sufficiently large c a "dead core" Ω_0 develops in Ω, i.e. a region where $u \equiv 0$. Problem (1.1) and various generalizations have been studied since by many different authors. There is no attempt made here of giving a survey of the literature on this problem and only the papers directly related to the results of this paper will be cited.

The qualities of interest for which estimates will be derived are:

- the critical value c_0 of c above which a dead core will exist,

- the location of the dead core,

- the "effectiveness factor" $\eta := \dfrac{\displaystyle\int_\Omega u^p \, dx}{|\Omega|}$.

The estimates to be derived complement the corresponding ones given in [3].

2   Estimates derived from optimal supersolutions

In the following an essential assumption made is that the mean curvature of $\partial\Omega$ be nonnegative everywhere. Most of the estimates will in general no longer hold if this assumption is dropped, as counterexamples show. Hence, this is not merely a purely technical assumption.

We now construct an optimal supersolution by combining the one-dimensional version of (1.1) with a suitable linear problem defined on Ω. This idea is essentially contained in [5] in a different form. It was extended then to more general elliptic problems in $\mathbb{R}^2$ in [6], to problems on a two-dimensional manifold in [9] and to the case of nonlinear boundary conditions in [10].

Let now $X(s)$ be the solution of

$$
(2.1) \qquad \begin{cases} X''(s) = c^2 f(X) \text{ in } (0, s_0) \\ X'(0) = 0, \; X(s_0) = 1, \end{cases}
$$

where at the moment the precise form of $f(X)$ is not yet relevant.

As a first choice of a linear problem consider the "torsion problem", i.e.

$$
(2.2) \qquad \begin{cases} \Delta\psi + 1 = 0 \text{ in } \Omega \\ \psi = 0 \text{ in } \partial\Omega. \end{cases}
$$

One then constructs a supersolution $\overline{u}(x)$ to (1.1) having the same level lines as the torsion function by setting

$$
(2.3) \qquad \overline{u}(x) = X(s(x)), \quad x \in \Omega,
$$

where

$$
(2.4) \qquad s(x) = \sqrt{2(\psi_m - \psi(x))}, \quad \psi_m = \max_\Omega \psi.
$$

The choice of $s(x)$ is suggested by the one-dimensional version of (2.2) when $s(x) = x$. In problem (2.1) we thus choose $s_0 = \sqrt{2\,\psi_m}$.

The main result from which the estimates follow can be stated as

Theorem 1 *Assume that the mean curvature of $\partial\Omega$ is nonnegative everywhere and* $f(0) \geq 0$, $f'(s) \geq 0$ *for* $s \geq 0$. *Then*

$$
\overline{u}(x) = X(s(x)) \text{ is a supersolution, i.e.}
$$
$$
\Delta\overline{u} \leq c^2 f(\overline{u}) \text{ in } \Omega
$$
$$
\overline{u} = 1 \text{ on } \partial\Omega.
$$

Proof: Calculate first

$$
\nabla s = -\frac{\nabla\psi}{s},
$$
$$
\Delta s = -\frac{\Delta\psi}{s} - \frac{|\nabla\psi|^2}{s^3} = \frac{1}{s} - \frac{|\nabla\psi|^2}{s^3},
$$

and then

$$\Delta \bar{u} = X' \cdot \Delta s + X'' \cdot |\nabla s|^2 = \frac{1}{s} X' \left(1 - \frac{|\nabla \psi|^2}{s^2} \right) + X'' \cdot \frac{|\nabla \psi|^2}{s^2} ,$$

from where one finds

$$\Delta \bar{u} - c^2 f(\bar{u}) = \left(\frac{1}{s} X' - c^2 f(X) \right) \left\{ 1 - \frac{|\nabla \psi|^2}{s^2} \right\} .$$

It was shown in [4] that under our assumption on $\partial \Omega$ one has

$$|\nabla \psi|^2 \le 2(\psi_m - \psi(x)) ,$$

which means that the term $\{\}$ is nonnegative. It remains to check the sign of the other factor. To this end consider

$$g(s) = X' - c^2 s f(X)$$

which satisfies $g(0) = 0$ and

$$g'(s) = X'(s) - c^2 f(X(s)) - c^2 \frac{df}{dX} \cdot X'(s) = -c^2 \frac{df}{dX} \cdot X'(s) \le 0 ,$$

since $\frac{df}{dX} \ge 0$ and $X' \ge 0$.

Therefore $g(s) \le 0$ for $s \ge 0$ so that $\bar{u}(x)$ satisfies the required differential inequality. If we select $s_0 = \sqrt{2\,\psi_m}$ in (2.1) the boundary condition is satisfied as well.

For the particular choice $f(u) = u^p$, $0 \le p < 1$ the usual properties of sub- or supersolutions still hold even if $f'(0)$ becomes unbounded (see [11]).

As a first application of Theorem 1 we note

Corollary 1.1 *The critical value satisfies*

(2.5)
$$c_0^2 \le \frac{p+1}{(1-p)^2 \psi_m} ,$$

with equality if Ω degenerates to an infinite slab of width $2\sqrt{2\psi_m}$.

Proof: For $f(u) = u^p$ the value c_0 in the one dimensional problem(2.1) is given by

(2.6)
$$c_0^2 = \frac{2(p+1)}{(1-p)^2} \cdot \frac{1}{s_0^2} .$$

Since $\bar{u}(x) = X(s(x))$ is a supersolution inequality (2.6) follows immediately.

Remarks:

a) In order to make inequality (2.5) more explicit we need a lower bound for ψ_m. It was shown in [4] that

$$\psi_m \ge \frac{A^2}{2L^2} \quad (A = |\Omega|, \ L = |\partial \Omega|)$$

with equality for a slab (taking appropriate limits) under our assumption on $\partial\Omega$. For a strictly convex plane domain it follows from inequality (3) of Webb [12] that even

$$\psi_m \geq \frac{A^2}{L^2} \cdot \frac{L - k_0\, A}{2L - 3k_0\, A} \; , \quad (\text{curvature } k \geq k_0 > 0)$$

with equality for a circle or an infinite strip.

b) It was shown in [3] that

$$(2.7) \qquad c_0^2 \geq \frac{2(p+1)}{(p-1)^2 \cdot \rho^2} \; , \quad \rho = \text{radius of largest ball in } \Omega \; ,$$

again with equality for a slab. Hence (2.5) is the optimal counterpart to (2.7) since $\psi_m \leq \frac{\rho^2}{2}$ noted in [4].

There is also information on the location and size of Ω_0 contained in Theorem 1, which may be stated as

Corollary 1.2 *The dead core Ω_0 contains the set*

$$\left\{ x \in \Omega \,|\, \psi(x) \geq d(p,c) \left[\sqrt{2\,\psi_m} - \frac{1}{2}\, d(p,c) \right] \right\} \; ,$$

where $d(p,c) = \dfrac{\sqrt{2(p+1)}}{(1-p)c}$.

Proof: In the one-dimensional problem (2.1) with $f(X) = X^p$ one can easily calculate the dead core as the interval $(0, \sigma(p,c))$ where

$$(2.8) \qquad \sigma(p,c) = s_0 - d(p,c) \; .$$

The level set $\sqrt{2(\psi_m - \psi(x))} = s_0 - d(p,c) = \sqrt{2\,\psi_m} - d(p,c)$ must be contained in Ω_0 since $X(s(x))$ is a supersolution. This implies the statement of Corollary 1.2.

Remarks:

a) It was shown in [3] that the dead core is contained in the set

$$\{ x \in \Omega \,|\, \text{dist}(x, \partial\Omega) \geq d(p,c) \} \; .$$

b) The torsion function $\psi(x)$ is only known explicitly in some special cases (e.g. ellipse, equilateral triangle). If $\psi(x)$ is not known explicitly Corollary 1.2 is still useful if one makes use of the monotonic behavior of $\psi(x)$ with respect to the domain: if $\tilde{\Omega} \subset \Omega$ then the corresponding solutions satisfy $\tilde{\psi}(x) \leq \psi(x)$ for any $x \in \tilde{\Omega}$.

As a third consequence of Theorem 1 one has

Corollary 1.3 *For given value of ψ_m the effectiveness factor η is a minimum for the slab of width $2\sqrt{2\,\psi_m}$. In particular, if $c \geq c_0$ one has*

$$(2.9) \qquad \eta \geq \frac{1}{c\sqrt{(p+1)\psi_m}} \; .$$

Proof: Since $\bar{u} = X(s(x))$ is a supersolution which satisfies the boundary condition one has

$$\frac{\partial u}{\partial n} \geq \frac{\partial \bar{u}}{\partial n} \quad \text{on } \partial\Omega$$

which implies

$$c^2 |\Omega| \, \eta \geq - \oint_{\partial\Omega} \frac{X'(s_0)}{s_0} \frac{\partial \psi}{\partial n} \, d\sigma = \frac{X'(s_0)}{s_0} |\Omega|, \quad (d\sigma = \text{element of } \partial\Omega)$$

that is

$$\eta \geq \frac{X'(s_0)}{c^2 \cdot s_0} = \text{effectiveness factor for slab of width } 2 \cdot s_0.$$

If $c \geq c_0$ one gets (see Aris [1], p. 146)

$$\eta = \sqrt{\frac{2}{c^2(p+1) \cdot s_0}} = \frac{1}{c\sqrt{(p+1)\psi_m}},$$

which completes the proof.

Remarks:

a) For $c < c_0$ the value of η in the one-dimensional case is determined from the relation (see [1], p. 148)

$$c \cdot s_0 = \sqrt{\frac{2}{1+p} \, (1 - u_0^{p+1})} \cdot F\left(1, \frac{p}{p+1}; \frac{3}{2}; 1 - u_0^{p+1}\right) =: h(u_0, p)$$

and then,

$$\eta = \frac{1}{F\left(1, \frac{1}{p+1}; \frac{3}{2}; 1 - u_0^{p+1}\right)}$$

there $F(a, b; c; z)$ is the hypergeometric function and $u_0 = u(0) = $ minimum value. The function $h(u_0, p)$ is monotonically decreasing in u_0 for any $p \in (0, 1)$. One can also prove that η is a decreasing function of s_0, so that an upper bound for ψ_m is needed in Corollary 1.3. A number of upper bounds for ψ_m are known (see e.g. [2], [4], [8]). One has e.g. $\psi_m \leq \frac{\rho^2}{2}$ under our assumptions on $\partial\Omega$.

b) It follows from Lemma 3.1 of [3] that for $c \geq c_0$

$$(2.10) \qquad \eta \leq \sqrt{\frac{2}{p+1}} \cdot \frac{L}{cA}$$

and the equality sign holds also in the limit if Ω degenerates into a slab.

A second choice of a linear problem is the fixed membrane problem on Ω i.e.

$$(2.11) \qquad \begin{cases} \Delta\varphi + \lambda\varphi = 0 & \text{in } \Omega \\ \varphi = 0 & \text{on } \partial\Omega. \end{cases}$$

In this case we replace $s(x)$ as defined in (2.4) by

$$(2.12) \qquad t(x) = \frac{1}{\sqrt{\lambda_1}} \cos^{-1} \left(\frac{\varphi(x)}{\varphi_m} \right),$$

where $\lambda_1 =$ first eigenvalue with associated eigenfunction $\varphi(x)$ and $\varphi_m = \max\limits_{\Omega} \varphi$. Also $s_0 = \frac{\pi}{2\sqrt{\lambda_1}}$ is now the length of the interval.

The analogue of Theorem 1 is now

Theorem 2 *Assume that the mean curvature of $\partial\Omega$ is nonnegative everywhere and $f(0) \geq 0$, $f'(s) \geq 0$ for $s \geq 0$. Then*

$$\begin{aligned}
\overline{u}(x) &= X(t(x)) &&\text{is a supersolution, i.e.} \\
\Delta\overline{u} &\leq c^2 f(\overline{u}(x)) &&\text{in } \Omega \\
\overline{u} &= 1 &&\text{on } \partial\Omega
\end{aligned}$$

Proof: A straightforward calculation gives now

$$\Delta\overline{u} - c^2 f(\overline{u}) = \left(\lambda_1 - \frac{|\nabla\varphi|^2}{\varphi_m^2 - \varphi^2} \right) \left\{ X' \cdot \cot(\sqrt{\lambda_1}\, t) \sqrt{\lambda_1} - c^2 f(X(t)) \right\}.$$

By a result of Payne & Stakgold [7] one has $\lambda_1 \geq \dfrac{|\nabla\varphi|^2}{\varphi_m^2 - \varphi^2}$ if the mean curvature of $\partial\Omega$ is non negative.

The term $\{\}$ is nonpositive as a similar reasoning as in the proof of Theorem 1 shows, now for $g(t) = X'\sqrt{\lambda_1} \cos(\sqrt{\lambda_1}\, t) - \sin(\sqrt{\lambda_1}\, t) f(X(t))$.

The counterparts of Corollaries 1.1 - 1.3 are now obvious:

Corollary 2.1 *The critical value c_0 satisfies*

$$(2.13) \qquad c_0^2 \leq \frac{8\lambda_1(p+1)}{(1-p)^2 \pi^2}.$$

Remarks: Since $\lambda_1 \geq \dfrac{\pi^2}{8\,\psi_m}$ as noted by Payne [5], (2.13) is weaker than (2.5).

The counterpart of Corollary 1.2 may still be useful in some cases (see e.g. Example 1). It now reads

Corollary 2.2 *The dead core is contained in the set*

$$\{x \in \Omega \,|\, \varphi(x) \geq \varphi_m \cdot \sin(\sqrt{\lambda_1}\, d(p,c))\}$$

It is interesting to see however that inequality (2.9) can now be improved. In fact we have

Corollary 2.3 *For given value of λ_1 the effectiveness factor η is a minimum for the slab of width $\dfrac{\pi}{\sqrt{\lambda_1}}$. In particular if $c \geq c_0$ one has*

$$(2.14) \qquad \eta \geq \frac{2}{c\pi} \sqrt{\frac{2\lambda_1}{p+1}}.$$

Examples

1. $\Omega =$ Rectangle of sides $a = 2$, $b = 1$, $p = \frac{1}{2}$. Inequalities (2.5) and (2.7) then yield

$$8 \leq \frac{(1-p)^2}{p+1} c_0^2 \leq 8.782 .$$

For $c \geq c_0$ (2.10) and (2.14) give

$$\sqrt{\frac{2}{p+1}} \cdot \frac{2.236}{c} \leq \eta \leq \sqrt{\frac{2}{p+1}} \cdot \frac{3}{c} .$$

For $p = \frac{1}{2}$ Corollary 1.2 and the estimate in [3] give the following pictures (the boundary of Ω_0 must lie in the shaded regions).

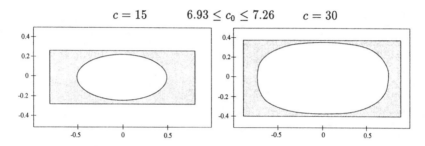

$$c = 15 \qquad 6.93 \leq c_0 \leq 7.26 \qquad c = 30$$

2. $\Omega =$ Equilateral triangle of height 1. One has $\psi_m = \frac{1}{27}$ and $\lambda_1 = 4\pi^2$ and (2.5) and (2.7) now give

$$18 \leq \frac{(1-p)^2}{p+1} c_0^2 \leq 27 .$$

A better lower bound in some cases (for plane domains) is $c_0^2 \geq \dfrac{4\pi}{(1-p) \cdot A}$ as given in Corollary 3.1 of [3].

For $c \geq c_0$ (2.10) and (2.14) show that

$$\sqrt{\frac{2}{p+1}} \cdot \frac{4}{c} \leq \eta \leq \sqrt{\frac{2}{p+1}} \cdot \frac{6}{c} .$$

For $p = 0.5$ one has therefore

$$9.33 \leq c_0 \leq 12.73 ,$$

and one obtains the following pictures

224

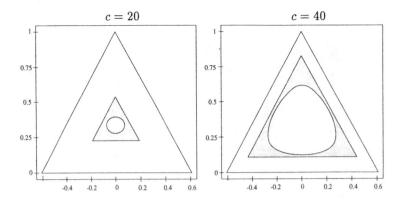

$c = 20$ $c = 40$

The boundary of Ω_0 lies in the shaded region.

References

[1] Aris R., *The mathematical theory of diffusion and reaction in permeable catalysts*, Vol. 1, Clarendon Press, Oxford 1975.

[2] Bandle C., *Isoperimetric inequalities and applications*, Pitman, London 1980.

[3] Bandle C., Sperb R. and Stakgold I., *Diffusion and reaction with monotone kinetics*, Nonlinear Analysis 8, (1984), 321-333.

[4] Payne L.E., *Bounds for the maximum stress in the Saint Venant torsion problem*, Indian J. Mech. Math., Special Issue 1968, pp. 51-59.

[5] Payne L.E., *Bounds for the solutions of a class of quasilinear elliptic eigenvalue problem in terms of the torsion function*, Proc. Roy. Soc. Edinborough 1581, pp. 51-59.

[6] Payne L.E. and Phillipin G., *Comparison theorems for a class of nonlinear elliptic problems*, Nonlinear Analysis 3, 1985, pp. 787-797.

[7] Payne L.E. and Stakgold I., *Nonlinear problems in nuclear reactor analysis*, Springer Lecture Notes in Math., Vol. 322, pp. 298-307.

[8] Sperb R., *Maximum principles and their applications*, Academic Press, New York 1981.

[9] Sperb R., *Optimal bounds for the critical value in a semilinear boundary value problem on a surface*, General inequalities 5, ISNM 80(1987), 391-400.

[10] Sperb R., *Optimal bounds in semilinear elliptic problems with nonlinear boundary conditions*, ZAMP 44, 1993, pp. 639-653.

[11] Stuart C.A., *Maximal and minimal solutions of elliptic differential equations with discontinuous nonlinearities*, Math. Z. 163 (1978), p. 239.

[12] Webb J.R.L., *Maximum principles for functionals associated with the solution of semilinear elliptic boundary value problems*, ZAMP 40, 1989, pp. 330-338.

GUIDED TM-MODES IN A SELF-FOCUSING ANISOTROPIC DIELECTRIC

C.A.STUART

Abstract. We present a new formulation of the equations for planar and cylindrical TM-modes in a uniaxial self-focusing dielectric medium. By using the amplitude of the magnetic field as the basic unknown, and by inverting the usual constitutive assumption, we obtain a single second order equation. The existence of guided planar TM-modes is the studied.

1. Introduction. The guidance of light is usually achieved by exploiting the effect of variations in the refractive index due to inhomogeneity of the medium through which the beam is propagating. As can be understood from Snell's law, the favorable configuration consists of a region of high refractive index surrounded by layers of material having a lower refractive index. However it is also well-known that nonlinear effects can be used to enhance,or even to produce, guidance. For example, no guidance occurs in a homogeneous linear medium (where the refractive index is constant) whereas it will occur, at least for sufficiently intense beams, in a homogeneous self-focusing medium (where the refractive index is an increasing function of the intensity of the light passing through it).

The mathematical discussion of this phenomenon amounts to the study of special solutions of Maxwell's equations in a homogeneous medium whose constitutive relation expresses the electric displacement field as a nonlinear function of the electric field. In a self-focusing material this dielectric response is an increasing function of the electric field strength. The special solutions which have been studied in this context are either TE (transverse electric field) or TM (transverse magnetic field) modes having a planar or cylindrical symmetry. For isotropic self-focusing materials the mathematical analysis of these problems has been undertaken in [1],[2] ,[3],[4],[5] and [6].

In this paper we deal with TM-modes in an anisotropic self-focusing medium. (For TE-modes in a uniaxial material propagating in the direction of the axis, the equation for the amplitude of the electric field is the same as that for an isotropic material.) It is generally recognized (see [20], [13], [12], [10], [17], [18], [19] for example) that, even for isotropic materials, the discussion of TM-modes is considerably more complicated than the corresponding treatment of TE-modes because the nonlinear effect is a function only of the electric field strength. In a TM-mode the electric field is composed of two orthogonal components which are $\pi/2$ out of phase and the problem is usually formulated as a system of equations for the amplitudes of these two components. In a TE-mode the electric field has a single harmonic component and the problem reduces to a second order differential equation for its amplitude. Using a mixture of approximations and numerics, TM-modes in anisotropic media have been discussed in [11], [12], [13], [15], [14], and [16]. Most of these contributions deal only with the special case of a Kerr nonlinearity and there are a number of other contributions dealing only with isotropic materials [8], [9], [16], [17], [18], [19].

Here we give a rigorous discussion based on a new, but equivalent, formulation of the problem which covers a wide range of constitutive laws for anisotropic materials, with or without saturation. For planar and cylindrical waveguides, we use the amplitude of the magnetic field to reduce the study of TM-modes to a single second order differential equation as in the case of TE-modes. To do so we must first invert the constitutive relation for a self-focusing anisotropic medium and this is the subject of the Section 2. The resulting equations have forms that are more complicated than the semilinear equations obtained for the corresponding TE-modes but, at least for planar modes, the equation does have a first integral. This is exploited in Section 3 to give necessary and sufficient conditions for the existence of planar guided TM-modes. The corresponding equation for a cylindrical waveguide is nonautonomous and it does not have a first integral. The discussion of guided modes in this case is less elementary and is the subject of [6].

2. Equations for TM-modes. We start with Maxwell's equations in the form

$$c(\nabla \wedge E) = -\partial_t B, c(\nabla \wedge H) = \partial_t D, \nabla.D = 0, \nabla.B = 0$$

where c is the speed of light in a vacuum,[7].

To introduce the constitutive assumption for a homogeneous uniaxial material we choose an orthonormal basis $\{e_i : i = 1, 2, 3\}$ with e_3 in the direction of the axis of the medium. All the e-m fields E, D, H and B that we consider can be expressed in the form

$$F(x, y, z, t) = F_T(x, y) \cos(kz - \omega t) + F_z(x, y) \sin(kz - \omega t)$$

where $F_T, F_z : \Re^2 \longrightarrow \Re^3$ are such that $F_T.e_3 = 0$ and $F_z.e_i = 0$ for $i = 1, 2$.

Thus the transverse and axial components of the fields are out of phase by a quarter of a cycle and the fields constitute monochromatic waves propagating in the direction of the material axis.

For fields of this type the constitutive relation which we adopt is expressed as follows.

(CR) $B = H$ and there exist two continuous functions $\varepsilon_i : [0.\infty)^2 \longrightarrow (0, \infty)$ such that

$$D_T(x, y) = \varepsilon_1(| E_T(x, y) |^2 /2, | E_z(x, y) |^2 /2) E_T(x, y)$$

$$D_z(x, y) = \varepsilon_2(| E_T(x, y) |^2 /2, | E_z(x, y) |^2 /2) E_z(x, y)$$

where $0 < A < \varepsilon_i(s_1, s_2)$ for $i = 1, 2$ and for $s_1, s_2 \geq 0$.

This means that with respect to the basis $\{e_i : i = 1, 2, 3\}$ the dielectric response tensor is represented by the diagonal matrix $\begin{pmatrix} \varepsilon_1 & 0 & 0 \\ 0 & \varepsilon_1 & 0 \\ 0 & 0 & \varepsilon_2 \end{pmatrix}$ where the elements are functions of the time-averages,$| E_T(x, y) |^2 /2$ and $| E_z(x, y) |^2 /2$, of the intensities of the transverse and axial components of the electric field E.

By a TM-mode we mean a solution of the above type for which $H_z \equiv 0$ and for such modes we expect to reduce Maxwell's equations and (CR) to a system of equations for H_T. Indeed since $\partial_t^2 D = -\omega^2 D$ we must have $-\omega^2 D = c\nabla \wedge \partial_t H$ and hence we set $D = -(c/\omega)\nabla \wedge H_T \sin(kz - \omega t)$. This ensures that $\nabla . D = 0$ and so it suffices to choose H_T in such a way that

$$\nabla . H = 0 \text{ and } c(\nabla \wedge E) = -\omega H_T \sin(kz - \omega t).$$

Setting $H_T.e_i = h_i$ for $i = 1, 2$ the first equation becomes $\partial_x h_1 + \partial_y h_2 = 0$ but to express the second equation using H_T we must express E using H_T. Since D has been given in terms of H_T this can be done by inverting the constitutive relation (CR). To show that this is possible we require some additional properties of the dielectric response.

According to (CR) the dielectric response of the medium is determined by a function $\varepsilon = (\varepsilon_1, \varepsilon_2) : [0, \infty)^2 \longrightarrow (0, \infty)^2$ about which we now make some further assumptions.

(V) There is a potential $\varphi \in C^1([0, \infty)^2)$ such that $\varepsilon = \nabla \varphi$.

This property of the dielectric response is used explicitly in [13] where it is assumed that $\partial_2 \varepsilon_1 = \partial_1 \varepsilon_2$ for all $s_1, s_2 \geq 0$. When (V) holds the potential, normalized so that $\varphi(0) = 0$ is given by

$$\varphi(s) = \int_0^1 \varepsilon(ts).s dt$$

and we define an auxiliary function Φ by

$$\Phi(s_1, s_2) = \varphi(s_1^2/2, s_2^2/2).$$

(C) We suppose that $\Phi \in C^2(\Re^2)$ and that $D^2\Phi(s)$ is positive definite for all $s \in \Re^2$.

The positive definitiveness of $D^2\Phi(s)$ implies that $\partial_i \left[\varepsilon_i(s_1^2/2, s_2^2/2)s_i\right] > 0$ for $s \in \Re^2$ and $i = 1, 2$ and that Φ is strictly convex. Since $\Phi(0) = 0$ and $\nabla \Phi(0) = 0$, it follows that $\Phi(s) > 0$ for all $s \in \Re^2 \setminus \{0\}$. By (CR), $\|\nabla\Phi(s_1, s_2)\| \geq A\sqrt{(s_1^2 + s_2^2)}$ for all $(s_1, s_2) \in \Re^2$ and so from the conclusions of Theorem 26.4 to Lemma 26.7 of[22] we find that $\nabla\Phi : \Re^2 \longrightarrow \Re^2$ is a diffeomorphism and $[\nabla\Phi]^{-1} = \nabla\Phi^*$ where Φ^* is the Legendre transform of Φ which can be defined by

$$\Phi^*(\tau) = \sup\left\{s.\tau - \Phi(s) : s \in \Re^2\right\} \text{ for } \tau = (\tau_1, \tau_2) \in \Re^2.$$

From this formula we deduce that $\Phi^*(0,0) = 0$ and $\Phi^*(s_1, s_2) = \Phi^*(|s_1|, |s_2|)$. Since ε_1 and $\varepsilon_2 > 0$ we also have that $\nabla\Phi$ maps the four quadrants and half-axes onto themselves. Of course $\nabla\Phi(s_1, s_2) = \begin{pmatrix} \varepsilon_1(s_1^2/2, s_2^2/2)s_1 \\ \varepsilon_2(s_1^2/2, s_2^2/2)s_2 \end{pmatrix}$ and we now show that $\nabla\Phi^*$ has a similar structure. From (C) we deduce that $\Phi^* \in C^2(\Re^2)$ and that $D^2\Phi^*(t)$ is positive definite for all $t \in \Re^2$. See Corollary 4.2.10 of [21].

Define $\psi : [0, \infty)^2 \longrightarrow \Re$ by

$$\psi(t_1, t_2) = \Phi^*(\sqrt{2t_1}, \sqrt{2t_2})$$

so that

$$\Phi^*(t_1, t_2) = \psi(t_1^2/2, t_2^2/2).$$

Clearly, $\psi \in C^2((0,\infty)^2)$ with $\partial_i \Phi^*(t_1, t_2) = \partial_i \psi(t_1^2/2, t_2^2/2)t_i$ and we define $\gamma = (\gamma_1, \gamma_2) \in C^1((0,\infty)^2, \Re^2)$ by $\gamma_i = \partial_i \psi$. Then

$$\begin{pmatrix} t_1 \\ t_2 \end{pmatrix} = \begin{pmatrix} \varepsilon_1(s_1^2/2, s_2^2/2)s_1 \\ \varepsilon_2(s_1^2/2, s_2^2/2)s_2 \end{pmatrix}$$

$$\Leftrightarrow t = \nabla\Phi(s) \Leftrightarrow \nabla\Phi^*(t) = s \Leftrightarrow$$

$$\begin{pmatrix} s_1 \\ s_2 \end{pmatrix} = \begin{pmatrix} \gamma_1(t_1^2/2, t_2^2/2)t_1 \\ \gamma_2(t_1^2/2, t_2^2/2)t_2 \end{pmatrix}.$$

It follows that for $i = 1, 2$ and $t_1, t_2 > 0$, $\gamma_i(t_1, t_2) =$

$$1/\varepsilon_i(\gamma_1(t_1, t_2)^2 t_1, \gamma_2(t_1, t_2)^2 t_2) =$$

$$1/\varepsilon_i\left(\left[\partial_1 \Phi^*(\sqrt{2t_1}, \sqrt{2t_2})\right]^2/2, \left[\partial_2 \Phi^*(\sqrt{2t_1}, \sqrt{2t_2})\right]^2/2\right).$$

Using this formula, and recalling that $\nabla\Phi^*$ is continuous on $\Re^2$, we see that γ can be extended by continuity to $[0,\infty)^2$.

Let us note an additional property of γ. Since $D^2\Phi^*(t)$ is positive definite for all $t \in \Re^2$, $0 < \partial_i^2 \Phi^*(t_1, t_2) = \partial_i \{\gamma_i(t_1^2/2, t_2^2/2)t_i\}$ for $i = 1, 2$ and all $(t_1, t_2) \in \Re^2$.

Summarizing these remarks we see that $\nabla\Phi^*(t_1, t_2) = \begin{pmatrix} \gamma_1(t_1^2/2, t_2^2/2)t_1 \\ \gamma_2(t_1^2/2, t_2^2/2)t_2 \end{pmatrix}$ where γ has the following properties.

(H) $\gamma \in C^1((0,\infty)^2, \Re^2) \cap C([0,\infty)^2, \Re^2)$ and for $i = 1, 2$,

$$0 < \gamma_i(t_1, t_2) < 1/A,$$

$$\gamma_i(t_1^2/2, t_2^2/2) = 1/\varepsilon_i(s_1^2/2, s_2^2/2) \text{ where } t_i = \varepsilon_i(s_1^2/2, s_2^2/2)s_i \text{ and }$$

$$\partial_i \left\{\gamma_i(t_1^2/2, t_2^2/2)t_i\right\} > 0 \text{ for all } (t_1, t_2) \in \Re^2.$$

As an example of the above procedure let us consider the Kerr law for a uniaxial material.

Example We suppose that there are four positive constants α_T, α_z, P and Q such that

$$\varepsilon_1(s_1, s_2) = \alpha_T + Ps_1 + Qs_2$$
$$\varepsilon_2(s_1, s_2) = \alpha_z + Qs_1 + Ps_2 \text{ for all } s_1, s_2 \geq 0.$$

Clearly $\varepsilon = \nabla\varphi$ where $\varphi(s_1, s_2) = \alpha_T s_1 + \alpha_z s_2 + Q s_1 s_2 + P(s_1^2 + s_2^2)/2$ and it is easy to check that $\Phi(s_1, s_2) = \varphi(s_1^2/2, s_2^2/2)$ satisfies the condition (C) provided that $Q/P \le 3$. Referring to [12] or to page 246 of [11], we see that the ratio Q/P is $\frac{1}{3}$ when the nonlinearity of the dielectric response is due to electronic distortion, whereas it is 1 when this nonlinearity is caused by electrostriction.

Returning to the general case we see that, if the function ε in the constitutive relation (CR) has the properties (V) and (C), then (CR) can be expressed in the following way where the function γ has the properties (H).

(CR)* $B = H$ and

$$E_T(x, y) = \gamma_1(\mid D_T(x, y) \mid^2 /2, \mid D_z(x, y) \mid^2 /2) D_T(x, y)$$

$$E_z(x, y) = \gamma_2(\mid D_T(x, y) \mid^2 /2, \mid D_z(x, y) \mid^2 /2) D_z(x, y)$$

where γ is obtained from ε by the above construction.

Returning to the problem of expressing the equations for TM-modes using H_T, and recalling that we have $D = -(c/\omega)\nabla \wedge H_T \sin(kz - \omega t)$ with $H_T = h_1 e_1 + h_2 e_2$, we find that

$$D_T = k(c/\omega)\{-h_2 e_1 + h_1 e_2\} \text{ and that } D_z = (c/\omega)\{\partial_x h_2 - \partial_y h_1\}.$$

Using (CR)* we can now express E in the equation $c(\nabla \wedge E) = -\omega H_T \sin(kz - \omega t)$ in terms of H_T. To write the resulting equations in a concise way it is convenient to introduce the following notation.
Set

$$g_i = \gamma_i \left((kc/\omega)^2 \left[h_1^2 + h_2^2\right]/2, (c/\omega)^2 [\partial_x h_2 - \partial_y h_1]^2 /2\right) \text{ for } i = 1, 2.$$

The magnetic field

$$H_T(x, y, z, t) = [h_1(x, y)e_1 + h_2(x, y)e_2]\cos(kz - \omega t)$$

satisfies Maxwell's equations and the constitutive relation (CR), where the dielectric response ε is supposed to have the properties (V) and (C), provided that (h_1, h_2) is a solution of the following system of equations.

$$\partial_x h_1 + \partial_y h_2 = 0$$

$$\partial_y \{g_2 [\partial_x h_2 - \partial_y h_1]\} + k^2 g_1 h_1 - (\omega/c)^2 h_1 = 0$$

$$-\partial_x \{g_2 [\partial_x h_2 - \partial_y h_1]\} + k^2 g_1 h_2 - (\omega/c)^2 h_2 = 0$$

There are two important cases where this system can be reduced to a single equation.

Planar TM-modes

There is a solution of the form

$$h_1(x, y) = 0 \text{ and } h_2(x, y) = (\omega/kc)u(kx) \tag{1}$$

provided that u satisfies the equation

$$\{g_2 u'(x)\}' - g_1 u(x) + \lambda u(x) = 0 \tag{2}$$

where now $g_i = \gamma_i(u(x)^2/2, u'(x)^2/2)$ for $i = 1, 2$ and $\lambda = (\omega/kc)^2$.

Cylindrical TM-modes

There is a solution of the form

$$h_1(x, y) = -(\omega/kc)u(r)y/r \text{ and } h_2(x, y) = (\omega/kc)u(r)u(r)x/r$$

where $r = \sqrt{x^2 + y^2}$, provided that u satisfies the equation

$$\{g_2 v(r)\}' - g_1 u(r) + \lambda u(r) = 0 \tag{3}$$

where now $g_i = \gamma_i(u(r)^2/2, v(r)^2/2)$ for $i = 1, 2$, $v(r) = u'(r) + u(r)/r$ and $\lambda = (\omega/kc)^2$.

3. Guided planar TM-modes. We suppose henceforth that the dielectric response function ε in (CR) has the properties (V) and (C).

By a guided planar TM-mode we mean a solution $u \not\equiv 0$ of equation (2) which has the properties that

$$\lim_{x \to \pm\infty} u(x) = \lim_{x \to \pm\infty} u'(x) = 0. \tag{4}$$

In this section we discuss the existence of such solutions. The analysis is based on the observation that the function I defined by

$$I(p, q) = \Phi^*(p, q) - q\partial_2\Phi^*(p, q) - \lambda p^2/2$$

is a first integral for (2). Indeed, recalling that $\nabla\psi = \gamma = (\gamma_1, \gamma_2)$, we see that

$$I(p, q) = \psi(p^2/2, q^2/2) - \gamma_2(p^2/2, q^2/2)q^2 - \lambda p^2/2$$

and hence it is easy to verify that if u satisfies (2) then

$$\{I(u(x), u'(x))\}' = u'(x)\left[g_1 u(x) - \lambda u(x) - \{g_2 u'(x)\}'\right] = 0.$$

Since $\Phi^*(0, 0) = 0$ and $0 < \gamma_2(p, q) < 1/A$, it follows that if u satisfies (2) and (4) then

$$I(u(x), u'(x)) = 0 \text{ for all } x \in \Re.$$

Hence the orbit of a guided TM-mode lies in the set $I^{-1}(0) \setminus \{(0, 0)\}$ and, in view of the symmetries of I, it is sufficient to discuss the set

$$C = \{(p, q) : I(p, q) = 0 \text{ with } p \geq 0 \text{ and } q \geq 0\}.$$

For this we introduce some additional hypotheses about the dielectric response.

(S) (a) $\varepsilon_1(s_1, 0)$ is a strictly increasing function of s_1 on $[0, \infty)$, and

(b) there exist $B > 0$ and $\alpha \geq 0$ such that $\varepsilon_2(s_1, s_2)/s_2^\alpha \longrightarrow B$ as $s_2 \longrightarrow \infty$, uniformly for s_1 in bounded subsets of $[0, \infty)$.

We observe that, in the example of a Kerr material which was discussed in the previous section, the condition (S) is satisfied with $\alpha = 1$. On the other hand for realistic constitutive laws which model saturation, the value $\alpha = 0$ in (S)(b) is appropriate. (See [11] for examples of such constitutive relations.)

These properties of ε imply some analogous behaviour of the function γ.

(a) Recalling that $\gamma_1(t_1, 0) = 1/\varepsilon_1(\gamma_1(t_1, 0)^2 t_1, 0)$, it follows from (H) that $\gamma_1(t_1, 0)$ is a strictly decreasing function of t_1 on $[0, \infty)$. Setting $\gamma_1(\infty, 0) = \lim_{t \to \infty} \gamma_1(t, 0)$, we have $0 \leq \gamma_1(\infty, 0) = 1/\lim_{s \to \infty} \varepsilon_1(s, 0) = 1/\varepsilon_1(\infty, 0)$.

(b) For $t_1, t_2 \in [0, \infty)$ let $\begin{pmatrix} s_1 \\ s_2 \end{pmatrix} = \begin{pmatrix} \gamma_1(t_1^2/2, t_2^2/2)t_1 \\ \gamma_2(t_1^2/2, t_2^2/2)t_2 \end{pmatrix}$.

Then $\begin{pmatrix} t_1 \\ t_2 \end{pmatrix} = \begin{pmatrix} \varepsilon_1(s_1^2/2, s_2^2/2)s_1 \\ \varepsilon_2(s_1^2/2, s_2^2/2)s_2 \end{pmatrix}$ and so $t_1 \geq As_1$. Hence if t_1 varies over a bounded subset of $[0, \infty)$ so does s_1. Also, setting $\beta = \alpha/(1 + 2\alpha)$, we have that

$$\gamma_2(t_1^2/2, t_2^2/2)[t_2^2/2]^\beta = 2^{-\beta} \left\{ \varepsilon_2(s_1^2/2, s_2^2/2)s_2 \right\}^{2\beta} / \varepsilon_2(s_1^2/2, s_2^2/2)$$

$$= 1/\left\{ \varepsilon_2(s_1^2/2, s_2^2/2)/(s_2^2/2)^\alpha \right\}^{1/(1+2\alpha)} \longrightarrow L \text{ as } t_2 \longrightarrow \infty,$$

uniformly for t_1 in bounded subsets of $[0, \infty)$, where $L = 1/B^{1/(1+2\alpha)}$.

LEMMA 3.1. *Let the dielectric response have the properties (CR),(V),(C) and (S). If $\lambda \notin (\gamma_1(\infty, 0), \gamma_1(0, 0))$, then $C \cap \{(p, 0) : p > 0\} = \emptyset$. If $\lambda \in (\gamma_1(\infty, 0), \gamma_1(0, 0))$, then there exist $p_\lambda > 0$ and $f \in C([0, p_\lambda]) \cap C^1((0, p_\lambda))$ such that*

$$f(0) = 0, \ f(p_\lambda) = 0, \ f(p) > 0 \text{ for all } p \in (0, p_\lambda)$$
and $C = \{(p, f(p)) : 0 \leq p \leq p_\lambda\}$.

Furthermore $\lim_{p \to 0} f'(p) = \sqrt{\dfrac{\gamma_1(0,0) - \lambda}{\gamma_2(0,0)}}$ *and* $\lim_{p \to p_\lambda} f'(p) = -\infty$.

Also, $p_\lambda \longrightarrow \begin{cases} 0 \\ \infty \end{cases}$ *as* $\lambda \longrightarrow \begin{cases} \gamma_1(0, 0) \\ \gamma_1(\infty, 0) \end{cases}$.

Proof. For $p > 0$,

$$I(p, 0) = p^2 \left\{ \Phi^*(p, 0)/p^2 - \lambda/2 \right\} \text{ and}$$

$$\Phi^*(p, 0)/p^2 = \int_0^1 \gamma_1(t^2p^2/2, 0)t \, dt \longrightarrow \begin{cases} \gamma_1(0, 0)/2 \\ \gamma_1(\infty, 0)/2 \end{cases} \text{ as } p \longrightarrow \begin{cases} 0 \\ \infty \end{cases}.$$

Furthermore, by (S)(a), $\gamma_1(t_1, 0)$ is a strictly decreasing function of t_1 on $[0, \infty)$ and so $\Phi^*(p, 0)/p^2$ is a strictly decreasing function of p on $[0, \infty)$. Thus, if $\lambda \notin (\gamma_1(\infty, 0), \gamma_1(0, 0))$, then $C \cap \{(p, 0) : p > 0\} = \emptyset$ and if $\lambda \in (\gamma_1(\infty, 0), \gamma_1(0, 0))$, then there exists a unique $p_\lambda > 0$ such that $I(p_\lambda, 0) = 0$.

Clearly $p_\lambda \longrightarrow \begin{cases} 0 \\ \infty \end{cases}$ as $\lambda \longrightarrow \begin{cases} \gamma_1(0, 0) \\ \gamma_1(\infty, 0) \end{cases}$.

From now on we suppose that $\lambda \in (\gamma_1(\infty, 0), \gamma_1(0, 0))$.

For all (p, q), $\partial_2 I(p, q) = -q\partial_2^2 \Phi^*(p, q)$ and so $\partial_2 I(p, q) < 0$ if $q > 0$. Hence for $p = 0$ and $p \geq p_\lambda$, $I(p, q) < 0$ for all $q > 0$.

Using (S)(b) we now show that $I(p, q) \longrightarrow -\infty$ as $q \longrightarrow \infty$. In fact,

$$\Phi^*(p, q) - q\partial_2\Phi^*(p, q) = \int_0^1 \partial_1\Phi^*(tp, tq)p + \partial_2\Phi^*(tp, tq)q \, dt - q\partial_2\Phi^*(p, q)$$

$$= \int_0^1 \gamma_1(t^2p^2/2, t^2q^2/2)tp^2 + \gamma_2(t^2p^2/2, t^2q^2/2)tq^2 \, dt - \gamma_2(p^2/2, q^2/2)q^2 \ .$$

Hence

$$q^{-2(1+\alpha)/(1+2\alpha)} \{\Phi^*(p, q) - q\partial_2\Phi^*(p, q)\} =$$

$$q^{-2(1+\alpha)/(1+2\alpha)} \int_0^1 \gamma_1(t^2p^2/2, t^2q^2/2)tp^2 dt + \int_0^1 \gamma_2(t^2p^2/2, t^2q^2/2)[t^2q^2/2]^\beta[t^2/2]^{-\beta}t \, dt$$

$$-\gamma_2(p^2/2, q^2/2)[q^2/2]^\beta 2^\beta$$

$$\longrightarrow 0 + 2^\beta L \left\{ \int_0^1 t^{1-2\beta} dt - 1 \right\} = -2^\beta L/2(1+\alpha) < 0 \text{ as } q \longrightarrow \infty, \text{ uniformly for } p$$

in bounded intervals.

It follows that, given $P > 0$, there exists $Q(P) > 0$ such that $I(p, q) < 0$ for all $p \in [0, P]$ and $q \geq Q(P)$.

Thus, for $p \in (0, p_\lambda)$, there is a unique $q = f(p)$ such that $I(p, f(p)) = 0$. Since $\partial_2 I(p, f(p)) < 0$, the implicit function theorem yields $f \in C^1((0, p_\lambda))$.

Furthermore, if $p_n \longrightarrow 0$ and $f(p_n) \longrightarrow q$ we have that $q \in [0, Q(p_\lambda)]$ and so $I(0, q) = 0$. This implies that $q = 0$ and we may conclude that $\lim_{p\to 0} f(p) = 0$.

Similarly, $\lim_{p\to p_\lambda} f(p) = 0$.

For $p \in (0, p_\lambda)$,

$$f'(p) = -\partial_1 I(p, f(p))/\partial_2 I(p, f(p)) = \partial_1 I(p, f(p))/f(p)\partial_2^2\Phi^*(p, f(p))$$

and

$$\partial_1 I(p, q) = \gamma_1(p^2/2, q^2/2)p - \partial_1\gamma_2(p^2/2, q^2/2)pq^2 - \lambda p.$$

Since $\partial_1 I(p, f(p))/p \longrightarrow \gamma_1(p_\lambda^2/2, 0) - \lambda < \gamma_1(0, 0) - \lambda < 0$ as $p \longrightarrow p_\lambda$, we see that $f'(p) \longrightarrow -\infty$ as $p \longrightarrow p_\lambda$.

On the other hand, $I(p, q) = 0$

$\Rightarrow \int_0^1 \gamma_1(t^2p^2/2, t^2q^2/2)tp^2 dt + \gamma_2(t^2p^2/2, t^2q^2/2)tq^2 dt - \gamma_2(p^2/2, q^2/2)q^2 - \lambda p^2/2 = 0$

$\Rightarrow \{\int_0^1 \gamma_2(t^2p^2/2, t^2q^2/2)tdt - \gamma_2(p^2/2, q^2/2)\}q^2 = -\{\int_0^1 \gamma_1(t^2p^2/2, t^2q^2/2)tdt - \lambda/2\}p^2$

$\Rightarrow \{f(p)/p\}^2 \longrightarrow \frac{\lambda/2 - \gamma_1(0,0)/2}{\gamma_2(0,0)/2 - \gamma_2(0,0)} = \frac{\gamma_1(0,0) - \lambda}{\gamma_2(0,0)}$ as $p \longrightarrow 0$.

Hence $f'(p) \longrightarrow \sqrt{\frac{\gamma_1(0,0) - \lambda}{\gamma_2(0,0)}}$ as $p \longrightarrow 0$.□

From these properties of C we can deduce the following information about guided TM-modes in a straightforward way so we only give a brief indication of the arguments.

Recall that $\gamma_1(0,0) = 1/\varepsilon_1(0,0)$ and that $\gamma_1(\infty,0) = 1/\varepsilon_1(\infty,0)$.

THEOREM 3.2. *Let the dielectric response have the properties (CR),(V),(C) and (S).*

If $\lambda \notin (\gamma_1(\infty,0), \gamma_1(0,0))$, then there is no guided TM-mode.

If $\lambda \in (\gamma_1(\infty,0), \gamma_1(0,0))$, then there exists a unique guided TM-mode u_λ such that $u_\lambda(0) = p_\lambda$ and $u'_\lambda(0) = 0$.

Furthermore, $u_\lambda(x) = u_\lambda(-x) > 0$ for all $x \in \Re$ and,

for any $\mu < \sqrt{\frac{\gamma_1(0,0)-\lambda}{\gamma_2(0,0)}}, \lim_{x \longrightarrow \infty} \exp(\mu x)\, u_\lambda(x) = 0$.

All guided TM-modes are of the form $\pm u_\lambda(x + \delta)$ for some $\delta \in \Re$ and some $\lambda \in (\gamma_1(\infty,0), \gamma_1(0,0))$.

Proof. If u is a guided TM-mode so is $\pm u(x + \delta)$ for all $\delta \in \Re$. Hence we may assume that $u(0) = \max u(x) > 0$ and that $u'(0) = 0$. Since $(u(0), u'(0)) \in C$, this shows that $\lambda \in (\gamma_1(\infty,0), \gamma_1(0,0))$.

For λ in this range we denote by u_λ the solution of (2) which satisfies the initial conditions $u(0) = p_\lambda$ and $u'(0) = 0$ where p_λ is given by the lemma. Note that in (2) we have that

$$\{g_2 u'(x)\}' = \partial_1 \gamma_2 (u(x)^2/2, u'(x)^2/2) u(x) u'(x)^2$$

$$+ \{\partial_2 \gamma_2 (u(x)^2/2, u'(x)^2/2) u'(x)^2 + \gamma_2(u(x)^2/2, u'(x)^2/2)\} u''(x)$$

Recalling that $\partial_2 \{\gamma_2(t_1^2/2, t_2^2/2) t_2\} > 0$, we see that the coefficient of $u''(x)$ is a C^1–function of u and u' which cannot vanish. It is now easy to show that u_λ is unique, is defined on all of $\Re$ and is a guided TM-mode.□

REFERENCES

[1] C.A.Stuart, Self-trapping of an electromagnetic field and bifurcation from the essential spectrum, Arch. Rational Mech.Anal.,113(1991), 65-96

[2] C.A.Stuart, Guidance properties of nonlinear planar waveguides, Arch. Rational Mech.Anal.,125(1993), 145-200

[3] J.B.McLeod, C.A.Stuart and W.C.Troy, An exact reduction of Maxwell's equations, in Proc.Gregynogg Conference, Birkhäuser, Basel, 1992

[4] H.-J.Ruppen, TE_n-modes for a planar self-focusing waveguide, preprint, 1993

[5] H.-J.Ruppen, TE_n-modes in a cylindrical self-focusing waveguide, preprint, 1994

[6] C.A.Stuart, Cylindrical TM-modes in a homogeneous self-focusing dielectric, preprint

[7] M.Born and E.Wolf, Principles of Optics, fifth edition, Pergamon Press, Oxford, 1975

[8] N.N.Akhmediev, Nonlinear theory of surface polaritons, Sov. Phys. JETP, 57(1983),1111-1116

[9] K.M.Leung, p-polarized nonlinear surface polaritons in materials with intensity-dependent dielectric functions, Phys. Rev. B, 32(1985), 5093-5101

[10] Q.Chen and Z.H.Wang, Exact dispersion relations for TM waves guided by thin dielectric films bounded by nonlinear media, Opt. Letters, 18(1993), 260-262

[11] D.Mihalache, G.I.Stegeman, C.T.Seaton, E.M.Wright, R.Zanoni, A.D.Boardman and T.Twardowski, Exact dispersion relations for transverse magnetic polarised guided waves at a nonlinear interface, Opt. Lett.,12(1987), 187-189

[12] K.Ogusu, TM waves guided by nonlinear planar waveguides, IEEE Trans. Microwave Th. and Tech.,37(1989), 941-946

[13] R.I.Joseph and D.N.Christodoulides, Exact field decomposition for TM waves in nonlinear media, Opt. Lett.,10(1987), 826-82.

[14] Y.Chen,TE and TM families of self-trapped beams, IEEE J. Quantum Elect.,27(1991), 1236-1241

[15] Y.Chen and A.W.Snyder, TM-type self-guided beams with circular cross-section, Electr. Lett.,27(1991), 565-56

[16] X.H.Wang and G.K.Cambrell, Full vectorial simulation of bistability phenomena in nonlinear optical channel waveguides, J. Opt. Soc. Am. B, 10(1993), 1090-1095

[17] M.Yokota, Guided transverse-magnetic waves supported by a weakly nonlinear slab waveguide, J. Opt. Soc. Ab. B, 10(1993), 1096-1101

[18] S.-W.Kang, TM modes guided by nonlinear dielectric slabs, J. Lightwave Tech., 13(1995), 391-395

[19] M.S.Kushwaha, Exact theory of nonlinear surface polatitons : TM case, Japanese J. Appl. Phys., 29(1995), 1826-1828

[20] G.I.Stegeman, Nonlinear guided wave optics, in Contemporary Nonlinear Optics, edited by G.P.Agrawal and R.W.Boyd, Academic Press, Boston, 1992

[21] J.B.Hiriart-Urruty and C.Lemaréchal,Convex Analysis and Minimisation Algorithms II, Springer-Verlag, Berlin, 1993

[22] R.T.Rockafellar,Convex Analysis, Princeton Univ. Press, Princeton, 1970

A boundary flux property for a general nonlinear diffusion reaction system *

Mabo Suzuki [†]   *Alex McNabb* [‡]   *Anthony J. Bracken* [§]

Abstract

Studies of chemical fluxes through biological membranes and diffusion-convection-reaction models for the transport processes have brought to light some intriguing reciprocal relationships between output fluxes for complementary experiments in which flow directions are reversed (Sten-Knudsen and Ussing, 1981: Bass and Bracken, 1983). In this note, a new dimension to these flux studies involving input fluxes and their asymptotic behaviour is explored. Simple relationships between time-lags associated with the asymptotic behaviour of input and output fluxes of the various interacting chemicals and the final chemical contents of the whole region are obtained.

Dedicated to Professor Ivar Stakgold

Submitted with warmest regards and best wishes in retirement for a colleague and friend who has made many great contributions to this field of nonlinear diffusion-reaction phenomena.

1 Introduction

Studies of chemical fluxes through biological membranes and analyses of various transport models for the processes have brought to light intriguing reciprocity properties of model equations (For example, Sten-Knudsen and Ussing, 1981: Bass and Bracken, 1983). These in turn have been a stimulus for experimental endeavours to explore the structure and transport properties of these complex systems. This work has primarily focussed on steady throughput fluxes, time-dependent output fluxes and comparisons of fluxes for reciprocal experiments. Such researches have brought to light many results concerning flux ratios and their surprising constancy for all time (Bass, Bracken and Hilden, 1986; McNabb *et al.*, 1990; Bracken *et al.*, 1988, 1994), and raises many questions concerning analogous results for input fluxes. It is evident that nothing much new would come from this if attention were confined to steady fluxes, but time-dependent input fluxes for reciprocal experiments have the potential to reveal aspects of the model structure that are not derivable from output fluxes. For many linear situations, output flux ratios are constant for all time. And therefore the initial values for the ratios coincide with the steady state flux ratios. On the other hand, the initial input fluxes are governed by the diffusivities in the neighbourhood of the input faces initially, and so the ratio of these initial fluxes for reciprocal experiments

---

*This research project was partially supported by the New Zealand Foundation for Research, Science and Technology under a grant from the Marsden Fund.

[†]Department of Mathematics, The University of Queensland, Brisbane, QLD.4072 Australia

[‡]Division of Science and Technology, The University of Auckland, Auckland, New Zealand

[§]Department of Mathematics, The University of Queensland, Brisbane, QLD.4072 Australia

reflects the ratio of the diffusivities near opposite faces of a membrane. General ratio results for all time for these input fluxes are illusive. However, a general result can be found for the asymptotic behaviour at large time. This concerns a relationship between the final membrane content of each conserved component of the system and the time-lags associated with the asymptotic fluxes of the diffusing components which contain the conserved entity.

2 General transport equations

Consider a general system of N diffusion equations incorporating convection and chemical reactions terms. Denote the concentration of substance i at each point $\mathbf{x}$ of a domain Ω at time t by $c_i(\mathbf{x}, t)$, its flux by $\mathbf{j}_i(\mathbf{x}, t)$ and the local rate of chemical generation of it by $f_i(\mathbf{x}, c_1, c_2, \cdots, c_N)$. The conservation equations that describe such a diffusion-convection-reaction system in the domain Ω may then be written as

$$
(1) \qquad \frac{\partial c_i(\mathbf{x}, t)}{\partial t} + \operatorname{div} \mathbf{j}_i(\mathbf{x}, t) = f_i(\mathbf{x}, c_1, c_2, c_3, \cdots c_N)
$$

$$
(\mathbf{x} \in \Omega,\ t \geq 0).
$$

$$
(2) \qquad \mathbf{j}_i(\mathbf{x}, t) = -D_i(\mathbf{x}) \operatorname{grad} c_i(\mathbf{x}, t) + \mathbf{v}_i(\mathbf{x}) c_i(\mathbf{x}, t).
$$

These equations originate from Fick's laws, where the flux expression of the ith component has diffusion coefficient $D_i(\mathbf{x})$ and drift velocity $\mathbf{v}_i(\mathbf{x})$. Both the diffusion coefficients D_i and the drift velocities $\mathbf{v}_i$ can be functions of space coordinates, and it is assumed $D_i(\mathbf{x}) > 0$ and the drift velocities $\mathbf{v}_i$ are conservative in the sense that they derive from potential functions ϕ_i and $\mathbf{v}_i = D_i \operatorname{grad} \phi_i$. (see McNabb and Bass, 1990). The functions $f_i(\mathbf{x}, c_1, c_2, c_3, \cdots, c_N)$ constitute source terms describing chemical interactions generating c_i, which may involve significant nonlinearities. The functions are assumed to satisfy such physical reality conditions as $f_i(\mathbf{x}, c_1, c_2, c_3, \cdots, c_N) \geq 0$ when $c_i = 0$, they all vanish when all the c_j are zero, and they are assumed to be continuous with bounded continuous derivatives in all the variables so that they have a linear approximation when all c_j are small. They are also constrained by chemical conservation principles. If the N components are all derived from M conserved constituents with local content $C_i(\mathbf{x}, t)$, which form fundamental molecular building blocks for all the ingredients c_i, there is an $M \times N$ matrix $\mathbf{K}$ of terms k_{ij} such that $\sum_{j=1}^{N} k_{ij} c_j = C_i$ and since there are no sources for these in the domain Ω, $\sum_{i=1}^{N} k_{ij} f_i(\mathbf{x}, c_1, c_2, c_3, \cdots, c_N) = 0$. In matrix notation, $\mathbf{Kc} = \mathbf{C}$ and $\mathbf{Kf} = 0$.

General initial conditions for the system are

$$
(3) \qquad c_i(\mathbf{x}, 0) = r_i(\mathbf{x}) \qquad (\mathbf{x} \in \Omega),
$$

but in our flux studies (following Sten-Knudsen and Ussing, 1981; Bass and Bracken, 1983), we assume $r_i(\mathbf{x}) = 0$.

Dirichlet boundary conditions of the form

$$
(4) \qquad c_i(\mathbf{x}, t) = y_i(t) \quad t \geq 0, \qquad (\mathbf{x} \in \partial\Omega_1, t \geq 0),
$$

$$
(5) \qquad c_i(\mathbf{x}, t) = z_i(t) \quad t \geq 0, \qquad (\mathbf{x} \in \partial\Omega_2, t \geq 0),
$$

$$
(6) \qquad \mathbf{j}_i(\mathbf{x}, t) \cdot \mathbf{n} = 0 \quad t \geq 0, \qquad (\mathbf{x} \in \partial\Omega_3, t \geq 0),
$$

where $\partial\Omega_1$, $\partial\Omega_2$ and $\partial\Omega_3$ are disjoint subsets of $\partial\Omega = \partial\Omega_1 \cup \Omega_2 \cup \Omega_3$, and $\mathbf{n}$ is the outward unit normal to $\partial\Omega_3$, are in general sufficient to specify a solvable boundary value problem with a unique solution.

Here we assume

(7) $$y_i(t) = a_i > 0 \qquad (t \geq 0),$$

(8) $$z_i(t) = 0 \qquad (t \geq 0),$$

where a_i are constant.

The boundary conditions (7)–(8) give rise to input fluxes $q_{i\text{-in}}(t)$ and output fluxes $q_{i\text{-out}}(t)$ of the i-th component through $\partial\Omega_1$ and $\partial\Omega_2$ respectively, given by the expressions

(9) $$q_{i\text{-in}}(t) = -\iint_{\partial\Omega_1} \mathbf{j}_i(\mathbf{x}, t) \cdot \mathbf{n} \, dS \geq 0 \qquad (t \geq 0),$$

(10) $$q_{i\text{-out}}(t) = \iint_{\partial\Omega_2} \mathbf{j}_i(\mathbf{x}, t) \cdot \mathbf{n} \, dS \geq 0 \qquad (t \geq 0).$$

These may be compared with the input and output fluxes $q^*_{i\text{-in}}$ and $q^*_{i\text{-out}}$ through $\partial\Omega_2$ and $\partial\Omega_1$ respectively, generated by the solutions c^*_i of the complementary boundary value problem for which y_i and z_i in the boundary conditions (7)–(8) are interchanged. Here $\mathbf{n}$ is an outward unit normal of the boundaries $\partial\Omega_1$ and $\partial\Omega_2$. The complementary input fluxes provide a new dimension for flux ratio analysis. In the next section, we make a small contribution to such a study.

3   Conservation and accumulated fluxes

At time t, the total content $Q_i(t)$ of the fundamental chemical building block C_i in the whole region Ω is given by

$$Q_i(t) = \int_\Omega C_i \, dv$$

(11) $$= \int_\Omega \sum_{j=1}^N k_{ij} c_i(x, t) dv$$

$$= \int_0^t \int_\Omega \sum_{j=1}^N k_{ij} \frac{\partial c_i(x, t)}{\partial t} \, dt \, dv.$$

From equation (1) we see

(12) $$Q_i(t) = -\sum_{j=1}^N k_{ij} \int_0^t \int_{\partial\Omega} \operatorname{div} \mathbf{j}_j \cdot n \, dv \, dt,$$

since $\sum_{j=0}^N k_{ij} f_j = 0$. From Gauss's theorem we see that

$$Q_i(t) = -\sum_{j=1}^N k_{ij} \int_0^t \iint_{\partial\Omega} \mathbf{j}_j \cdot \mathbf{n} \, dS \, dt$$

$$(13) \qquad = -\sum_{j=1}^{N} k_{ij} \int_{0}^{t} \iint_{\partial\Omega_1} \mathbf{j}_j \cdot \mathbf{n} \, dS \, dt$$

$$= -\sum_{j=1}^{N} k_{ij} \int_{0}^{t} \iint_{\partial\Omega_2} \mathbf{j}_j \cdot \mathbf{n} \, dS \, dt, -\sum_{j=1}^{N} k_{ij} \int_{0}^{t} \iint_{\partial\Omega_3} \mathbf{j}_j \cdot \mathbf{n} \, dS \, dt,$$

and hence

$$Q_i(t) = \sum_{j=1}^{N} k_{ij} \left[\int_{0}^{t} q_{i\text{-in}}(\mathbf{x}, s) ds - \int_{0}^{t} q_{i\text{-out}}(\mathbf{x}, s) ds \right]$$

$$(14) \qquad = \sum_{j=1}^{N} k_{ij} \left[q_{i\text{-in}}(\infty)t - q_{i\text{-out}}(\infty)t \right.$$

$$\left. + \int_{0}^{t} \{q_{i\text{-out}}(\infty) - q_{i\text{-out}}(s)\} ds - \int_{0}^{t} \{q_{i\text{-in}}(\infty) - q_{i\text{-in}}(s)\} ds \right]$$

Since $q_{i\text{-out}}(\infty) = q_{i\text{-in}}(\infty)$ and the integrands converge to zero exponentially as $s \to \infty$, we can write in the limit as time t goes to infinity,

$$(15) \qquad Q_i(t) \sim \sum_{j=1}^{N} k_{ij} q_{j\text{-in}}(\infty)(\tilde{t}_{j\text{-out}} - \tilde{t}_{j\text{-in}}),$$

where $\tilde{t}_{j\text{-out}}$ and $\tilde{t}_{j\text{-in}}$ are the time-lags describing the intercepts of the linear asymptotes for $q_{j,\text{out}}(t)$ and $q_{j,\text{in}}(t)$ with the time axis given by the expressions

$$(16) \qquad q_{j\text{-out}}(\infty)\tilde{t}_{j\text{-out}} = \int_{0}^{t} \{q_{j\text{-out}}(\infty) - q_{j\text{-out}}(s)\} ds$$

$$(17) \qquad q_{j\text{-in}}(\infty)\tilde{t}_{j\text{-in}} = \int_{0}^{t} \{q_{j\text{-in}}(\infty) - q_{j\text{-in}}(s)\} ds$$

4 Remarks

Direct measurements of instantaneous values of fluxes and concentrations in diffusion systems are experimentally difficult to make. On the other hand, accumulated quantities are more readily measurable. Membrane transport problems are complicated and difficult to explore internally. They are essentially 'black box problems', and the basic investigative techniques are in the nature of system response analyses. Past researches have focused on output responses to imposed boundary conditions. The results of this paper encourage consideration and further analyses of the input flux responses.

Acknowledgements

We would like to thank Associate Professor Jørgen Sand for useful discussions during his visit to the University of Queensland. Special thanks go also to Ms. Y. McNabb for her cheerful support and encouragement during this collaboration.

References

[1] Parshotam, A., McNabb, A. and Wake, G. (1993). Comparison theorems for coupled reaction-diffusion equations in chemical reactor analysis. *Journal of Mathematical analysis and Applications*, **178**, No. 1, pp.196–220.

[2] Barrer, R. M. (1941). *Diffusion in and through solids*, Chapter 1. Cambridge University Press., New York.

[3] Bass, L. and Bracken, A. J. (1983). The flux–ratio equation under nonstationary boundary conditions. *Mathematical Biosciences*, **66**, pp.87-92.

[4] Bass, L., Bracken, A. J. and Hilden, J. (1986). Flux ratio theorems for nonstationary membrane transport with temporary capture of tracer. *Journal of Theoretical Biology*, **118**, pp.327–338.

[5] Bracken, A. J., McNabb, A., and Suzuki, M., (1994). Ussing's flux ratio theorem for nonlinear diffusive transport with chemical interactions. *proceedings of 1994 Korean Automatic Control Conference* (South Korea), **9**, pp.747–752, October, 1994.

[6] Carr, J. and Pego, R. L. (1989). Metastable patterns in solution of $u_t = \epsilon^2 u_{xx} - f(u)$. *Communications on Pure and Applied Mathematics*, **42**, pp.523-576.

[7] Daynes, H. A. (1920). The process of diffusion through a rubber membrane, *Proceedings of Royal Society, London*, **97A**, pp.286–307.

[8] Brown, K. J., Lacey, A. A., (1990) *Reaction–diffusion systems*, Oxford University Press, London.

[9] McNabb, A. (1962). On the flux of diffusing matter through boundaries, *Proceedings of American Mathematical Society*, **13**, pp.170–174.

[10] McNabb, A. and Bass, L. (1991). A diffusion-reaction model for the cellular uptake of protein-bound ligands. *SIAM Journal of Applied Mathematics*, **51**, No. 1, pp.124–149.

[11] McNabb, A. (1986). Comparison theorems for differential equations. *Journal of Mathematical analysis and Applications*, **119**, No. 1/2, pp.417–428.

[12] McNabb, A., Bass, L. and Suzuki, M., (1995). Ussing flux ratio theorem for albumin enhanced fatty acid fluxes across a lipid water interface. *SIAM Journal of Applied Mathematics*, **8**, No.3, pp.63–68.

[13] McNabb, A. and Wake, G. C. (1991) Heat conduction and finite measures for transition times between steady states. *IMA Journal of Applied Mathematics*, **47**, pp.193–206.

[14] Schuss, Z. (1980). *Theory and Applications of Stochastic Differential Equations*. Wiley-Interscience, New York.

[15] Sten-Knudsen, O. and Ussing, H. H. (1981). The flux ratio equation under nonstationary conditions. *Journal of membrane biology*, **63**, pp.233–242.

[16] Ussing, H. H. (1978). Interpretation of tracer fluxes. In: *Membrane Transport in Biology* (Giebisch, G., Tosteson, D. C. and Ussing, H. H. Eds.), Vol. 1, pp.115–40, Springer, New York.

[17] Weisiger R. A., Pond S. M. and Bass, L. (1989). Albumin enhanced unidirectional fluxes of fatty acid across a lipid-water interface: theory and experiments. *American Journal of Physiology*, **257**, pp.G904–G916.

[18] Weisiger, R. A., Pond, S. M. and Bass, L. (1991). Hepatic uptake of protein-bound ligands: extended sinusoidal perfusion model. *American Journal of Physiology*, **261**, pp.G872–G884.

Different kinds of singular solutions
of nonlinear parabolic equations [*]

Juan Luis Vazquez [†]   Laurent Véron [‡]

dedicated to I. Stakgold on his birthday

1 Introduction

The aim of the qualitative theory of partial differential equations is in its general formulation to describe the complete set of solutions of the different initial and boundary-value problems of interest, to show how they depend on the data and further to analyse the main features of such solutions. This is usually a formidable task. A great help in this endeavour, stressed since old times by the experts in continuum mechanics, is the observation that the most interesting phenomena admit a rather simple example in the form of a *special solution* with particular properties that make it easier to be calculated. One of the most apparent examples is the heat equation; if we consider the large time behaviour of any nonnegative solution with finite initial energy (i.e., finite L^1 norm) then this behaviour is given in first approximation by the so-called *fundamental* solution

$$(1) \qquad\qquad V(x,t) = \frac{1}{(4\pi t)^{N/2}} e^{-\frac{x^2}{4t}} .$$

This is an example of *singular solution*, since it solves the heat equation in the classical sense in $Q = \mathbf{R}^N \times (0, \infty)$, while it takes the following singular initial data

$$(2) \qquad\qquad V(x,0) = 0 \quad \text{for } x \neq 0, \quad V(0,0) = \infty.$$

Moreover, the singularity takes the form of a Dirac delta (obtained as a limit of V when $t \to 0$). For a linear equation, the fundamental solutions allow to derive in principle the structure and properties of all other solutions. This is today a well-known part of the partial differential equations curriculum.

The situation is quite different in the realm of nonlinear equations. The nonlinearity prevents the direct representation of arbitrary solutions in terms of fundamental or other special solutions. On the other hand, nonlinear equations exhibit a number of curious features, like free boundaries, dead cores or blow-up singularities, which are absent in linear problems and usually contain the information of greater interest for the applied scientist. Prof. Stakgold has contributed many enlightening pages to these subjects. It has been gradually discovered and verified that different kinds of special solutions play a crucial

[*]Partially supported by HCM Project CHRX-CT94-0618.

[†]Depto. Matemáticas, Univ. Autónoma de Madrid, 28049 Madrid, Spain.

[‡]Dépt. de Mathématiques, Fac. des Sciences, Univ. de Tours, 37200 Tours, France

role in establishing the qualitative theory and describing the special phenomena. These solutions are characterized in various ways. Two typical ways are through their group invariance (self-similarity) and through the classification of their singularities. Nonlinear equations admit new and sometimes unusual kinds of such solutions. Several problems are then posed, namely which special solutions of a certain type exist, which are interesting, how to obtain them and which properties do they describe.

In this paper we will present a rather unusual feature of the theory of nonlinear parabolic equations as they appear for instance in problems of heat propagation in nonlinear media. Much progress has been done in this area, cf. e.g. [9] or [10] and their references. Our purpose is to describe the different types of singular solutions that may occur in nonlinear parabolic equations of the form

$$(3) \qquad u_t - \Delta\phi(u) + \psi(u) = 0,$$

where ϕ and ψ are continuous and nondecreasing real functions, normalized so that $\phi(0) = \psi(0) = 0$. We are interested in solutions defined in the whole domain $\overline{Q} = \mathbf{R}^N \times [0, \infty)$ but for certain singular points or sets. These solutions will have a direct influence on the description of the large-time behaviour of general classes of solutions.

The most common type of such solutions with singularities is given by the already mentioned fundamental solutions, i.e. solutions $V_c(x, t)$ of (3) in $Q = \mathbf{R}^N \times (0, \infty)$ such that

$$(4) \qquad V_c(x, t) \to c\delta(x) \qquad \text{as} \quad t \to 0.$$

Here $c \in \mathbf{R}$ and $\delta(x)$ stands for Dirac's delta function (centered at $x = 0$). This is the singularity encountered in the linear theory. In the simplest linear heat equation $u_t = \Delta u$ it is well-known the set of such solutions is given by the formula

$$(5) \qquad V_c(x, t) = cV(x, t),$$

with V as in (1). If we now let $c \to \infty$ we find a uniform limit

$$(6) \qquad \lim_{c \to \infty} V_c(x, t) = \infty \quad \text{everywhere in } Q.$$

This seems to preclude further developments in the direction of the existence of new, interesting solutions with singularities. However, the theory of quasilinear equations of the type (3) has produced a number of new possiblities. We will be interested here in the types of solutions that can appear in the limit $c \to \infty$ of the the fundamental solutions. *Our main result states that there are exactly 3 types of singular limits apart from the trivial limit (6).* These types will be called *very singular solution, flat solution* and *razor blade.*

In the sequel we will present them through the most characteristic occurrences, prove the above result and briefly discuss the properties of the different solutions. While the first two types seem to be closely connected with the limit of fundamental solutions, there can be lots of razor blades which have nothing to do with the limit. Therefore, the problem is posed of identifying the razor blades which are limits.

2 Types of singular solutions

2.I. The simplest type of singularity to be considered is the *isolated singularity:* we have a solution of the equation in $Q = \mathbf{R}^N \times (0, \infty)$ which takes standard initial data for $x \neq 0$, say $u(x, 0) = 0$, but which tends to ∞ as $(x, t) \to (0, 0)$. The fundamental solutions are examples of solutions with an isolated singularity. In the 80's a new type of solutions with an isolated singularity was discovered in nonlinear parabolic equations. Thus, Brezis, Peletier and Terman [2] and Galaktionov, Kurdyumov and Samarski [5] found one such solution V_* in the equation with power-like nonlinearities

$$(7) \qquad\qquad u_t = \Delta u^m - u^p$$

in the range $1 < m < p < m + (2/N)$. This singularity is of a stronger nature than the fundamental solution since

$$(8) \qquad\qquad \lim_{t \to 0} \int_{\mathbf{R}^N} V_*(x, t) \, dx \to \infty,$$

and was consequently called a *very singular solution* (VSS for short). Similar 'large' singularities had been found before in elliptic equations arising in the so-called Emden equations and the Thomas-Fermi model in atomic theory, cf. Fowler [3] and Brezis-Lieb [1]. They received the name of solutions with a *strong singularity*, see [12], [11].

2.II. There are two other types of limits in equations of the form (7), though they do not exhibit an isolated singularity. Thus, in the range $1 < p < m$ Kamin, Peletier and Vazquez [6] have proved that the limit of fundamental solutions gives as result

$$(9) \qquad\qquad \lim_{c \to \infty} V_c(x, t) = c_* t^{-\frac{1}{p-1}},$$

where $c_* = (p - 1)^{-\frac{1}{p-1}}$. This solution can be characterized as an x-homogeneous solution which is singular at $t = 0$. Indeed, we can find such type of solutions for equations of the form (3) by assuming that u is only a function of t. A solution of the form $u(x, t) = U(t)$ is called a *flat solution*. It is clear that as a function only of time a flat solution is a solution of the ODE

$$(10) \qquad\qquad u_t + \psi(u) = 0,$$

whose analysis is easy. A singular flat solution takes on initial value $U(0) = \infty$.

LEMMA 2.1. *A singular flat solution U is either identically infinite or finite for all times $t > 0$. The latter happens if and only if $1/\psi(s)$ is integrable as $s \to \infty$, i.e.,*

$$(11) \qquad\qquad \int^{\infty} \frac{ds}{\psi(s)} < \infty.$$

In this case U is implicitly given by

$$(12) \qquad\qquad t = \int_u^{\infty} \frac{ds}{\psi(s)}.$$

The proof is easy. If for instance ψ is a power function, $\psi(s) = s^p$ for $s \geq 0$, then (11) holds whenever $p > 1$. The explicit formula is then given by (9).

2.III. Let us now take the purely diffusive equation

$$(13) \qquad\qquad u_t = \Delta u^m$$

in the *fast diffusion range* $m < 1$. We find that for $m > m_c = (N-2)/N$ there exist fundamental solutions given explicitly by the formula

$$(14) \qquad\qquad V_c(x,t) = t^{-k}\left(A + B\frac{|x|^2}{t^{2k/N}}\right)^{-\frac{1}{1-m}}$$

with

$$(15) \qquad\qquad B = \frac{k(1-m)}{2mN}, \quad k = ((2/N) + m - 1)^{-1},$$

while $A > 0$ is a free constant which can be uniquely determined as a function of the mass $c = \int u(x,t)\,dx$, i.e., $A = A(c,m,N)$. The limit $c = \infty$ corresponds to putting $A = 0$, which gives a new type of singular solution

$$(16) \qquad\qquad V_{\#}(x,t) = \lim_{c\to\infty} V_c(x,t) = a(m,N)\left(t/|x|^2\right)^{\frac{1}{1-m}}$$

where

$$(17) \qquad\qquad a^{1-m} = 2m\left(\frac{2}{1-m} - N\right),$$

which is well-defined and positive precisely for $m > m_c$. This is a solution of the equation for all $x \neq 0$ and $t > 0$ with a singularity which stays at $x = 0$ for all time. The initial values are 0 for every $x \neq 0$. Looking for a name we have decided to give this kind of solution the name of *razor blade*, because of the way the singularity advances straight ahead along the t-axis. Razor blades have a *standing singularity*.

A similar situation happens when we take the diffusion-absorption case

$$(18) \qquad\qquad u_t = \Delta u^m - u, \quad m_c < m < 1.$$

There is then a simple change of variable that transforms this equation into the purely diffusive case (13), namely

$$(19) \qquad\qquad v(x,\tau) = e^t u(x,t), \quad \tau = \frac{1}{1-m}e^{(1-m)t}.$$

Using the previous results (with the origin of time $\tau = 1/(1-m)$) we easily check that the limit of fundamental solutions is given by

$$(20) \qquad\qquad V_{\#}(x,t) = c\frac{(1 - e^{-(1-m)t})^{1/(1-m)}}{|x|^{\frac{2}{1-m}}},$$

with $c = a\,(1-m)^{-1/(1-m)}$. The interesting novelty is that the razor blade now stabilizes towards a *singular stationary profile* as $t \to \infty$, while in the previous case it blew up.

3   Study of the limit of fundamental solutions

In this central section we assume that the fundamental solutions V_c exist and are unique and study the special singular solution defined as

$$(21) \qquad\qquad U(x,t) = \lim_{c \to \infty} V_c(x,t).$$

Conditions on ϕ and ψ to have these properties of the fundamental solutions as well as the Maximum Principle are given in the literature and need not concern us here. In particular, when ϕ and ψ are powers like in (7), this happens when $p < m + (2/N)$. For larger values of p there exist no solutions with a singularity at $(0,0)$.

In order to perform the study we take into account the following properties, derived from the Maximum Principle:

$V_c(x,t)$ is nonnegative, radially symmetric and decreasing as a function of x (for t and c fixed), increasing as a function of c, and $V_c(0,t)$ is decreasing in t.

By the monotonicity of the family $\{V_c\}$ in c the limit (21) exists in $\mathbf{R} \cup \{+\infty\}$, i.e. it can be finite or infinite. We can also define for every $t > 0$ the singular set of U (also called *blow-up set*)

$$(22) \qquad\qquad S(t) = \{x \in \mathbf{R^N} : U(\mathbf{x,t}) = \infty\}.$$

From the monotonicity of the solutions in x we conclude that $S(t)$ is either empty or a ball with center 0 and radius $a(t)$ or the whole space $\mathbf{R^N}$ (i.e., $a(t) = \infty$). Our analysis will show that there are four possibilities.

Case 1. $S(t) = \mathbf{R^N}$ for every t. This corresponds to a limit $U \equiv \infty$. This is the case which occurs in the linear theory. Example: take linear constitutive functions $\phi(s) = as$, $\psi(s) = bs$, $a, b > 0$.

Case 2. $S(t)$ is empty for all positive t and the initial conditions $U(x,0) = 0$ for $x \neq 0$ are taken, i.e. $S(0) = \{0\}$. This corresponds to the Very Singular Solution, with an isolated singularity at $x = 0$, $t = 0$.

Case 3. $S(t)$ is empty for all positive t and $S(0) = \mathbf{R^N}$. This corresponds to the Flat Solution, which is singular for all $x \in \mathbf{R^N}$ if $t = 0$.

Case 4. $S(t) = \{0\}$ for all $t > 0$. The initial values are 0 for $x \neq 0$. This is a new kind of singular solution, so-called the razor blade (or "lame de couteau", as we called it originally in French).

Our main result can be stated as follows

THEOREM 3.1. *For the limit $U(x,t)$ defined by (21) one of these 4 possibilities holds.*

Let us establish the result step by step. We consider different options according to the behaviour of U near $t = 0$. A first option is to admit instantaneous blow-up, which we formulate as follows:

(H1): There exists $x_0 \neq 0$ such that the function $U(x_0,t)$ is not bounded near $t = 0$.

We then have

LEMMA 3.1. *Assume that (H1) holds. Then $U(x,t)$ is independent of x, $U(x,t) = U(t)$.*

Proof. Under hypothesis (H1) there exist sequences $t_n \to 0$ and $M_n \to \infty$ such that $U(x, t_n) \geq M_n$ for $|x| \leq |x_0|$. Therefore, taking k fixed and n large enough we have that at $t = t_n$ the function $U(x, t_n)$ is bounded below by M_n times the characteristic function of the ball $B_{n,k}$ with radius k/M_n and center $x_n = x_0 - (k/M_n)^{1/N} e_0$, where $e_0 = x_0/|x_0|$. By the Maximum Principle we get for $t > 0$

$$(23) \qquad U(x, t + t_n) \geq u_{n,k}(x, t),$$

where $u_{n,k}$ is the solution of (3) with initial data $M_n \chi(B_{n,k})$. We pass to the limit $n \to \infty$, the function $u_{n,k}$ must converge to the fundamental solution $V_{\omega k}(x - x_0, t)$ for some constant $\omega = \omega(N)$. Hence, in the limit

$$(24) \qquad U(x, t) \geq V_{\omega k}(x - x_0, t),$$

and this holds for any x_0 and any $k > 0$. Take now limit in (24) as $k \to \infty$

$$U(x, t) \geq U(x - x_0, t).$$

Iterating this process implies that

$$(25) \qquad U(x, t) \geq U(x - lx_0, t)$$

for any $l \in \mathbf{N}$ and any $(x, t) \in Q$. This and the monotonicity of U in $|x|$ allows to conclude that U must be constant in x. ☐

We examine now the cases where (H1) does not hold.

LEMMA 3.2. *If $S(t_0)$ is empty for some $t_0 > 0$, so is $S(t)$ for all $t \geq t_0$.*

Proof. If $S(t)$ is empty at some time $t = t_0$, then $U(x, t_0)$ is bounded, hence by the Maximum Principle so is $U(x, t)$ for $x \in \mathbf{R}^{\mathbf{N}}$ and $t \geq t_0$, so that $S(t)$ is empty for all $t > t_0$. ☐

LEMMA 3.3. *Assume that for every $x_0 \neq 0$ the function $U(x_0, t)$ is bounded in a neighbourhood of $t = 0$. Then either $S(t)$ is empty for all times $t > 0$ or, otherwise, $a(t)$ is finite for t small and $a(t) \to 0$ as $t \to 0$. Moreover, the initial conditions $U(x, 0) = 0$ for $x \neq 0$ are taken continuously.*

Proof. We take a point $x_0 \neq 0$. By hypothesis there exists $t_0 = t(x_0)$ such that $U(x_0, t)$ is bounded, say by M, in the interval $0 < t < t_0$. By standard comparison in the region $R = \{x : |x| \geq |x_0|, t \in (0, t_0)\}$ we conclude that $V_c(x, t)$, $c > 0$, and $U(x, t)$ are bounded above by M and this region. The standard regularity theory implies then that U is continuous and takes the initial data $U(x, 0) = 0$ for $|x| \geq |x_0|$. Clearly, we also have $a(t) \leq |x_0|$ for $0 < t < t_0$. ☐

LEMMA 3.4. *The radius of $S(t)$, $a(t)$, satisfies*

$$(26) \qquad a(t_1) + a(t_2) \leq a(t_1 + t_2)$$

for every $t_1, t_2 > 0$. In particular, if $a(t)$ is finite at time t_1 then it is also for all $t_2 \geq t_1$.

Proof. We will assume that $N = 1$ to make the writing simpler. By a construction we did in Lemma 3.1 we have for $x > a(t_1)$

$$U(x, t + t_1) \geq U(x - a(t_1), t),$$

hence

$$U(a(t_1) + a(t_2) - \epsilon, t_1 + t_2) \geq U(a(t_2) - \epsilon, t_2).$$

Since $U(a(t_2) - \epsilon, t_2) = \infty$, we get $U(a(t_1) + a(t_2) - \epsilon, t_1 + t_2) = \infty$. ☐

As a consequence we have

COROLLARY 3.1. *The function $a(t)/t$ is nondecreasing. In particular, if $a(t)$ is finite at some time t_0, it is finite for $0 < t < t_0$.*

LEMMA 3.5. *$a(t)$ is a constant function with value either 0 or ∞.*

Proof. For $0 < l < 1$ we can easily check that

$$(27) \qquad V_{c,l}(x,t) = V_c(lx, l^2 t)$$

is a supersolution for equation (3) with initial mass c/l^N. By the Maximum Principle

$$(28) \qquad V_{c,l}(x,t) \geq V_{c/l}(x,t),$$

which in the limit $c \to \infty$ implies that the function U_l defined by the same scaling (27) satisfies

$$(29) \qquad U_l(x,t) = U(lx, l^2 t) \geq U(x,t).$$

Putting $t = 1$ we get

$$U(l(a(1) - \epsilon)\mathbf{e}, l^2) \geq U((a(1) - \epsilon)\mathbf{e}, 1) = \infty.$$

Here $\mathbf{e}$ is any unitary vector. This means that $la(1) \leq a(l^2)$, i.e.

$$a(t) \geq \sqrt{t}\, a(1), \quad 0 < t < 1.$$

If $a(1)$ is finite and nonzero we have a contradiction with formula (26). If $a(1)$ is infinite, it implies that $a(t)$ is also infinite for all $0 < t < 1$. The same happens starting from any other time instead of 1. ☐

Summing up. the analysis of Option 2, when (H1) is not satisfied, leads to two different possibilities:

(i) when $S(t)$ is empty for all t, so that the solution has an isolated singularity at $(0,0)$, what we call a Very Singular Solution, and

(ii) when $S(t) = \{0\}$ for all t. We have a *razor blade*.

4 Properties of the Very Singular Solutions

In order to give the reader an idea of the relative behaviour of these singular solutions we state a number of results for the very singular solutions (VSS).

PROPOSITION 4.1. *If there exists a VSS then it is bigger than every FS. In that case the limit of fundamental solutions (FS) is a VSS.*

Proof. The main idea can be seen in [6]. Observe also that the limit of weak solutions with a uniform bound from above is a weak solution. ☐

The VSS obtained as limit of FS is then the *minimal* element in the set of very singular solutions.

PROPOSITION 4.2. *If there exists a VSS and a flat solution then the VSS is smaller than the flat solution.*

PROPOSITION 4.3. *If there exists a flat solution then the limit of FS is either the flat solution or a VSS, if this one exists.*

PROPOSITION 4.4. *For the power-like equations (7) there exist VSS precisely for the range*

$$(30) \qquad p > 1, \quad m < p < m + \frac{2}{N}.$$

The VSS is unique, selfsimilar and bounded for all $t > 0$ and can be obtained as the limit of fundamental solutions as in (21).

The uniqueness of the VSS for equation (7) was proved by Kamin and Véron [8].

5 Construction of Razor Blades

There are many razor blades, most of them having nothing to do with the limits of fundamental solutions, like the following.

PROPOSITION 5.1. *Let us consider the heat equation. For every function $c(t) \geq 0$ there is a unique solution of the problem*

$$u_t = \Delta u - au + c(t)\delta(x) \qquad \text{in } Q$$
$$u(x, 0) = 0 \qquad \text{for } x \in \mathbf{R}^N$$

Each of these solutions is a different razor blade.

For the equation with power nonlinearities (7) we can describe the cases where razor blades appear as limits of fundamental solutions.

PROPOSITION 5.2. *The limit (21) of the family of fundamental solutions is a razor blade if $m_c < m < 1$ and $p \leq 1$. For $p < 1$ the formula is selfsimilar,*

$$(31) \qquad U(x, t) = t^\alpha f(|x|t^\beta),$$

with $\alpha = 1/(1 - p)$ and $\beta = (p - m)/2(p - 1)$. The profile f is a positive and decreasing function of $\xi \geq 0$ with $f(0) = \infty$.

The following result should hold in general for equations of the type (3) though we only know it for special cases.

PROPOSITION 5.3. *If a razor blade has infinite mass for all $t > 0$ then (i) it is bigger than the fundamental solutions, (ii) if there are VSS it is bigger than the minimal VSS, (ii) it is increasing in time.*

6   Some remarks on the elliptic problems

When we neglect the time dependence, i.e., when we consider the stationary solutions of the previous equations, we obtain a nonlinear elliptic equation. The study of singular solutions is then different but closely related. Without loss of generality we can write the equation in the form

$$(32) \qquad\qquad\qquad -\Delta u + f(u) = 0.$$

This leads to the study of existence and uniqueness of the fundamental solutions, which in the stationary case are defined as solutions of (32) in the space minus a point, say $\mathbf{R^N} - \{0\}$, which can be interpreted as distribution solutions of the equation $-\Delta u + f(u) = c\delta(x)$. Fundamental solutions are the typical solutions having an isolated singularity. But in the nonlinear case there are, as we have said, new isolated singularities, the so-called strong singularities. For power nonlinearities they turned out to be in limits as $c \to \infty$ of fundamental solutions, [12]. However, this is not always the case. Thus, in [11] the authors have analysed examples of nonlinearities f for which fundamental solutions do not exist, but strong singularities exist. Let us finally remark that, when viewed as stationary solutions for the evolution problem (3), the isolated singularities of (32) are a kind of razor blades, with the difference that they do not take the values $u(x,0) = 0$ for $x \neq 0$.

7   Further developments and generalizations

In the case where the nonlinearities are power-like we can make a complete study of the existence of the different types of solutions and their relative interplay. A complete classification for general ϕ and ψ is however out of question, since a casuistic discussion ensues involving different integral expressions of the relative strength of both nonlinearities.

The above results can be extended inside the area of nonlinear parabolic equations in a number of directions. The first step is to include explicit dependence of the coefficients on x, t. More generally, we can consider equations of the form

$$(33) \qquad\qquad u_t = \nabla \cdot \mathbf{A}(x, t, u, \nabla u) + B(x, t, u, \nabla u).$$

One particular instance worth mentioning is the so-called *p-Laplacian* case where

$$(34) \qquad\qquad \mathbf{A}(x, t, u, \nabla u) = |\nabla u|^{p-2} \nabla u.$$

The existence and properties of very singular solutions when there is a power-type zero-order term is studied in [7]. In the stationary case the singular solutions of this type of equations without zero-order term are classified in [4].

A detailed account of the properties and applications of the singular solutions here described, with proof of the results of Sections 4 and 5, as well as the generalizations just mentioned, will be given elsewhere.

References

[1] H. Brezis, E. Lieb, *Long-range atomic potentials in Thomas-Fermi theory*, Comm. Math. Phys., 65 (1980), pp. 231-246.

[2] H. Brezis, L. A. Peletier, D. Terman, *A very singular solution of the heat equation with absorption*, Arch. Rat. Mech. Anal., 95, (1986), pp. 185-209.

[3] R. H. Fowler, *Further studies on Emden's and similar differential equations*, Quart. Jour. Math., 2 (1931), pp. 259-288.

[4] A. Friedman, L. Véron, *Singular solutions of some quasilinear elliptic equations*, Archive Rat. Mech. Anal., 96 (1986), pp. 359-387.

[5] V. A. Galaktionov, S. P. Kurdyumov, A. A. Samarskii, *On asymptotic "eigenfunctions" of the Cauchy problem for a nonlinear parabolic equation*, Mat. Sbornik, 126 (1985), pp. 435-472 (English transl.: Math. USSR Sbornik, 54 (1986), pp. 421-455).

[6] S. Kamin, L. A. Peletier, J. L. Vazquez, *Classification of singular solutions of a nonlinear heat equation*, Duke Math. Jour., 58 (1989), pp. 601-615.

[7] S. Kamin, J. L. Vazquez, *Singular solutions of some nonlinear parabolic equations*, Jour. Anal. Math., 59 (1992), pp. 51-74.

[8] S. Kamin, L. Véron, *Existence and uniqueness of the very singular solution for the porous medium equation with absorption*, Jour. Anal. Math., 51 (1988), pp. 245-258.

[9] A. A. Samarskii, V. A. Galaktionov, S. P. Kurdyumov and A. P. Mikhailov, "Blow-up in Problems for Quasilinear Parabolic Equations", Nauka, Moscow, 1987 (in Russian); English translation: Walter de Gruyter, Berlin, 1995.

[10] J. L. Vazquez, *Asymptotic behaviour of nonlinear parabolic equations. Anomalous exponents*, in "Degenerate Diffusions", IMA Volumes in Mathematics **47**, Springer Verlag (1993), pp. 215-228.

[11] J. L. Vazquez, L. Véron, *Isolated singularities of some semilinear elliptic equations*, Jour. Diff. Eqns., 60 (1985), pp. 301-321.

[12] L. Véron, *Singular solutions of some nonlinear elliptic equations*, Nonlinear Anal., 5 (1981), pp. 225-242.

Uncertainty in epidemic models

S.D.Watt *   G.C.Wake †

Abstract

When uncertainty is present in a biological model, there will be a significant effect on the outcome. By using stochastic calculus, it is possible to quantify this effect. The tool used is the Fokker-Planck equation which is an equation which the probability distribution of solutions of the stochastic differential equation satisfies. Via this approach, it is possible to find explicitly the expected values and variances of the stochastic populations in terms of the amplitude of the noise.

Tribute

This contribution is part of a tribute to Professor Ivar Stakgold, previously President of the Society of Industrial and Applied Mathematics. We note here his two extensive visits to New Zealand in 1981 and 1987, the second as an invited speaker at the Annual Applied Mathematics Conference under the auspices of what is now ANZIAM.

1 Introduction

The deterministic model of the spread of bovine tuberculosis in possums has been well studied [3, 4], and it is relatively straight forward to use the standard tools of dynamics systems. However, when randomness is introduced into the model, the analysis changes from differential calculus to stochastic calculus.

The method we have chosen to introduce the randomness is to add a small random process to one of the parameters. We have chosen to perturb the disease transmission parameter, β, in the SI model in view of this being the most difficult parameter to measure. Once this is done, the model changes from an ordinary differential equation (ODE) to a stochastic differential equation (SDE). By solving this equation numerically, it is possible to find sample solutions. Given enough of these samples, we can find statistical quantities, for example, the expected values and variances, and in principle the other moments.

However, this method assumes the numerical algorithm is accurate. Even though the numerics give fair answers, there are problems not just with the accuracy of the method, but a lot of emphasis on the generation of random numbers. By looking at the solutions in a different framework, these problems do not arise.

As a consequence of diffusion having a parallel to a random process, the probability distribution of solutions of an SDE can be a solution of a diffusion-type equation. This equation is the Fokker-Plank equation, or sometimes called the forward Kolmogorov equation. Using this equation, it is possible to calculate the moments, for example the variance. By using the method of moments, as described in Section 4, the Fokker-Planck equation need not necessarily be solved explicitly.

---

*Mathematics Department,Massey University, Private Bag 11-222, Palmerston North, New Zealand

†Industrial and Applied Mathematics, Tamaki Campus, University of Auckland, Private Bag 92-019, Auckland, New Zealand

1.1 The IN model and its stochastic analogue

The model we wish to analyse is the spatially uniform population model described in [3]. The model is

$$\dot{N} = (a - b)N - \alpha I$$
$$\dot{I} = \beta I(N - I) - (\alpha + b)I$$

where

- a is the natural birth rate

- b is the natural death rate

- α is the death rate due to the disease

- β is the disease transmission rate

We now assume that β is a stochastic parameter

$$\beta \to \beta(1 + \epsilon\eta(t))$$

where $\eta(t)$ is white noise with zero mean and unit variance.

If we now rescale time $t \to t/\beta$, we get the system

(1)
$$\dot{N} = CN - DI$$

(2)
$$\dot{I} = I(N - I) - EI + \epsilon\eta(t)I(N - I)$$

where we define the dimensionless parameters

$$C = \frac{a - b}{\beta}$$
$$D = \frac{\alpha}{\beta}$$
$$E = \frac{\alpha + b}{\beta}$$

2 The Fokker-Planck equation

The analysis of stochastic differential equations, and more generally stochastic processes, is helped by the connection to diffusion processes. When the stochastic variability is white noise, the probability distribution of the solutions satisfy an advection-diffusion type equation. In one dimension, this is

$$\frac{\partial P}{\partial t} = -\frac{\partial}{\partial x}(M(x)P) + \frac{1}{2}(V(x)P)$$

where P is the probability distribution of the solutions.

Consider a one dimensional SDE

$$\dot{x} = F(x, \eta(t))$$

where $\eta(t)$ is white noise. Then $M(x)$ and $V(x)$ are defined as

$$M(x) = E[F(x, \eta(t))]$$
$$V(x) = E[(F(x, \eta(t)) - M)^2]$$

where the averages are taken over the white noise.

For example, if the SDE is of the form

$$\dot{x} = f(x) + g(x)\eta(t)$$

then $M(x) = f(x)$ and $V(x) = g(x)^2$.

For the more general case, consider a system of m SDE's

$$\dot{x}_i = f_i(\boldsymbol{x}) + \sum_{j=1}^{n} g_{ij}(\boldsymbol{x})\eta_j(t)$$

where $\boldsymbol{\eta}$ is an n-dimensional white noise process.

The corresponding diffusion process satisfies

$$\frac{\partial P}{\partial t} = -\sum_{i=1}^{m} \frac{\partial}{\partial x_i}[f_i P] + \frac{1}{2}\sum_{i=1}^{m}\sum_{j=1}^{m}\frac{\partial^2}{\partial x_i \partial x_j}[a_{ij}P]$$

where a_{ij} are the components of $A = \boldsymbol{g}\boldsymbol{g}^T$.

This equation is known as the Fokker-Planck equation or the forward Kolmogorov equation.

The initial condition of the distribution is

$$P(t = 0) = \prod_{i=1}^{m} \delta(x_i - x_i^0)$$

where $\boldsymbol{x}^0$ is the initial condition for the SDE, $\boldsymbol{x}(t = 0) = \boldsymbol{x}^0$.

3   Linearised moments

The problem of finding the evolution of the specific statistical quantities, namely the expectation values and variances of the populations, is not straight forward. Even the problem of finding the equilibrium quantities is not easily solved. However, by considering the equilibrium solutions of the SDE near the stable equilibrium of the deterministic ODE, that is $\epsilon = 0$, some progress can be made. The following analysis is similar to that found in May [5].

The deterministic model has the stable equilibrium

$$(N^*, I^*) = \left(\frac{DE}{D - C}, \frac{CE}{D - C}\right)$$

We now transform coordinates to

$$\begin{aligned} x &= N - N^* \\ y &= I - I^* \end{aligned}$$

Thus the model (1-2) becomes

$$\begin{aligned} \dot{x} &= Cx - Dy \\ \dot{y} &= y(x - y) + I^*(x - y) \\ &+ \epsilon\eta(t)[y(x - y) - \frac{E(2C - D)}{D - C}y + I^*x + EI^*] \end{aligned}$$

If we assume that x and y are of order ϵ, then to first order in ϵ, the equations are

$$\dot{x} = Cx - Dy$$
$$\dot{y} = I^*(x - y) + \epsilon\eta(t)EI^*$$

The corresponding Fokker-Planck equation, as in Section 2, is

(3) $$\frac{\partial P}{\partial t} = -\frac{\partial}{\partial x}[(Cx - Dy)P] - \frac{\partial}{\partial y}[I^*(x - y)P] + \frac{1}{2}(\epsilon EI^*)^2\frac{\partial^2 P}{\partial y^2}$$

As we are concerned with the equilibrium probability distribution, we set the LHS to zero. Due to the linearity of the coefficients in the Fokker-Planck equation, the solution is of the form of a Gaussian distribution

$$P = C\exp[-c_{20}x^2 + c_{11}xy - c_{02}y^2]$$

where $C = (2\pi)^{-1}\sqrt{4c_{20}c_{02} - c_{11}^2}$ is a normalising constant.

Substituting this into (3), we find

$$c_{20} = \frac{\Delta_1\Delta_2(C + E)}{\epsilon^2 E^4 D^2}$$
$$c_{11} = \frac{2\Delta_1\Delta_2}{\epsilon^2 E^4 D}$$
$$c_{02} = \frac{\Delta_1\Delta_2}{\epsilon^2 E^4 C}$$

where $\Delta_1 = D - C$ and $\Delta_2 = E - D + C$.

From this, we find the small-noise limit of the statistical quantities

$$<N> = N^*$$
$$<I> = I^*$$
$$\sigma_N^2 = \frac{D^2 E^3}{2\Delta_1\Delta_2}\epsilon^2$$
$$\sigma_I^2 = \frac{(C + E)CE^3}{2\Delta_1\Delta_2}\epsilon^2$$

These results are used later as a second order approximation.

4   Method of moments

One of the main objectives of this research is to get a fair description of the expectation values and variances of the stochastic populations. The method of moments allows the possibility to find these quantities, which are derived from the first and second moments, directly without having to find the solution of the Fokker-Planck equation.

We define a moment as

$$M_{n,m} =<x^ny^m> = \int_{-\infty}^{\infty}\int_{-\infty}^{\infty}x^ny^m P\,dx\,dy$$

where the kth order moments are all the moments where $n + m = k$.

We will assume that P satisfies a general time-dependent Fokker-Planck equation

(4) $$\frac{\partial P}{\partial t} = -\frac{\partial}{\partial x}[f_1 P] - \frac{\partial}{\partial y}[f_2 P] + \frac{1}{2}\frac{\partial^2}{\partial y^2}[g^2 P]$$

with $P = \delta(x - x_0)\delta(y - y_0)$ at $t = 0$.

Taking the derivative of $M_{n,m}$ with respect to time yields

$$\dot{M}_{n,m} = \int_{-\infty}^{\infty} \int_{-\infty}^{\infty} x^n y^m \frac{\partial P}{\partial t} \, dx \, dy$$

Substituting the RHS of (4) for $\frac{\partial P}{\partial t}$, we get

$$\dot{M}_{n,m} = \int_{-\infty}^{\infty} \int_{-\infty}^{\infty} x^n y^m \left[-\frac{\partial}{\partial x}[f_1 P] - \frac{\partial}{\partial y}[f_2 P] + \frac{1}{2}\frac{\partial^2}{\partial y^2}[g^2 P] \right] \, dx \, dy$$

By integrating this by parts, we get

(5) $\qquad \dot{M}_{n,m} = n < x^{n-1} y^m f_1 > + m < x^n y^{m-1} f_2 > + \frac{1}{2} m(m-1) < x^n y^{m-2} g^2 >$

As a consequence of the initial condition for P, we have

$$M_{n,m} = x_0^n y_0^m \text{ at } t = 0$$

By using this method, it is possible to find the evolution equations of the moments directly. However, in most cases, the moment equation (5) will not be closed. That is, the evolution equations for the kth order moments will involve higher order moments. Some methods of trying to resolve this problem is discussed in [1].

When the equilibrium is sought, there is a greater chance of finding closure. This is due to the possible existence of two other relations.

By considering the equilibrium solutions, we set the LHS of (5) to zero to give

(6) $\qquad 0 = n < x^{n-1} y^m f_1 > + m < x^n y^{m-1} f_2 > + \frac{1}{2} m(m-1) < x^n y^{m-2} g^2 >$

There are also two more possible relations

(7) $\qquad\qquad 0 = \lim_{n \to 0} \frac{\dot{M}_{n,0}}{n} \; = \; < x^{-1} f_1 >$

(8) $\qquad\qquad 0 = \lim_{m \to 0} \frac{\dot{M}_{0,m}}{m} \; = \; < y^{-1} f_2 > - \frac{1}{2} < y^{-2} g^2 >$

The existence of these equations depends on the form of f_1, f_2 and g.

4.1 The calculation of moments for the IN model

By using the method of moments, it is possible to find approximations to the expectation values and variances. In the IN model, we assign the functions

$$
\begin{aligned}
f_1 &= CN - DI \\
f_2 &= I(N - I) - EI \\
g &= \epsilon I(N - I)
\end{aligned}
$$

Substituting these in (6), we have the relations

(9) $0 = C < N > - D < I >$

(10) $0 = < IN > - < I^2 > - E < I >$

(11) $0 = 2C < N^2 > - 2D < IN >$

(12) $0 = C < IN > - D < I^2 > + < IN^2 > - < I^2 N > - E < IN >$

(13) $0 = < I^2 N > - < I^3 > - E < I^2 > + \epsilon^2 < I^2 N^2 > - 2\epsilon^2 < I^3 N > + \epsilon^2 < I^4 >$

As (12) and (13) introduce third and fourth order moments, we will disregard them.

Due to the form of f_2 and g, we can use (8) to get

$$0 = < N > - < I > - E - \frac{\epsilon^2}{2} < N^2 > + \epsilon^2 < IN > - \frac{\epsilon^2}{2} < I^2 >$$

Thus we have four equations for five unknowns.

Assume perturbation expansions

$$< N > = \sum_{n=0}^{\infty} < N >_n \epsilon^{2n}$$

$$< I > = \sum_{n=0}^{\infty} < I >_n \epsilon^{2n}$$

$$< N^2 > = \sum_{n=0}^{\infty} < N^2 >_n \epsilon^{2n}$$

$$< IN > = \sum_{n=0}^{\infty} < IN >_n \epsilon^{2n}$$

$$< I^2 > = \sum_{n=0}^{\infty} < I^2 >_n \epsilon^{2n}$$

We then get a hierarchy of equations

$$
\begin{aligned}
0 &= C < N >_n - D < I >_n \\
0 &= < IN >_n - < I^2 >_n - E < I >_n \\
0 &= 2C < N^2 >_n - 2D < IN >_n \\
0 &= < N >_n - < I >_n - E - \frac{1}{2} < N^2 >_{n-1} + < IN >_{n-1} - \frac{1}{2} < I^2 >_{n-1}
\end{aligned}
$$

By using the fact that the variance is zero in the noise-free case ($\epsilon = 0$) and also the results of the linearised moments, we can find the mean population and infected population to fourth order in the noise, ϵ

$$< N > \approx \frac{DE}{\Delta_1} + \frac{DE^2}{2\Delta_1}\epsilon^2 - \frac{DE^3(\Delta_1\Delta_2 + DE)}{4\Delta_1^2\Delta_2}\epsilon^4$$

$$< I > \approx \frac{CE}{\Delta_1} + \frac{CE^2}{2\Delta_1}\epsilon^2 - \frac{CE^3(\Delta_1\Delta_2 + DE)}{4\Delta_1^2\Delta_2}\epsilon^4$$

If we assume that these are exact means and then substitute these into (9-11) and (4.1), we can find find the variances to fourth order in ϵ also

$$\sigma_N^2 \approx \frac{D^2E^2}{2\Delta_1\Delta_2}\epsilon^2 + \frac{DE^4(4D\Delta_1^2 - D^2\Delta_1 - C^2\Delta_2 - 5DE\Delta_1)}{4\Delta_1^3\Delta_2}\epsilon^4$$

$$\sigma_I^2 \approx \frac{CE^3(C + E)}{2\Delta_1\Delta_2}\epsilon^2 + \frac{CE^4(D^2\Delta_1 + C^2\Delta_2 - 2D^2E)}{4\Delta_1^3\Delta_2}\epsilon^4$$

5 Numerical results

When given an SDE, the Fokker-Planck approach, as discussed in previous sections, is an analytical tool and quite a theoretical approach. Another, and more tangible, is to solve the equation numerically. This approach is straightforward, yet not. As stochastic calculus does not behave in exactly the same way as differential calculus, the plethora of numerical algorithms for ODE's do not necessarily translate to stochastic calculus.

There have been much research into this area, and the topic is well described in Kloeden & Platen [2]. Here we have chosen a relatively simple Euler-type scheme.

Consider a set of SDE's

$$
\begin{align}
\dot{x} &= f_1(x, y) \tag{14} \\
\dot{y} &= f_2(x, y) + \eta(t)g(x, y) \tag{15}
\end{align}
$$

where $\eta(t)$ is white noise with zero mean and unit variance.

We then discretise in time and use the scheme [7]

$$
\begin{align}
x_{j+1} &= x_j + f_1(x_j, y_j)\Delta \tag{16} \\
y_{j+1} &= y_j + f_2(x_j, y_j)\Delta + g(x_j, y_j)\sqrt{\Delta}\zeta_j \tag{17}
\end{align}
$$

where $t = \Delta j$ and ζ_j are independent random variables with zero mean and unit variance. The random number generator we used was DRNOR [6].

Using the numerical scheme, we can calculate an array of solutions. By averaging these solutions, we can get results for the expectation values and variances.

5.1   Comparison of the numerical and analytical results

We can compare the results given for the expectation values and variances by the numerical scheme and the Fokker-Planck approach.

The technique we use is to integrate the model (1-2) over time. We form an array of solutions, with the same initial condition. We integrate for long enough so that the probability distribution of the solutions is stationary. Using this array of solutions, we can calculate the expectation values and variances.

Assume that we calculate m independent solutions, denoted $(N_j(t), I_j(t))$ for $j = 1 \ldots m$, and integrated till time T. We then define

$$
< N > = \frac{1}{m} \sum_{n=1}^{m} N_n(T) \tag{18}
$$

$$
< I > = \frac{1}{m} \sum_{n=1}^{m} I_n(T) \tag{19}
$$

$$
\sigma_N^2 = \frac{1}{m} \sum_{n=1}^{m} N_n(T)^2 - \frac{1}{m^2} \left(\sum_{n=1}^{m} N_n(T) \right)^2 \tag{20}
$$

$$
\sigma_I^2 = \frac{1}{m} \sum_{n=1}^{m} I_n(T)^2 - \frac{1}{m^2} \left(\sum_{n=1}^{m} I_n(T) \right)^2 \tag{21}
$$

For the comparison of the numerical and analytical results, we have made the assumptions that : $C = 1$, $D = 2$, $E = 2$, $T = 40$, $\Delta = 10^{-2}$, $m = 1000$ and $epsilon \in [0, 0.1]$. The results are shown in Figures 1 and 2. The two results are in good agreement, which suggests that both methods are good approximations.

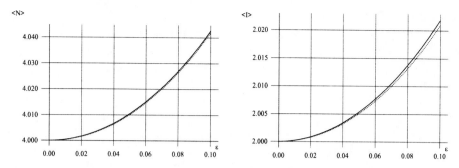

FIG. 1. *Comparison of (a) the mean population and (b) the mean infected population via numeric (—) and analytic (. . .) methods*

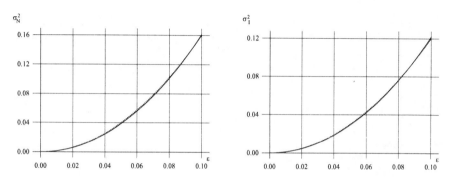

FIG. 2. *Comparison of (a) the population variance and (b) the infected population variance via numeric (—) and analytic (. . .) methods*

6   Conclusion

We have considered the problem of uncertainty of parameters in a biological system. The changing of the model from a set of ODE's to a set of SDE's creates the need for a new analysis.

One of the useful tools of stochastic calculus is the possibility of placing the solution distribution in a seemingly simple framework via the Fokker-Planck equation. An advantage of this approach is that as long as the stochasticity appears linearly, the corresponding equation for the solution distribution is a linear PDE, regardless if the SDE is nonlinear or not.

However, the Fokker-Planck equation is a relatively hard equation to solve, as the coefficients are non-constant. The solution is not necessarily needed though, via the calculation of moments. The method of moments, as described in Section 5, allows the possible calculation of the statistically significant expectation values and variances.

Using this method, we calculated the expectation values and variance. As would be expected, if a parameter is changed by a small amount, the effect on the system is small. In fact, the change on the system is of the order of the perturbation squared.

As for further research, we would like to look at more general models. For example, the *SI* model described in [4].

Acknowledgements

The support (SDW) of a Massey University/University of Auckland Postdoctoral Fellowship to support this work and the support of a subcontract from the Foundation of Research Science and Technology via New Zealand Pastoral Research Ltd., Contract C10 420, are gratefully acknowledged.

References

[1] D.C.C. Bover "Moment equation methods for nonlinear stochastic systems". *J. Math. Anal. and Appl.* **65** pp. 306–320 (1978).

[2] P.E. Kloeden and E. Platen "Numerical solution of stochastic differential equations". Springer-Verlag (1992).

[3] K. Louie, M.G. Roberts and G.S. Wake. "Thresholds and stability analysis of models for the spatial spread of a fatal disease". *IMA J. Math. App. in Med. and Bio.* **10** pp. 207–226 (1993).

[4] M.R. Roberts. "The dynamics of Bovine Tuberculosis in possum populations, and it eradication or control by culling or vaccination". *Preprint* (1995).

[5] R. May. "Stability and complexity in model ecosystems". Princeton University Press (1974).

[6] D. Kahaner, C. Moler and S. Nash. "Numerical methods and software". Prentice Hall. (1988).

[7] D. Zwillinger "Handbook of differential equations" Wiley (1992).

Energy Estimates in Thermoelasticity*

R. J. Weinacht[†]   B. F. Esham[‡]

Dedicated to: Ivar Stakgold

Abstract

Higher order, ϵ-dependent energy estimates are given for solutions of an initial-boundary value problem in linear thermoelasticity (in which the square of ϵ is the inertial constant).

1 Introduction

In this paper we present energy estimates for an initial-boundary value problem in thermoelasticity. Although we consider a linear problem, it is precisely these types of estimates (but for variable coefficients rather than the constant coefficient case considered here) that prove useful in approaching the corresponding nonlinear problem via à priori estimates of the linearized PDEs (see Slemrod [11] for one-dimensional nonlinear thermoelasticity and Esham/Weinacht [4] and the references therein for nonlinear hyperbolic-parabolic problems in which ϵ dependence is investigated).

For definiteness consider the following initial-boundary value problem (in non-dimensional variables) in linear thermoelasticity for a homogeneous isotropic medium. The problem consists of the PDE system

$$(1.1) \qquad \partial_t \theta - \Delta\theta + \gamma \operatorname{div} \partial_t \mathbf{u} = \Phi$$

$$(1.2) \qquad \epsilon^2 \partial_t^2 \mathbf{u} - E[\mathbf{u}] + \gamma \operatorname{grad} \theta = \mathbf{F}$$

for (x,t) in $\Omega \times (0,\infty)$ where Ω is a bounded region in $I\!\!R^n$ and where E is the elastostatic operator

$$E[\mathbf{u}] := \operatorname{grad} \operatorname{div} \mathbf{u} + \nu \Delta \mathbf{u}$$

with initial conditions

$$
\begin{aligned}
\theta(x,0;\epsilon) &= \phi(x;\epsilon) \\
\mathbf{u}(x,0;\epsilon) &= \mathbf{f}(x;\epsilon) \\
\partial_t \mathbf{u}(x,0;\epsilon) &= \mathbf{g}(x;\epsilon)
\end{aligned}
$$
(1.3)

and homogeneous Dirichlet boundary conditions

$$\theta\Big|_{\partial\Omega\times(0,\infty)} = 0$$

(1.4)

$$\mathbf{u}\Big|_{\partial\Omega\times(0,\infty)} = 0.$$

---

* The work of R. J. Weinacht was supported in part by GNAFA-CNR of the Consiglio Nazionale delle Ricerche of Italy.

[†] Professor, University of Delaware, Department of Mathematical Sciences, Newark, DE 19716.

[‡] Associate Professor, Department of Mathematics, State University of New York, Geneseo, New York.

259

In fact, we will treat in detail the case of data $\{\Phi, \mathbf{F}, \phi, \mathbf{f}$ and $\mathbf{g}\}$ of special form, especially with regard to their ϵ-dependence. In Section 2 we indicate explicitly the special choices with the reason thereof.

We refer to [3, 5, 6], and references therein for a derivation of (1.1) - (1.2), including the scaling leading to the non-dimensional form. Here we only mention that θ is the (deviation from the reference) temperature and $\mathbf{u}$ is the displacement. The constant γ is the "coupling constant" which measures thermal-elastic interaction and ϵ^2 is the "inertial constant"; ν is a ratio of Lamé constants. A major point of interest is the behavior of the solution $\{\theta, \mathbf{u}\}$ of (1.1) - (1.4) as ϵ tends to zero from above. For $\epsilon = 0$ (1.1) - (1.2) become the PDEs for the coupled/quasi-static approximation [3, 5, 6].

The system (1.1) - (1.2), roughly speaking, is a parabolic (heat) PDE coupled with a hyperbolic (wave) PDE but this is somewhat misleading since it is precisely the coupling which gives the special features of the system. For this reason the standard estimates for hyperbolic systems or parabolic systems cannot be applied directly to this case. Of course, it is the strength of the Energy Estimate approach (as emphasized especially by Friedrichs [8]) that it does not depend on the type (elliptic/parabolic/hyperbolic) of the system. Energy Estimates in thermoelasticity were used implicitly by Dafermos [1] in considering "finite energy" solutions in inhomogeneous anisotropic linear thermoelasticity (see also their explicit use in Dafermos/Nohel [2], Slemrod [11] in one-dimensional nonlinear thermoelasticity, Day in a series of papers on one-dimensional linear thermoelasticity [see his monograph [3] and references therein] and by Esham/Weinacht [5] in investigating ϵ-dependence in one-dimensional linear thermoelasticity).

In [1] Dafermos proved existence and uniqueness of "finite energy" weak solutions for an initial-boundary value problem which includes (1.1) - (1.4) for any positive ϵ. He further gave regularity results (under additional smoothness and compatibility assumptions) leading to the existence of classical smooth solutions. The estimates of the present paper can be regarded as applying to the classical smooth (to a certain order) solution of (1.1) - (1.4) provided by the results of [1]. Alternatively we can view our estimates to be valid for elements of the closure (in appropriate norms) of spaces of smooth functions defined on the closure of $\Omega \times (0, T)$ for any fixed $T > 0$. Such estimates are then available to prove existence of solutions of (1.1) - (1.4). We omit explicit assumptions on the data for the latter purposes, referring instead to [1]. It should be mentioned that in using the estimates to prove uniform asymptotic validity of certain perturbation solutions as ϵ tends to zero (see [5, 6]) much more regularity is required than for the existence of a solution of (1.1) - (1.4) for fixed positive ϵ (because derivatives of such a solution are used in the construction).

In a concluding remark at the end of this paper we mention an analogue to the problem (1.1) - (1.4) for a system of evolution equations in Hilbert space which corresponds to the case of an inhomogeneous anisotropic medium.

In addition to standard notations (for partial differentiation: ∂_t^k for time derivatives, both multi-index notation D^α and subscripts $\theta_{,\ell}$, $u^j_{,\ell}$, etc. along with the summation notation for space derivatives) we use the following special notations for dealing with scalar-valued functions such as θ and vector-valued functions such as $\mathbf{u}$: For scalar-valued θ, χ we denote the $L^2(\Omega)$ inner product by $\langle \cdot, \cdot \rangle$

$$\langle \theta, \chi \rangle := \int_\Omega \theta(x,t)\chi(x,t)dx$$

with corresponding norm (without subscript) $\| \cdot \|$

$$\|\theta\| := \langle \theta, \theta \rangle^{\frac{1}{2}}$$

while for vector-valued $\mathbf{u}, \mathbf{w}$ we denote the $\mathbf{L}^2(\Omega)$ inner product by $[\cdot, \cdot]$

$$[\mathbf{u}, \mathbf{w}] := \int_\Omega \mathbf{u}(x, t) \cdot \mathbf{w}(x, t) dx$$

with corresponding norm $\| \cdot \|$

$$\|\mathbf{u}\| = [\mathbf{u}, \mathbf{u}]^{\frac{1}{2}}.$$

The norms in the Sobolev space $H^s(\Omega)$ and the corresponding Sobolev space $\mathbf{H}^s(\Omega)$ of functions with values in $\mathbb{R}^n$ will be denoted by $\| \cdot \|_s$ and $\| \cdot \|_s$ respectively.

2 A Remainder IBVP

Beginning in the next section estimates will be given for the smooth solution $\{\theta, \mathbf{u}\}$ of the initial-boundary value problem (IBVP) in $\Omega \times (0, \infty)$

(2.1)
$$\partial_t \theta - \Delta \theta + \gamma \operatorname{div} \partial_t \mathbf{u} = \epsilon^m \rho(x, t; \epsilon)$$

(2.2)
$$\epsilon^2 \partial_t^2 \mathbf{u} - E[\mathbf{u}] + \gamma \operatorname{grad} \theta = \epsilon^{m+1} \mathbf{r}(x, t; \epsilon)$$

(2.3)
$$\theta(x, 0; \epsilon) = 0$$
$$\mathbf{u}(x, 0; \epsilon) = 0$$
$$\partial_t \mathbf{u}(x, 0; \epsilon) = \epsilon^m \mathbf{z}(x; \epsilon)$$

(2.4)
$$\theta\Big|_{\partial\Omega \times (0, \infty)} = 0$$

$$\mathbf{u}\Big|_{\partial\Omega \times (0, \infty)} = 0$$

where m is a given positive integer. The IBVP (2.1) - (2.4) is merely (1.1) - (1.4) when the data $\{\Phi, \mathbf{F}, \phi, \mathbf{f}, \mathbf{g}\}$ take the particular form indicated.

The main reason for this choice of data, especially the powers of ϵ which appear, is that the data are precisely of this form in certain "remainder problems" in investigating the asymptotics for small ϵ of the solution $\{\theta, \mathbf{u}\}$ of (1.1) - (1.4) (see Esham/Weinacht [5, 6]). Hence we call (2.1) - (2.4) a remainder IBVP. In such remainder problems it is vital to get estimates which show ϵ-dependence of various derivatives of θ and $\mathbf{u}$. Throughout all estimates are for ϵ tending to zero from above. If ϵ-dependence is not of interest for (1.1) - (1.4) the estimates obtained in this paper are still of value and for such purposes one can merely set $\epsilon = 1$ throughout. In that case an $L^2(\Omega)$ estimate such as (3.1) below

$$\int_\Omega |\partial_t^p \theta(x, t; \epsilon)|^2 dx = O\left(\epsilon^{2(m-p)}\right)$$

which is uniform in t for t in $[0, T]$ and which depends only on the data is a useful $L^2(\Omega)$ bound (uniform in t) for $\partial_t^p \theta$; we could indicate exactly which derivatives of the data appear in each such bound but this would become very cumbersome and that information is easily obtained by merely following the procedure which we give.

The choice of homogeneous initial data $\{\phi, \mathbf{f}\}$ can be, of course, obtained by reduction from the general case (under certain regularity assumptions).

In addition to the usual compatibility assumptions on the data $\{\rho, \mathbf{r}, \mathbf{z}\}$ of (2.1) - (2.4) which will allow smooth (up to a certain order) solutions $\{\theta, \mathbf{u}\}$ on the closure of $\Omega \times (0, T)$, we require that

(2.5)
$$\begin{aligned}
\partial_t^p D^\alpha \rho(x, t; \epsilon) &= \epsilon^{-p} \rho_{\alpha p}(x, t; \epsilon) \\
\partial_t^p D^\alpha \mathbf{r}(x, t; \epsilon) &= \epsilon^{-p} \mathbf{r}_{\alpha p}(x, t; \epsilon) \\
D^\alpha \mathbf{z}(x; \epsilon) &= \mathbf{z}_\alpha(x; \epsilon)
\end{aligned}$$

where

(2.6)
$$\int_0^T \|\rho_{\alpha p}(\cdot, t; \epsilon)\|^2 dt = H_{\alpha p} < \infty$$

$$\int_0^T \|\mathbf{r}_{\alpha p}(\cdot, t; \epsilon)\|^2 dt = N_{\alpha p} < \infty$$

$$\|\mathbf{z}_\alpha(\cdot; \epsilon)\| = M_\alpha < \infty$$

with $H_{\alpha p}$, $N_{\alpha p}$ and M_α independent of ϵ. The above restrictions which might seem very special are also dictated by actual circumstances in certain remainder IBVPs.

3 Estimates by Energy Inequalities

In this section we show that for any given non-negative integer $p < m$ (for the m appearing in (2.1) - (2.3)).

(3.1)
$$\left.\begin{aligned}
\|\partial_t^p \theta(\cdot, t; \epsilon)\| \\
\|\epsilon \partial_t^{p+1} \mathbf{u}(\cdot, t; \epsilon)\| \\
\|\partial_t^p u^j_{,\ell}(\cdot, t; \epsilon)\|
\end{aligned}\right\} = O(\epsilon^{m-p})$$

as well as

(3.2)
$$\int_0^t \|\partial_s^p \operatorname{grad} \theta(\cdot, s, \epsilon)\|^2 ds = O(\epsilon^{2(m-p)})$$

uniformly in t for t in $[0, T]$.

The above estimates will be obtained by energy inequalities by considering $E_p(t)$ defined by

$$\begin{aligned}
E_p(t) : = \frac{1}{2} \int_\Omega \{ (\partial_t^p \theta(x, t; \epsilon))^2 + \epsilon^2 |\partial_t^{p+1} \mathbf{u}(x, t; \epsilon)|^2 \\
+ (\partial_t^p \operatorname{div} \mathbf{u}(x, t; \epsilon))^2 + \nu \, \partial_t^p u^j_{,\ell}(x, t; \epsilon) \partial_t^p u^j_{,\ell}(x, t; \epsilon) \} dx
\end{aligned}$$

which we call the p-th energy at time t of the (smooth) solution $\{\theta, \mathbf{u}\}$ of the initial-boundary value problem (2.1) - (2.4).

By applying ∂_t^p (the p-th derivative with respect to t) to (2.1) and (2.2), then multiplying the first of these by $\partial_t^p \theta$ and taking the dot product of the second with $\partial_t^{p+1} \mathbf{u}$ and subsequently integrating the sum over $\Omega \times (0,t)$ we arrive at the Energy Identity

$$E_p(t) + \int_0^t \|\partial_s^p \operatorname{grad} \theta(\cdot, s; \epsilon)\|^2 ds = E_p(0) + \epsilon^m \int_0^t \{\langle \partial_s^p \theta, \partial_s^p \rho \rangle$$

(3.3)
$$+ [\epsilon \partial_s^{p+1} \mathbf{u}, \partial_s^p \mathbf{r}]\} ds$$

after appropriate integration by parts and use of the given boundary conditions (2.4) as well as the resulting "t-differentiated" boundary conditions. Applying the arithmetic-geometric mean inequality and the Gronwall inequality to (3.3) yields the Energy Inequality

$$E_p(t) + \int_0^t \|\partial_s^p \operatorname{grad} \theta(\cdot, s; \epsilon)\|^2 ds$$

(3.4)
$$\leq e^T \left(E_p(0) + \frac{\epsilon^{2m}}{2} \int_0^T \{\|\partial_t^p \rho(\cdot, t; \epsilon)\|^2 + \|\partial_t^p \mathbf{r}(\cdot, t; \epsilon)\|^2\} dt \right).$$

As agreed upon previously $\langle \cdot, \cdot \rangle$ denotes the inner product in $L^2(\Omega)$ and $[\cdot, \cdot]$ the inner product in $\mathbf{L}^2(\Omega)$; the corresponding norms are indicated by $\| \cdot \|$ and $\|\hspace{-0.3em}\| \cdot \|\hspace{-0.3em}\|$, respectively.

Inequality (3.4) gives estimates for exactly the quantities appearing in (3.1) and (3.2), all in terms of the RHS of (3.4), which involves only data of the problem (2.1) - (2.4). It is important that these estimates be uniform in t for t in $[0, T]$ but we will refrain from repeating that fact in each instance.

It should be noted that the higher order energy inequalities "in time" given by (3.4) via (3.3) do not have counterparts "in space" (corresponding to differentiation with respect to the space variables) because there are boundary terms left over in the integration by parts leading to the analogue of (3.3). Nonetheless we want to estimate $\partial_t^p D^\alpha \theta$ and $\partial_t^p D^\alpha \mathbf{u}$ for non-negative integers p and multi-indices $\alpha = (\alpha_1, \ldots, \alpha_n)$ of interest. This will be done in the following section.

Our next step in establishing (3.1) and (3.2) is to estimate the RHS of (3.4). It follows immediately from the assumptions (2.5) - (2.6) on $\rho, \mathbf{r}$ that the second term (the integral term) on the RHS of (3.4) is $O(\epsilon^{2(m-p)})$. If we show also the (more favorable) estimate that

(3.5)
$$E_p(0) = O(\epsilon^{2(m+1-p)})$$

then the result (3.1), (3.2) is a direct consequence.

Thus it remains only to prove (3.5). For this it is sufficient to show that for $|\alpha| = 1$

(3.6)
$$\left. \begin{array}{r} \|\partial_t^k \theta(\cdot, 0; \epsilon)\| \\ \|D^\alpha \partial_t^k \mathbf{u}(\cdot, 0; \epsilon)\| \end{array} \right\} = O(\epsilon^{m+1-p})$$

and

(3.7)
$$\|\hspace{-0.3em}\|\partial_t^{p+1} \mathbf{u}(\cdot, 0; \epsilon)\|\hspace{-0.3em}\| = O(\epsilon^{m-p})$$

for which $p = -1$ is permitted. From the initial conditions (2.3) the estimates (3.6) for $p = 0$ and (3.7) for $p = -1, 0$ are trivial. For larger, general values of p the estimates (3.6) - (3.7) are most easily established by examining the form of $\partial_t^p \theta(x, 0; \epsilon)$ and $\partial_t^{p+1} \mathbf{u}(x, 0; \epsilon)$ for $p \geq 1$ as we do immediately below.

For ease of expression here and below we will refer to the PDE resulting from the application of $D^\alpha \partial_t^p$ to (2.1) (resp. (2.2)) as Equation $(\alpha, p) - 1$ (resp. Equation $(\alpha, p) - 2$) and the corresponding system of PDEs as System (α, p).

With regard to form we claim that each derivative $\partial_t^p \theta(x, 0; \epsilon)$ is a sum of two types of terms. The first is given explicitly by

$$\epsilon^m \partial_t^{p-1} \rho(x, 0; \epsilon)$$

which is clearly $O(\epsilon^{m+1-p})$ in the $\| \cdot \|$ norm. The second type of term is a sum of terms each of which has the property that it and its space derivatives through order two is also $O(\epsilon^{m+1-p})$ in the $\| \cdot \|$ norm. Similarly we assert that each derivative $\partial_t^{p+1} \mathbf{u}(x, 0; \epsilon)$ is the sum of the terms which are $O(\epsilon^{m-p})$ in the $\| \cdot \|$ norm. The first term in the sum is given explicitly by

$$\epsilon^{m-1} \partial_t^{p-1} \mathbf{r}(x, 0; \epsilon)$$

and each of the remaining terms in the sum has the property that it and its space derivatives through order two are $O(\epsilon^{m-p})$ in the $\| \cdot \|$ norm.

For example, for $p = 1$ from System $(0, 0)$

$$\begin{align}
(3.9) \qquad \partial_t \theta(x, 0; \epsilon) &= \epsilon^m \{\rho(x, 0; \epsilon) - \gamma \operatorname{div} z(x; \epsilon)\} \\
\partial_t^2 \mathbf{u}(x, 0; \epsilon) &= \epsilon^{m-1} \mathbf{r}(x, 0; \epsilon)
\end{align}$$

and for $p = 2$ from System $(0, 1)$

$$\begin{align}
(3.10) \qquad \partial_t^2 \theta(x, 0; \epsilon) &= \epsilon^{m-1} \{\epsilon[(\Delta + \partial_t)\rho(x, 0; \epsilon) - \gamma \Delta \operatorname{div} \mathbf{z}(x; \epsilon)] \\
&\quad - \gamma \operatorname{div} \mathbf{r}(x, 0; \epsilon)\} \\
\partial_t^3 \mathbf{u}(x, 0; \epsilon) &= \epsilon^{m-2} \{E[\mathbf{z}(x; \epsilon)] + \gamma^2 \operatorname{grad} \operatorname{div} \mathbf{z}(x; \epsilon) \\
&\quad + \epsilon \partial_t \mathbf{r}(x, 0; \epsilon) - \gamma \operatorname{grad} \rho(x, 0; \epsilon)\}.
\end{align}$$

The proof of the assertion about the form of $\partial_t^p \theta(x, 0; \epsilon)$ and $\partial_t^{p+1} \mathbf{u}(x, 0; \epsilon)$ is easily carried out by induction on p. For $p = 1$ the result is immediate as seen in (3.9) (and the conditions of (2.5) - (2.6)). For the induction step from p to $p + 1$ we merely use System $(0, p)$ and the induction hypothesis. With the form of $\partial_t^p \theta(x, 0; \epsilon)$ and $\partial_t^{p+1} \mathbf{u}(x, 0; \epsilon)$ thus established the estimates (3.6) - (3.7) follow directly and hence also (3.1).

It should be noted that the RHS of (3.9) and (3.10) consist entirely of terms depending on the data alone and, moreover, that successively larger p will place greater demands on the regularity of the data to ensure smoothness of the solution $\{\theta, \mathbf{u}\}$ on the closure of $\Omega \times (0, T)$.

Let us note here that the same argument which led to (3.6) also yields for $|\alpha| = 1$

$$(3.11) \qquad \|\partial_t^p D^\alpha \theta(\cdot, 0; \epsilon)\| = O(\epsilon^{m+1-p})$$

which will be useful in the following section.

4   Estimates for other derivatives

In this section we obtain the L^2 estimates

$$(4.1) \qquad \|D^\alpha \partial_t^p \theta(\cdot, t; \epsilon)\| = O\left(\epsilon^{m-p-\left[\frac{|\alpha|+1}{2}\right]}\right)$$

$$(4.2) \qquad \|D^\alpha \partial_t^p \mathbf{u}(\cdot, t; \epsilon)\| = O\left(\epsilon^{m-p-\left[\frac{|\alpha|}{2}\right]}\right)$$

for all (α, p) such that $|\alpha| + p < m$. The square bracket in the exponents of (4.1) and (4.2) denotes the greatest integer function. These estimates are uniform in t for t in $[0, T]$. The results (4.1) - (4.2) have been established in the previous section for some of the values of (α, p), namely those appearing in (3.1) which stem from the Energy Inequality (3.4): all pure time derivatives $\partial_t^p \theta$ and $\partial_t^{p+1} \mathbf{u}$ as well as $D^\alpha \partial_t^p \mathbf{u}$ with $|\alpha| = 1$. We proceed now to examine derivatives not included in (3.1).

For $\mathbf{u}$ itself ($\alpha = 0$, $p = 0$) the assertion (4.2) follows directly from the (scalar) Poincaré inequality

$$\langle u^j, u^j \rangle \leq C \langle u^j_{,\ell}, u^j_{,\ell} \rangle$$

(at first with no sum on j) and from the case $|\alpha| = 1$, $p = 0$ previously established in (3.1).

For all derivatives of the form $\partial_t^{p-1} \theta_{,\ell}$ we have

$$\partial_t^{p-1} \theta_{,\ell}(x, t; \epsilon) = \partial_t^{p-1} \theta_{,\ell}(x, 0; \epsilon) + \int_0^t \partial_s^p \theta_{,\ell}(x, s; \epsilon) ds$$

and hence by the Schwarz and arithmetic-geometric mean inequalities

$$\|\partial_t^{p-1} \theta_{,\ell}(\cdot, t; \epsilon)\|^2 \leq 2\{\|\partial_t^{p-1} \theta_{,\ell}(\cdot, 0; \epsilon)\|^2 + T \int_0^t \|\partial_s^p \theta_{,\ell}(\cdot, s; \epsilon)\|^2 ds\}$$

so that

$$\|\partial_t^{p-1} \theta_{,\ell}(\cdot, t; \epsilon)\| = O(\epsilon^{m-p})$$

because of (3.2) and (3.11).

Estimates for $\partial_t^p D^\alpha \theta$ and $\partial_t^p D^\alpha \mathbf{u}$ for $|\alpha| \geq 2$ can be obtained by elliptic regularity estimates. Instead of a formal induction proof of these results we prefer to make the idea clear by providing the key steps in the critical cases of $|\alpha| \leq 4$, starting now with $|\alpha| = 2$.

For fixed t (and ϵ) with t in $[0, T]$ we have from (2.1) that $\theta(\cdot, t; \epsilon)$ is the solution of a Dirichlet problem for the Poisson equation

$$(4.3) \qquad \Delta \theta = \Psi := \partial_t \theta + \gamma \operatorname{div} \partial_t \mathbf{u} - \epsilon^m \rho$$

with homogeneous Dirichlet boundary condition

$$\theta(\cdot, t; \epsilon) \Big|_{\partial\Omega} = 0.$$

By our estimates above on derivatives of θ and $\mathbf{u}$ with at most one space differentiation the RHS of (4.3) satisfies

$$\|\Psi(\cdot, t; \epsilon)\| = O(\epsilon^{m-1})$$

and so from the elliptic regularity estimate (see e.g. [7])

$$(4.4) \qquad \|\theta\|_2 \leq C\{\|\Psi\| + \|\theta\|\}$$

with the constant C independent of t and ϵ it follows that

$$\|\theta(\cdot, t; \epsilon)\|_2 = O(\epsilon^{m-1}).$$

Similarly from (2.2) we have that $\mathbf{u}(\cdot, t; \epsilon)$ is the solution of the displacement boundary value problem of elastostatics

$$E[\mathbf{u}] = \mathbf{H} : = \epsilon^2 \partial_t^2 \mathbf{u} + \gamma \operatorname{grad} \theta - \epsilon^{m+1} \mathbf{r}$$

$$\mathbf{u}(\cdot, t; \epsilon)\Big|_{\partial\Omega} = 0$$

with $\mathbf{H}$ on the RHS satisfying

$$\|\mathbf{H}(\cdot, t; \epsilon)\| = O(\epsilon^{m-1})$$

and thus

$$\|\mathbf{u}(\cdot, t; \epsilon)\|_2 = O(\epsilon^{m-1})$$

by the corresponding elliptic regularity estimate for the elastostatic system (see e.g. [10]).

By replacing θ by $\partial_t^p \theta$ and $\mathbf{u}$ by $\partial_t^p \mathbf{u}$ for the boundary-value problems just considered we obtain

$$\left.\begin{array}{c} \|\partial_t^p \theta(\cdot, t; \epsilon)\|_2 \\ \|\partial_t^p \mathbf{u}(\cdot, t; \epsilon)\|_2 \end{array}\right\} = O(\epsilon^{m-p-1}).$$

The next step ($|\alpha| = 4$) makes clear how the procedure can be continued to arbitrarily high order (provided the requisite regularity and compatibility conditions are satisfied). Applying div to (2.2) yields

$$\epsilon^2 \partial_t^2 \operatorname{div} \mathbf{u} - (\nu + 1)\Delta \operatorname{div} \mathbf{u} + \gamma \Delta \theta = \epsilon^{m+1} \operatorname{div} \mathbf{r}$$

from which we conclude

$$\|\Delta \operatorname{div} \mathbf{u}(\cdot, t; \epsilon)\| = O(\epsilon^{m-1})$$

while the Laplacian Δ applied to (2.1) gives

$$\Delta\Delta\theta = \Psi^* := \partial_t \Delta\theta + \gamma \operatorname{div} \partial_t \Delta\mathbf{u} - \epsilon^m \Delta\rho$$

with Ψ^* on the RHS satisfying

$$\|\Psi^*(\cdot, t; \epsilon)\| = O(\epsilon^{m-2}).$$

Thus $\theta(\cdot, t; \epsilon)$ is the solution of the biharmonic boundary-value problem

$$\Delta\Delta\theta = \Psi^*$$

$$\theta(\cdot, t; \epsilon)\Big|_{\partial\Omega} = 0$$

$$\frac{\partial\theta}{\partial n}(\cdot, t; \epsilon)\Big|_{\partial\Omega} = \gamma_1 \theta$$

where γ_1 denotes the trace operator of order one. The elliptic regularity estimate for this biharmonic problem (see e.g. [9])

$$\|\theta\|_4 \leq C\{\|\Phi^*\| + \|\theta\| + \|\theta\|_2\}$$

then gives

$$\|\theta(\cdot, t; \epsilon)\|_4 = O(\epsilon^{m-2}).$$

For the corresponding result for $\mathbf{u}$ we first apply D^α div with $|\alpha| = 1$ to (2.2) to get

$$\epsilon^2 \partial_t^2 (\operatorname{div} \mathbf{u})_{,j} - (\nu + 1)(\operatorname{div} \Delta \mathbf{u})_{,j} + \gamma(\Delta\theta)_{,j} = \epsilon^{m+1}(\operatorname{div} \mathbf{r})_{,j}$$

and so

$$\|\operatorname{div} \Delta \mathbf{u}_{,j}\| = O(\epsilon^{m-2}).$$

Then application of the Laplacian Δ to (2.2) gives

$$\nu \Delta \Delta u^j = H_*^j := \epsilon^2 \partial_t^2 \Delta u^j - \operatorname{div} \Delta \mathbf{u}_{,j} + \gamma \Delta \theta_{,j} - \epsilon^{m+1} \Delta r^j$$

with H_*^j on the RHS satisfying

$$\|H_*^j(\cdot, t; \epsilon)\| = O(\epsilon^{m-2}).$$

Thus u^i is the solution of the biharmonic boundary-value problem

$$\begin{aligned}
\Delta \Delta u^i &= H_*^i \\
u^i \Big|_{\partial\Omega} &= 0 \\
\frac{\partial u^i}{\partial n}\Big|_{\partial\Omega} &= \gamma_1 u^i
\end{aligned}$$

and hence by the elliptic regularity estimate for this problem (for each component of $\mathbf{u}$)

$$\|u(\cdot, t; \epsilon)\|_4 = O(\epsilon^{m-2}).$$

The same procedure applied to $\partial_t^p \theta$ in place of θ and $\partial_t^p \mathbf{u}$ in place of $\mathbf{u}$ therefore results in

$$\left. \begin{aligned}
\|\partial_t^p \theta(\cdot, t; \epsilon)\| \\
\|\partial_t^p \mathbf{u}(\cdot, t; \epsilon)\|
\end{aligned} \right\} = O(\epsilon^{m-p-2}).$$

Continuing according to this pattern we arrive at (4.1) - (4.2) for the values indicated. It should be noted that from the L^2 estimates (4.1) - (4.2) uniform estimates on the closure of $\Omega \times (0, T)$ can be obtained by use of the Sobolev embedding theorems.

5 Concluding Remark

The above estimates for solutions of (2.1) - (2.4) have exact analogues for the initial value problem for the system of evolution equations

(5.1) $$\dot\theta + \Lambda\theta - \gamma C \dot u = \Phi \qquad (= \epsilon^m \rho)$$
(5.2) $$\epsilon^2 \ddot u + L u + \gamma \Gamma \theta = F \qquad (= \epsilon^{m+1} r)$$

for θ in a Hilbert space N with inner product $< \cdot, \cdot >$ (and induced norm $\|\cdot\|$) and u in a Hilbert space H with inner product $[\cdot, \cdot]$ (and induced norm $\|\|\cdot\|\|$). The positive symmetric operators Λ and L are densely defined on N and H, respectively while the densely defined C (respectively Γ) maps from its domain in H into the space N (resp. from its domain in N into the space H). The pair of operators $\{C, \Gamma\}$ are dual in the sense that

$$\langle Cu, \theta \rangle = [u, \Gamma\theta]$$

for $\{\theta, u\}$ in the appropriate domains.

The pth energy is then

$$E_p(t) := \frac{1}{2}\{\langle \partial_t^p \theta,\ \partial_t^p \theta \rangle + \epsilon^2 [\partial_t^{p+1} u,\ \partial_t^{p+1} u] + \ell(\partial_t^p u,\ \partial_t^p u)\}$$

where the bilinear form ℓ is generated by L

$$\ell(u, v) := [Lu,\ v].$$

The energy inequality corresponding to (3.4) is

$$E_p(t) + \int\limits_0^t \lambda(\partial_s^p \theta, \partial_s^p \theta) ds \le e^T \{E_p(0) + \frac{\epsilon^{2m}}{2} \int\limits_0^T \{\langle \partial_t^p \rho, \partial_t^p \rho \rangle + [\partial_t^p r, \partial_t^p r]\} dt\}$$

where λ is the bilinear form generated by Λ

$$\lambda(\theta, \chi) = \langle \Lambda \theta, \chi \rangle.$$

A system of the form (5.1) - (5.2) is mentioned in a semi-group approach to thermo-elasticity [10; p. 359], and corresponds to the case of an inhomogeneous anisotropic medium.

6 Acknowledgment:

The work of R. J. Weinacht was supported in part by GNAFA-CNR while the author was a Visiting Professor at the Dipartimento di Matematica "Ulisse Dini" of the Università degli Studi di Firenze. He would like to express his gratitude to the Dipartimento for its hospitality, especially Professors Magnanini, Pucci and Talenti.

References

[1] C. M. Dafermos, *On the existence and the asymptotic stability of solutions to the equations of linear thermoelasticity*, Arch. Mech. Rat. Anal., 29, (1968), pp. 241-271.

[2] C. M. Dafermos and J. A. Nohel, *Energy methods for nonlinear hyperbolic Volterra integrodifferential equations*, Comm. in Partial Differential Equations, 4, (1979), pp. 219-278.

[3] W. A. Day, *Heat conduction within linear themoelasticity*, in Springer Tracts in Natural Philosophy, 30, Springer-Verlag, Berlin, 1985.

[4] B. F. Esham, Jr. and R. J. Weinacht, *Hyperbolic-parabolic singular perturbations for quasilinear equations*, SIAM J. Math. Anal., 20, (1989), pp. 1344-1365.

[5] B. F. Esham, Jr. and R. J. Weinacht, *Singular perturbations and the coupled/quasi-static approximation in linear thermoelasticity*, SIAM J. Math. Anal., 25 (1994), pp. 1521-1536.

[6] B. F. Esham, Jr. and R. J. Weinacht, *The coupled/quasi-static approximation in multidimensional linear thermoelasticity*, (in preparation).

[7] A. Friedman, *Partial Differential Equations*, Holt, Rinehart and Winston, New York, 1969.

[8] K. O. Friedrichs, *Symmetric positive linear differential equations*, Comm. Pure Appl. Math., 11, (1958), pp. 333-418.

[9] J. L. Lions and E. Magenes, *Non-homogeneous boundary value problems and applications*, Vol. I, Springer-Verlag, Berlin, 1972.

[10] J. E. Marsden and T. J. R. Hughes, *Mathematical Foundations of Elasticity*, Prentice-Hall, Englewood Cliffs, 1983.

[11] M. Slemrod, *Global existence, uniqueness and asymptotic stability of classical smooth solutions in one-dimensional non-linear thermoelasticity*, Arch. Rat. Mech. Anal., 76, (1981), pp. 97-133.

The Wigner-Poisson and Schrödinger-Poisson Systems[*]

P. F. Zweifel[†]   T. M. Morris, III [†]

1   The Wigner Transform

The Wigner Method in quantum mechanics was introduced originally[1] as an easy way of computing expectation values. It is based on the following transformation: Let $A(X, P)$ be an operator depending on the position operator X and the momentum operator $P = -i\hbar\nabla_X$. The Wigner transform of A is

$$(1) \qquad A_w(X, P) = \sum_{k,l} A_{kl} \int_{\mathbb{R}^d} e^{\frac{ip \cdot z}{\hbar}} u_k(x - \frac{z}{2})\bar{u}_l(x + \frac{z}{2}) \, dz,$$

$$(2) \qquad A_{kl} = (u_k, Au_l)_{L^2(\mathbb{R}^d)} \; .$$

Here u_k is an arbitrary basis in $L^2(\mathbb{R}^d)$, the quantum state space; $x = X_w$ and $p = P_w$ are interpreted as classical phase-space coordinates.

The following propositions are easy to prove.[2]

PROPOSITION 1.1. *For two operators A and B,*

$$(3) \qquad \text{Tr } AB = (2\pi\hbar)^{-d} \int_{\mathbb{R}^d \times \mathbb{R}^d} A_w(x, p) B_w(x, p) \, dx \, dp.$$

Note that neither A nor B need be trace class. For example,
$\text{Tr } X = (2\pi\hbar)^{-d} \int_{\mathbb{R}^d \times \mathbb{R}^d} x \, dx \, dp = 0$ *in an improper sense.*

PROPOSITION 1.2. *The following transformation rules hold:*

$$(4) \qquad (a) \quad \text{Let } A = A(X), \quad \text{Then } A_w = A(x).$$
$$(5) \qquad (b) \quad \text{Let } B = B(P), \quad \text{Then } B_w = B(p).$$

PROPOSITION 1.3. *For two operators A and B,*

$$(6) \qquad (AB)_w = A_w(x + \frac{i\hbar}{2}\nabla_p, p - \frac{i\hbar}{2}\nabla_x) B_w(x, p)$$

$$(7) \qquad = B_w(x - \frac{i\hbar}{2}\nabla_p, p + \frac{i\hbar}{2}\nabla_x) A_w(x, p).$$

The Wigner transform and the Weyl transform are studied in the book of Folland.[3] We shall not need the Weyl transform in our work, but it is we note that this mapping of phase-space functions of x and p to operators is the inverse of the Wigner transform.[2]

---
[*]The research was supported by NSF Grant No. DMS-940348.
[†]Center for Transport Theory and Mathematical Physics, Virginia Tech, Blacksburg, VA

2 The Wigner-Poisson System

A practical and important operator is the density matrix of quantum mechanics.[4] Any positive trace class operator ρ is a density matrix (usually one normalizes $\text{Tr}\,\rho = 1$). Since such an operator has a purely discrete spectrum, it can always be written as

$$(8) \qquad \rho = \sum_n \lambda_n P_n$$

with eigenvalues λ_n and spectral projections P_n. Writing

$$(9) \qquad P_n f = \psi_n(\psi_n, f)$$

one calculates from (1) that

$$(10) \qquad \rho_w = \sum_n \lambda_n \int_{\mathrm{IR}^d} e^{\frac{ip \cdot z}{\hbar}} \psi_n(x - \frac{z}{2}) \overline{\psi}_n(x + \frac{z}{2})\, dz.$$

In quantum mechanics,[4] expectation values are calculated as

$$(11) \qquad <A> = \text{Tr}\,\rho A.$$

From Proposition 1.1, we see that for the Wigner function:

$$(12) \qquad w(x, p) = (2\pi\hbar)^{-d} \rho_w(x, p),$$

$$(13) \qquad <A> = \int_{\mathrm{IR}^d \times \mathrm{IR}^d} w(x, p) A_w(x, p)\, dx\, dp.$$

Thus, the Wigner method reduces the quantum mechanical trace calculation to phase-space integrals. The Wigner function is, in a sense, analogous to a classical phase-space distribution, such as the Boltzmann distribution; however, it is not necessarily positive. It is easy, however, to show w to be real. The density

$$(14) \qquad n(x) = \int_{\mathrm{IR}^d} w(x, p)\, dp = \sum_n \lambda_n \mid \psi \mid^2$$

is manifestly positive and, in fact, gives the correct quantum-mechanical density.

The evolution equation for ρ depends upon the Hamiltonian $H(X, P)$ which we assume can be expressed as[4, 5]

$$(15) \qquad H = -\frac{\hbar^2}{2}\Delta + V(X),$$

where Δ is the d-dimensional Laplacian, and we have taken all masses to be unity. Then

$$(16) \qquad H_w = \frac{1}{2}p^2 + V(x).$$

The evolution equation for $w(x, p)$ can be obtained from that of ρ:[4, 5]

$$(17) \qquad i\hbar \partial_t \rho = H\rho - \rho H^*.$$

(If V is real, then $H = H^*$; for now we will assume this, but see Sec. 5.)

Performing the Wigner transform of (17) leads to an evolution equation for w:

$$(18) \qquad \partial_t w(x,p,t) + p \cdot \nabla_x w - \frac{i}{\hbar} \Theta(V) w \;=\; 0$$

where $\Theta(V)$ is the pseudo-differential operator with symbol

$$(19) \qquad \mathrm{Sym}\Theta(V) = V(x + \frac{\tau}{2}) - V(x - \frac{\tau}{2}).$$

Here we have used Propositions 1.2 and 1.3. The solution to (18) has initial condition:

$$(20) \qquad w(x,p,0) = w_I(x,p) \;=\; \sum_n \lambda_n \int\limits_{\mathbb{R}^d} e^{\frac{ip \cdot z}{\hbar}} \, \phi_n(x - \frac{z}{2}) \hat{\phi}_n(x + \frac{z}{2}) \, dz$$

with each ϕ_n simply the initial value of the ψ_n above, which obey:

$$(21) \qquad i\hbar \partial_t \psi_n \;=\; H\psi_n.$$

As expected, it can be shown[6] that the Wigner equation (18) is equivalent to the system of Schrödinger equations given in (21) which has the same dimension as the number of terms in (8).

The mean-field approximation to the Wigner equation gives the quantum Vlasov equation, in which V is approximated by the "self-consistent" field:[7]

$$(22) \qquad V(x) = \int\limits_{\mathbb{R}^d} n(y)v(y - x) \, dy.$$

where v is the two body potential and n is the density given in (14). (Then V depends on $\Psi = (\psi_n)$, making (21) nonlinear.) Taking v to be the standard Coulomb potential gives the Wigner-Poisson system, in which V obeys:

$$(23) \qquad \Delta V = -\epsilon n.$$

Here we can choose ϵ to be the normalized ($\epsilon = \pm 1$) electric charge. Positive ϵ gives the repulsive case, while negative ϵ gives the attractive case.

Since the Wigner-Poisson (WP) system is equivalent to the corresponding Schrödinger system, called the Schrödinger-Poisson (SP) system, we can consider WP to be a statistical version of the single particle mean-field Schrödinger equation with a Coulomb two-body potential (Hartree-Fock).[4, 5] In most of the analysis we consider first the SP system and then lift our results to the WP system. We can then utilize many existing results for the non-linear Schrödinger equation[8] and the Hartree-Fock system.[9] Alternatively, WP can be considered a quantum version of the Vlasov-Poisson system.[2] This is seen by expanding (18) in powers of $\hbar$ and formally tending $\hbar \to 0$. This process yields the classical Vlasov equation. Hereafter, for convenience, we set $\hbar = 1$.

3 Existence and Uniqueness in $\mathbb{R}^3$

To deal with the SP system we need the following direct sum Hilbert spaces:

$$(24) \qquad X^k := \{\Gamma = (\gamma_n)_{n \in \mathbb{N}} \mid \gamma_n \in H^k, \|\Gamma\|^2_{X^k} = \sum_n \lambda_n \|\gamma_n\|^2_{H^k}\}.$$

To prove global existence and uniqueness for WP and SP, we first show that a mild local in time solution to the SP system exists. We need

PROPOSITION 3.1. *The mapping* $J : X^2 \to X^2$, $J(\Psi)_m = V(\Psi)\psi_m$, *for* V *the Coulomb potential satisfying equation (23) is locally Lipschitz in* X^2.

This is proved by showing, $\forall \Gamma_1, \Gamma_2 \in X^2$

(25) $$\|V(\Gamma_1) - V(\Gamma_2)\|^2_{L^\infty} \leq C\|\Gamma_1 - \Gamma_2\|^2_{X^2}$$
(26) $$\|\nabla V(\Gamma_1) - \nabla V(\Gamma_2)\|^2_{L^2} \leq C\|\Gamma_1 - \Gamma_2\|^2_{X^2}$$
(27) $$\|\Delta V(\Gamma_1) - \Delta V(\Gamma_2)\|^2_{L^2} \leq C\|\Gamma_1 - \Gamma_2\|^2_{X^2}$$

where C is a constant depending on the X^2 norms of Γ_1 and Γ_2. These inequalities can be shown through use of the Minkowski, Hölder, and Gagliardo-Nirenberg inequalities.[10] A classical result on the Cauchy problem[11] then shows that there actually exists a unique local *strong* solution Ψ of the SP system for some $T > 0$ (i.e. $\Psi \in C([0,T], X^2) \cap C^1([0,T], X^0)$) such that $\Psi(x,0) = \Phi(x)$ for some $\Phi(x) \in X^2$. To proceed further, we need a conservation law.[23]

PROPOSITION 3.2. *For any mild, X^2-valued solution Ψ of SP on $[0,T]$ (which must also be a strong X^0-valued solution on $[0,T]$) with the initial datum Φ, the following hold:*

(28) $$\|\psi_m(t)\|^2_{L^2} = \|\phi_m\|^2_{L^2} = 1 \qquad \forall m$$
(29) $$\| \Psi(t) \|^2_{X^0} = \|\Phi\|^2_{X^0} = 1$$
(30) $$\|\nabla \Psi(t)\|^2_{X^0} + \epsilon\|\nabla V(t)\|^2_{L^2} = \|\nabla \Phi\|^2_{X^0} + \epsilon\|\nabla V_0\|^2_{L^2}$$

where V_0 and n_0 are given by

(31) $$n_0 = \sum \lambda_m |\phi_m|^2$$

(32) $$\Delta V_0 = -\epsilon n_0.$$

Note that (30) is not an equation for conservation of energy (which would have a factor of $\frac{1}{2}$ in the first term on each side), but, rather, an expression for conservation of the "action"

(33) $$Q = 2 <T> + <V>.$$

Here T is the kinetic energy and V is the potential energy.[12] This conservation law leads to a global bound on Ψ via Gronwall's inequality:

LEMMA 3.1. *Let $\Phi \in X^0$ and $\Psi(t)$ be a mild X^2-valued solution of the SP system on $[0,T]$. Then there is a C, depending only on Φ and T, such that*

(34) $$\|\Psi(t)\|^2_{X^2} \leq C \ \forall t \in [0,T].$$

yielding the desired unique global strong solution of SP.[23]

4 Floquet Solutions

For our purposes it is sufficient to seek Floquet solutions to WP and SP on the cube $Q = [0,1]^3$. Schrödinger Floquet states, states for which $\Phi(x + \alpha) = e^{2\pi i \alpha} \Phi(x), x, \alpha \in \mathbb{R}^3$, are easily seen to lead to periodic Wigner functions with the transformation $p \to p - \alpha$. The Floquet parameter α is proportional to the current through the system. In this case, the momentum is quantized: $p_n = 2\pi n, n \in \mathbf{Z}$, and the integrals over p are replaced by sums. The SP system to be considered obeys:

(35) $$\partial_t \Psi(x,t) = -\frac{1}{2}\Delta \Psi + V(\Psi)\Psi$$

$$(36) \qquad \Psi \;=\; (\psi_m)_{m \in \mathbb{N}}$$

$$(37) \qquad n(x,t) \;=\; \sum_{m=1}^{\infty} \lambda_m |\psi_m(x,t)|^2 - n^*$$

$$(38) \qquad \Delta V \;=\; -\epsilon n$$

where n^* is the background charge density. This forces the total charge to be zero:

$$(39) \qquad \int_Q n(x,t)\,dx = 0.$$

The main tool we shall use in solving the time-dependent problem is the Galerkin sequence $(\psi^{(N)})_{N \in \mathbb{N}}$, defined as:

$$(40) \qquad \psi_m^{(N)}(x,t) \;=\; \sum_{|k| \leq N} d_{m,k}^{(N)}(t) h_k(x)$$

$$(41) \qquad h_k(x) \;=\; e^{2\pi i(k+\alpha) \cdot x}, \; k \in \mathbf{Z}^3.$$

Each $\psi_m^{(N)}$ obeys

$$(42) \qquad (i\partial_t \psi_m^{(N)} + \frac{1}{2}\Delta \psi_m^{(N)} - V^{(N)}\psi_m^{(N)}, h_k) = 0 \qquad (|k| \leq N)$$

$$(43) \qquad \psi_m^{(N)}(x,0) \;=\; \sum_{|k| \leq N} (\phi_m, h_k) h_k$$

Now it is possible to derive a first-order (ordinary) nonlinear differential system for the $d_{m,l}^{(N)}$. By standard methods[13] it can be shown that the system has a unique solution in the space Y^2, where

$$(44) \qquad Y^k = \{ \Psi = (\psi_m) \mid \psi_m \text{ Floquet on } Q, \; \psi_m \in H_{loc}^k(\mathbb{R}^3) \} \;.$$

As before, conservation laws are important in obtaining the global solution. (These laws are valid for any order N of the Galerkin approximation.) It is necessary to obtain bounds like those in Lemma 3.1 for each step in the Galerkin sequence, uniform in N. Then we can use these uniform bounds to take the limit $N \to \infty$, using compactness.

In addition, error estimates for the Galerkin sequence can be obtained.[14] The convergence rate is proporional to $N^{-\frac{1}{4}}$

Now, we seek solutions to the SP system of the form

$$(45) \qquad \Psi(x,t) = e^{-iEt}\Phi(x)$$

where $\mathbf{E}$ is a matrix (in general, an operator in l^2). The corresponding SP system is time-dependent and contains the eigenmatrix as a generalization of the usual energy eigenvalue of elementary quantum mechanics. This problem, with $\mathbf{E}$ constant, was studied in [15] and [16]. [17] and [18] study the problem for diagonal $\mathbf{E}$. Here we will adopt a more standard approach.

If (45) is substituted into the SP system, the exponential factor will be homogeneous if the matrix $\mathbf{E}$ is self-adjoint in the Hilbert space

$$(46) \qquad \{\xi = (\xi_m)_{m \in \mathbb{N}} \mid \sum \lambda_m |\xi_m|^2 < \infty \}$$

274

since in this case

(47)
$$n(\Psi) = n(\Phi)$$

where

(48)
$$n(\Theta) = \sum \lambda_m |\theta_m|^2 \ .$$

The stationary system then becomes

(49)
$$-\frac{1}{2}\Delta\Phi + V(\Phi)\Phi \ + \ V^*\Phi = \mathbf{E}\,\Phi$$

(50)
$$-\Delta V(\Phi) \ = \ n_\Phi - n^* \ .$$

One again, $\Phi = (\phi_m)_{m\in\mathbb{N}}$. Here, V^* is a (possible) external field. Now, $\mathbf{E}$ will be self-adjoint if and only if

(51)
$$\lambda_k E_{km} = \lambda_m \overline{E_{mk}}$$

which can be recognized as a detailed balance condition. [19] Also, since $\mathbf{E}$ is diagonalizable by a unitary transformation, we assume it to be in diagonal form and label its eigenvalues μ_m. The system (49)-(50) reduces to

(52)
$$-\frac{1}{2}\Delta\phi_m + V(\Phi)\phi_m \ + \ V^*\phi_m = \mu_m\,\phi_m$$

(53)
$$-\Delta V(\Phi) \ = \ n_\Phi - n^*$$

(54)
$$n_\Phi \ = \ \sum \lambda_m(\mu_m)|\phi_m|^2 \ .$$

The essential difference for the time-dependent problem is that the λ_m depend explicitly upon the eigenvalues μ_m. One example of this is the Fermi distribution (λ_m is the probability of the state ϕ_m):

(55)
$$\lambda_m = \frac{\alpha_m}{e^{\beta(\mu_m - \mu_F)} + 1};$$

as before $\sum \lambda_m = 1$ (α_m and β are physical constants, and μ_F is the Fermi level). Other distributions are possible as long as certain technical conditions are satisfied.

The solvability of the stationary system can be proved using a Schauder fixed-point argument. We denote the solution $(\Psi, M), M = (\mu_m)_{m\in\mathbb{N}}$. Here, we will work in the Fréchet space

(56)
$$Y^k = l(H_{per}^k(Q)) \times l(\mathbb{R})$$

with metric

(57)
$$d([\Phi, M], [\tilde{\Phi}, \widetilde{M}]) = \sum_j \frac{1}{2^j} \left[\frac{||\phi_j - \tilde{\phi}_j||_{H^k}}{1 + ||\phi_j - \tilde{\phi}_j||_{H^k}} + \frac{|\mu_j - \tilde{\mu}_j|}{1 + |\mu_j - \tilde{\mu}_j|} \right] \ .$$

Define an operator T on Y^2 by

(58)
$$T(\Phi, M) = (\tilde{\Phi}, \widetilde{M})$$

where $\tilde{\Phi}$ and $\widetilde{M}$ are solutions of the linear eigenvalue problem corresponding to (52)-(54) (i.e. V fixed.)

The following lemma is proved in [20] using methods adapted from [21].

LEMMA 4.1. *Let $V \in L_{per}^\infty(Q)$ and real. Then the linear eigenvalue system*

(59)
$$-\frac{1}{2}\Delta\phi_m + V\phi_m + V^*\phi_m = \mu_m\,\phi_m \ (m \in \mathbb{N})$$

has a solution $(\Phi, M) \in Y^2$ *where* $\Phi = (\phi_m)_{m \in \mathbb{N}}$ *forms an orthonormal system in* $L^2_{per}(Q)$; *the eigenvalues* $M = (\mu_m)$ *are real, arranged in increasing order, and fulfill* $\mu_m \to \infty$ *as* $m \to \infty$. *Furthermore, there are positive generic constants* $C^{(1)}, C^{(2)} > 0$ *such that the estimate*

$$(60) \qquad - \|V\|_{L^\infty} + C^{(1)} m^{2/3} \leq \mu_m \leq C^{(2)} m^{2/3} + \|V\|_{L^\infty}$$

is valid.

(The major tools in proving this lemma are the min-max and comparison principles.[21])

Using this lemma, it is possible to prove that T maps the following set into itself

$$(61) \qquad S_\gamma = \{ (\Phi, M) \in Y^1 \,|\, \|\phi_m\|_{L^2} \leq 1, \ \|\nabla \phi_m\|_{L^2} \leq K_m, \ \mu_m \geq \mu_m^*(\gamma) \}$$

where $K_m = (\gamma + 2C^{(2)} m^{2/3})^{\frac{1}{2}}$ and $\mu_m^*(\gamma) = C^{(1)} m^{2/3} - \gamma$. In addition, T is continuous in the topology of Y^1 and maps the closed, convex set S_γ into $T(S_\gamma)$, relatively compact in this same space. This is sufficient to prove the existence of solutions in Y^2 (not necessarily unique) for the eigenmatrix problem.

5 Dissipative Systems

In some approximations, the magneto fluid-dynamics of quantum plasmas can be described by adding to the usual Hamiltonian terms representing dissipation (or viscosity). Here we shall add only linear dissipation, so that the nonlinearity is still only that induced by the mean field approximation.

We deal with the usual SP system (42) – (43) with (42) replaced by

$$(62) \qquad i\partial_t \psi_m = -\frac{1}{2} \Delta \psi_m + V \psi_m + i H_1 \psi_m$$

where

$$(63) \qquad H_1 = \alpha \Delta + \beta x^2 + \gamma.$$

The first term models viscous friction, the second loss of energy to a reservoir and the third kinetic friction. Overall, iH_1 gives a diffusive character to the Hamiltonian equation. The SP system leads to the Wigner equation (18) with

$$(64) \qquad \partial_t w(x, p, t) + p \cdot \nabla_x w - \Theta(V) w = \alpha(2p^2 - \frac{1}{2}\Delta_x)w + \beta(2x^2 - \frac{1}{2}\Delta_p)w + 2\gamma w.$$

(See Prop.2.1,[22]). Real terms in the right-hand side of (64), involving even order derivatives in v will lead to dissipative (skew adjoint) Hamiltonians if they are separable. Thus, the treatment in this section is an example of how one would treat more general cases. We shall need the space

$$(65) \qquad \tilde{X} = \{ \Psi = (\psi_m) \,|\, \Psi \in X^2, \ \|x^2 \Psi\|_{X^2}^2, \ \|x \otimes \nabla \Psi\|_{X^2}^2 < \infty \}$$

where

$$(66) \qquad \|x \otimes \nabla \Psi\|_{X^2}^2 = \sum \lambda_m \sum_{j,k=1}^{3} \|x_j \partial_{x_k} \psi_m\|_{L^2}^2$$

and

$$(67) \qquad \tilde{Z} = X^2 \cap \tilde{X}.$$

The existence proof of this system is similiar to that of [23] where the unitary group generated by the free Hamiltonian $-\frac{1}{2}\Delta$ is used as an " integrating factor"; however, here we use $-\frac{1}{2}\Delta + iH_1$ as the free Hamiltonian. Because of the presence of i, this cannot generate a unitary group, but instead generates a contraction semigroup on X for $\alpha \geq 0, \beta \leq 0, \gamma \leq 0$. (Lemma 3.1, [22]. The proof follows by the Hille-Yosida theorem. From this, local existence of solution to SP (and hence WP) follows (the Lipshitz proof for $V(\Psi)$ is the same as in[23], since V is the same).

Normally, global existence follows from conservation laws. Here, because the Hamiltonian contains dissipation terms, there is no conservation. All is not lost, since we can define a Liapunov functional:

$$(68) \qquad E(t, \Psi) = \int_{\mathbb{R}^3} \{|\nabla \Psi|^2 + |\nabla V|^2 + \delta |\Psi|^2\} dx \,.$$

One shows
$$(69) \qquad \frac{\partial E}{\partial t} \leq 0$$

for the signs of α, β, γ noted above. From this inequality one attains the *a priori* bound

$$(70) \qquad \|\Psi(t)\|_{\tilde{Z}} \leq C \,.$$

which leads to Theorem 4.3 of [22], which states the existence of a unique strong global solution of SP (and hence WP) with

$$(71) \qquad \Psi \in C([0, \infty); \tilde{Z}) \cap C^1([0, \infty); X^2)$$

and
$$(72) \qquad w \in C([0, \infty); L^2(\mathbb{R}^3 \times \mathbb{R}^3)) \cap L^\infty([0, \infty); L^s(\mathbb{R}^3 \times \mathbb{R}^3)) \,, \ 2 \leq s \leq \infty \,.$$

6 The Quasi-linear SP System

Here we study the periodic SP system (42) on the unit cube; however, we modify the Poisson equation to the following quasilinear equation, an approximation for moderately large fields. In particular the dielectric constant is written $\epsilon_0 + \epsilon_1 |\nabla V|^2$. Setting $\epsilon_0 = 1$ and $\epsilon_0 / \epsilon_1 = \theta$, the Poisson equation we consider is

$$(73) \qquad -\nabla \cdot (1 + \theta |\nabla V|^2)\nabla V = \epsilon(n - 1)$$

where, as before, $\epsilon = \pm 1$ is the normalized charge. Without loss of generality we set $\int_Q V \, dx = 0$. As usual, we require charge neutrality: $\int_Q n \, dx = 1$.

In the two and three dimensional case, we are able to prove existence of a strong solution to SP.[20] For the three (or two) dimensional case, we proceed as follows. First we study the weak solution of the Poisson equation (73). The weak formulation can be written as:

$$(74) \qquad \int_Q (1 + |\nabla V|^2)\nabla V \cdot \nabla \phi \, dx - \int_Q (n - 1)\phi \, dx = 0 \quad (\forall \phi \in C^\infty_{per}(Q)).$$

This equation is derived from the following Lagrangian functional

$$(75) \qquad L(V) = \frac{1}{2} \int_Q |\nabla V|^2 dx + \frac{1}{4} \int_Q |\nabla V|^4 \, dx - \int_Q (n - 1)V \, dx.$$

We seek a weak solution $V \in H^{1,4}_{per}$. The corresponding n must be in $(H^{1,4}_{per})^* = H^{-1,\frac{4}{3}}$. This follows if
$$(76) \qquad \sup_{\|\phi\|_{H^{1,4}_{per}} \leq 1} |<n, \phi>| < \infty.$$

Existence of the weak solution, V, follows from a Theorem of Struwe, [24] which requires following conditions:

(i) $L(V) \to \infty$ as $\|V\| \to \infty$, $V \in M$ (a weakly closed subset of a reflexive Banach space)

(ii) $\forall V \in M$ and any sequence $(V_m) \in M$ s.t. $V_m \to V$ weakly in M, we have $L(V) \leq \lim_{m \to \infty} \inf L(V_m)$.

If these conditions are satisfied, L is bounded below and attains its infimum on M. Condition (i) is verified for the reflexive Banach space $H_{per}^{1,4}(Q)$ with $M = \{V \in H_{per}^{1,4} | \int_Q V \, dx = 0\}$. The fact that $L(V) \to \infty$ as $\|V\| \to \infty$ follows from Poincare's, Young's, and Hölder's inequality, by which it is shown that

$$(77) \qquad L(V) \geq C_1 \|V\|_{H_{per}^{1,4}}^4 - C_2 .$$

For condition (ii), we first show that $n \in L^2$ implies $n \in H^{-1,\frac{4}{3}}$ ($n \in L^2$ comes from the analysis of the Schrödinger part of SP). Now we can deal with the term $\int_Q (n-1)V \, dx$ in $L(V)$ and proceed to show lower semicontinuity. Now we have shown the existence of a weak solution. That this is actually a strong solution and is unique follows from Theorems 5.1, 5.2, and 8.10 Chapter IV of [25]. The details are found in [26]. In proving existence for SP, the modified conservation law

$$(78) \qquad \int \left\{ |\nabla \psi|^2 + |\nabla V|^2 + \frac{3}{2}\theta |\nabla V|^2) \right\} dx = const.$$

is used. Using the Galerkin approximation, the conservation law is also shown to hold for every order, with a constant uniform in N. Similarly, a uniform H^2 bound for $\Psi^{(N)}$ is obtained, and from this one proceeds to prove existence (but not uniqueness) of a weak solution.

For one dimension, a unique strong solution of the quasilinear Poisson equation can be constructed by quadrature. [26]

7 Quantum BGK Modes

Any function, ρ_0, composed exclusively of observables which commute with the Hamiltonian, generates a solution to the stationary Wigner equation. This can be seen by substitution of ρ_0 into (17). Here we require that ρ_0 be positive, trace class, with $\text{Tr} \, \rho_0 = 1$ The solution thus generated, written $w = [\rho_0]_w$, is the quantum analog to the classical BGK modes,[27] which we will hereafter refer to as a quantum BGK (QBGK) mode. The following theorem is proven using the existence of a complete set of commuting operators:[28]

THEOREM 7.1. *Consider the self-adjoint Hamiltonian $H = \frac{1}{2}P^2 + V(x)$, and its associated stationary Wigner equation*

$$(79) \qquad p \cdot \nabla_x w(x,p) - \frac{i}{\hbar}\Theta(V)w(x,p) = 0$$

with $\Theta(V)$ the pseudo-differential operator as given in (19). Any solution to this equation which is a Wigner function is also a QBGK mode.

The non-uniqueness of the solution to (79), even with Floquet boundary conditions, causes problems in numerical studies of the WP system which are still not solved.

References

[1] E. Wigner, Physics Rev. **40** (1932), 749.

[2] P. F. Zweifel, Trans. Theor. and Stat. Phys. **22** (1993), 459.

[3] Gerald B. Folland, *Harmonic Analysis in Phase Space,* Annals of Mathematics Studies, Princeton University Press, 1989.

[4] Claude Cohen-Tannoudji, Bernard Diu, Franck Laloë, *Quantum Mechanics* Hermann and Vol1, p 295, John Wiley & Sons. Inc; Paris, 1977.

[5] Albert Messiah, *Quantum Mechanics,* Vol1, p 310, p 331, John Wiley & Sons; New York, 1958.

[6] P. A. Markowich, Math. Meth. Appl. Sci. **11** (1989), 459.

[7] R. Illner, Trans. Theor. and Stat. Phys. **22** (1993), 459

[8] T. Cazenave, *An Introduction to Nonlinear Schrödinger Equations,* extos de Methodos Mathematicos 22 Rio de Janerio, 1989.

[9] J. P. Dias and M. Figueira, C.R. Acad Sci. Paris **290**, (1980), 889; J. Math. Anal. Appl. **84** (1981), 486.

[10] A. Friedman, *Partial Differential Equations,* Rinehart and Winston, New York, 1969.

[11] A. Pazy, *Semi-groups Linear Operators and Applications to P. D. E.,* Springer Verlag, Berlin, 1983.

[12] H. Lange, B. Toomire, P. F. Zweifel, Trans. Theor. and Stat. Phys. **23** (1994), 731.

[13] H. Lange, B. Toomire, P. F. Zweifel, *Periodic Solutions to Wigner–Poisson Equation,* Nonlinear Analysis TMA, (to be published).

[14] S. Bohum R. Illner, H. Lange, and P. F. Zweifel, *Error Estimates for Galerkin Approximations to the Periodic Schrödinger–Poisson System,* ZAMM (to be published).

[15] O. Kavian and H. Lange,*Stationary States in the Schrödinger-Poisson System,* Math. Inst. , Univ. Köln, (1994) (preprint).

[16] O. Kavian, H. Lange, P. F. Zweifel, ZAMM, **74** (1994), T606-8.

[17] F. Nier, *A Variational Formulation of Schrödinger–Poisson Systems in Dimensions $d \leq 3$.* Preprint CMAP Ecole Polytechnique, Palaiseau (1992).

[18] G. Albinus, H.-C. Kaiser, J. Rehberg, *On Stationary Schrödinger-Poisson Equations,* Preprint No.66 IAAS (1993).

[19] C. Cercignani, *Theory and Applications of the Boltzmann Equation,* Elsevier, New York, 1975.

[20] H. Lange, B. Toomire, P. F. Zweifel, *Eigenmatrix Problem for Schrödinger–Poisson System,* preprint, Center for Transport Theory and Math. Physics (1994) (in preparation).

[21] R. Temam, *Infinite-Dimensional Dynamical Systems in Mechanics and Physics,* Springer–Verlag, Berlin, 1983.

[22] H. Lange, P. F. Zweifel, J. Math. Phys. **35** (1994), 1513.

[23] R. Illner, H. Lange, and P. F. Zweifel, Math. Meth. Appl. Sci. **17** (1994), 349.

[24] M. Struwe, *Variational Methods,* Springer–Verlag, Berlin, 1990.

[25] O. Ladyshenkaya and N. Ural'tseva, *Linear and Quasilinear Elliptic Equations,* Academic Press, New York, 1968.

[26] R. Illner *et.al.*, *Quasi-linear Schrödinger–Poisson System,* Center for Transport Theory and Math. Physics, (submitted).

[27] I. B. Bernstein, J. M. Greene,M. D. Kruskal, Phys. Rev. **108**, (1957) 546. The quantum case was first considered by M. Buchanan (unpublished).

[28] H. Lange, B. Toomire, P. F. Zweifel, *Quantum BGK Modes for the Wigner–Poisson System,* preprint, Center for Transport Theory and Math. Physics (1995) (in preparation).